Lecture Notes in Mathematics 1967

Editors:
J.-M. Morel, Cachan
F. Takens, Groningen
B. Teissier, Paris

T0217640

Benoit Fresse

Modules over
Operads and Functors

 Springer

Benoit Fresse
UFR de Mathématiques
Université des Sciences et Technologies de Lille
Cité Scientifique - Bâtiment M2
59655 Villeneuve d'Ascq Cedex
France
Benoit.Fresse@math.univ-lille1.fr

ISBN: 978-3-540-89055-3   e-ISBN: 978-3-540-89056-0
DOI: 10.1007/978-3-540-89056-0

Lecture Notes in Mathematics ISSN print edition: 0075-8434
                        ISSN electronic edition: 1617-9692

Library of Congress Control Number: 2008938192

Mathematics Subject Classification (2000): Primary: 18D50; Secondary: 55P48, 18G55, 18A25

Cover design: SPi Publisher Services

Printed on acid-free paper

9 8 7 6 5 4 3 2 1

springer.com

Preface

The notion of an operad was introduced 40 years ago in algebraic topology in order to model the structure of iterated loop spaces [6, 47, 60]. Since then, operads have been used fruitfully in many fields of mathematics and physics.

Indeed, the notion of an operad supplies both a conceptual and effective device to handle a variety of algebraic structures in various situations. Many usual categories of algebras (like the category of commutative and associative algebras, the category of associative algebras, the category of Lie algebras, the category of Poisson algebras, ...) are associated to operads.

The main successful applications of operads in algebra occur in deformation theory: the theory of operads unifies the construction of deformation complexes, gives generalizations of powerful methods of rational homotopy, and brings to light deep connections between the cohomology of algebras, the structure of combinatorial polyhedra, the geometry of moduli spaces of surfaces, and conformal field theory. The new proofs of the existence of deformation-quantizations by Kontsevich and Tamarkin bring together all these developments and lead Kontsevich to the fascinating conjecture that the motivic Galois group operates on the space of deformation-quantizations (see [35]).

The purpose of this monograph is to study not operads themselves, but *modules over operads* as a device to model functors between categories of algebras as effectively as operads model categories of algebras.

Modules over operads occur naturally when one needs to represent universal complexes associated to algebras over operads (see [14, 54]).

Modules over operads have not been studied as extensively as operads yet. However, a generalization of the theory of Hopf algebras to modules over operads has already proved to be useful in various mathematical fields: to organize Hopf invariants in homotopy theory [2]; to study non-commutative generalizations of formal groups [12, 13]; to understand the structure of certain combinatorial Hopf algebras [38, 39]. Besides, the notion of a module over an operad unifies and generalizes classical structures, like Segal's notion of a

Γ-object, which occur in homological algebra and homotopy theory. In [33], Kapranov and Manin give an application of the relationship between modules over operads and functors for the construction of Morita equivalences between categories of algebras.

Our own motivation to write this monograph comes from homotopy theory: we prove, with a view to applications, that functors determined by modules over operads satisfy good homotopy invariance properties.

Acknowledgements

I am grateful to Geoffrey Powell and the referee for a careful reading of a preliminary version of this work and helpful observations. I thank Clemens Berger, David Chataur, Joan Millès, John E. Harper for useful discussions, and for pointing out many mistakes in the first version of this monograph. Cet ouvrage n'aurait été mené à son terme sans l'atmosphère amicale de l'équipe de topologie de Lille. Je remercie mes collègues, et plus particulièrement David Chataur, Sadok Kallel et Daniel Tanré, pour toutes sortes de soutien durant l'achèvement de ce livre.

I thank the referee, the editors and the Springer-Verlag lecture notes team for their efficient processing of the manuscript.

The author is supported by the "Laboratoire Paul Painlevé", UMR 8524 de l'Université des Sciences et Technologie de Lille et du CNRS, by the GDR 2875 "Topologie Algébrique et Applications" du CNRS, and this research is supported in part by ANR grant 06-JCJC-0042 "Opérades, Bigèbres et Théories d'Homotopie".

Contents

Index and Glossary of Notation

Introduction

Main Ideas and Objectives

The background of the theory of operads is entirely reviewed in the first part of the monograph. The main characters of the story appear in a natural generalization of the symmetric algebra $S(X)$, the module spanned by tensors $x_1 \otimes \cdots \otimes x_n \in X^{\otimes n}$ divided out by the symmetry relations

$$x_{w(1)} \otimes \cdots \otimes x_{w(n)} \equiv x_1 \otimes \cdots \otimes x_n,$$

where w ranges permutations of $(1, \ldots, n)$. Formally, the symmetric algebra is defined by the expansion $S(X) = \bigoplus_{n=0}^{\infty} (X^{\otimes n})_{\Sigma_n}$, where the notation $(-)_{\Sigma_n}$ refers to a module of coinvariants under the action of the symmetric group of n-letters, denoted by Σ_n. The theory of operads deals with functors $S(M)$: $X \mapsto S(M, X)$ of generalized symmetric tensors

$$S(M, X) = \bigoplus_{n=0}^{\infty} (M(n) \otimes X^{\otimes n})_{\Sigma_n}$$

with coefficients in objects $M(n)$ equipped with an action of the symmetric groups Σ_n. The structure formed by the coefficient sequence $M = \{M(n)\}_{n \in \mathbb{N}}$ is called a Σ_*-object (or, in English words, a symmetric sequence or a symmetric object). The definition of $S(M, X)$ makes sense in the setting of a symmetric monoidal category $\mathcal{E}$. The map $S(M) : X \mapsto S(M, X)$ defines a functor $S(M) : \mathcal{E} \to \mathcal{E}$.

In this book we study a generalization of this construction with the aim to model functors on algebras over operads. For an operad P, we use the notation $\mathsf{P}\mathcal{E}$ to refer to the category of P-algebras in $\mathcal{E}$. We aim to model functors $F : \mathsf{R}\mathcal{E} \to \mathcal{E}$ from a category of algebras over an operad R to the underlying category $\mathcal{E}$, functors $F : \mathcal{E} \to \mathsf{P}\mathcal{E}$ from the underlying category $\mathcal{E}$ to a category of algebras over an operad P, as well as functors $F : \mathsf{R}\mathcal{E} \to \mathsf{P}\mathcal{E}$ from a category of algebras over an operad R to another category of algebras over an operad P.

To define functors of these types we use left and right modules over operads, the structures formed by Σ_*-objects equipped with left or right operad actions. For a right R-module M and an R-algebra A, we use a coequalizer to make the right R-action on M agrees with the left R-action on A in the object $S(M, A)$. This construction returns an object $S_R(M, A) \in \mathcal{E}$ naturally associated to the pair (M, A) and the map $S_R(M) : A \mapsto S_R(M, A)$ defines a functor $S_R(M) : {}_R\mathcal{E} \to \mathcal{E}$ naturally associated to M. For a left P-module N the map $S(N) : X \mapsto S(N, X)$ defines naturally a functor $S(N) : \mathcal{E} \to {}_P\mathcal{E}$. For a P-R-bimodule N, a right R-module equipped with a left P-action that commutes with the right R-action on N, the map $S_R(N) : A \mapsto S_R(N, A)$ defines naturally a functor $S_R(N) : {}_R\mathcal{E} \to {}_P\mathcal{E}$.

We study the categorical and homotopical properties of functors of these form.

Not all functors are associated to modules over operads, but we check that the categories of modules over operads are equipped with structures that reflect natural operations on functors. As a byproduct, we obtain that usual functors (enveloping operads, enveloping algebras, Kähler differentials, bar constructions, ...), which are composed of tensor products and colimits, can be associated to modules over operads.

In homotopy theory, operads are usually supposed to be cofibrant in the underlying category of Σ_*-objects in order to ensure that the category of algebras over an operad has a well defined model structure. In contrast, the category of right modules over an operad R comes equipped with a natural model structure which is always well defined if the operad R is cofibrant in the underlying symmetric monoidal category. Bimodules over operads form model categories in the same situation provided we restrict ourself to connected Σ_*-objects for which the constant term $N(0)$ vanishes. Thus for modules over operads we have more homotopical structures than at the algebra and functor levels. As a result, certain homotopical constructions, which are difficult to carry out at the functor level, can be realized easily by passing to modules over operads (motivating examples are sketched next). On the other hand, we check that, for functors associated to cofibrant right modules over operads, homotopy equivalences of modules correspond to pointwise equivalences of functors. In the case where algebras over operads form a model category, we can restrict ourself to cofibrant algebras to obtain that any weak-equivalence between right modules over operads induce a pointwise weak-equivalence of functors. These results show that modules over operads give good models for the homotopy of associated functors.

We use that objects equipped with left operad actions are identified with algebras over operads provided we change the underlying symmetric monoidal category of algebras. Suppose that the operad R belongs to a fixed base symmetric monoidal category $\mathcal{C}$. The notion of an R-algebra can be defined in any symmetric monoidal category $\mathcal{E}$ acted on by $\mathcal{C}$, or equivalently equipped with a symmetric monoidal functor $\eta : \mathcal{C} \to \mathcal{E}$.

The category of Σ_*-objects in $\mathcal{C}$ forms an instance of a symmetric monoidal category over $\mathcal{C}$, and so does the category of right R-modules. One observes that a left P-module is equivalent to a P-algebra in the category of Σ_*-objects and a P-R-bimodule is equivalent to a P-algebra in the category of right R-modules, for any operads P, R in the base category $\mathcal{C}$.

Because of these observations, it is natural to assume that operads belong to a base category $\mathcal{C}$ and algebras run over any symmetric monoidal category $\mathcal{E}$ over $\mathcal{C}$. We review constructions of the theory of operads in this relative context. We study more specifically the functoriality of operadic constructions with respect to the underlying symmetric monoidal category. We can deduce properties of functors of the types $S(N) : \mathcal{E} \rightarrow {}_P\mathcal{E}$ and $S_R(N) : {}_R\mathcal{E} \rightarrow {}_P\mathcal{E}$ from this generalization of the theory of algebras over operads after we prove that the map $S_R : M \mapsto S_R(M)$ defines a functor of symmetric monoidal categories, like $S : M \mapsto S(M)$. For this reason, the book is essentially devoted to the study of the category of right R-modules and to the study of functors $S_R(M) : {}_R\mathcal{E} \rightarrow \mathcal{E}$ associated to right R-modules.

Historical Overview and Prospects

Modules over operads occur naturally once one aims to represent the structure underlying the cotriple construction of Beck [3] and May [47, §9]. As far as we know, a first instance of this idea occurs in Smirnov's papers [56, 57] where an operadic analogue of the cotriple construction is defined. This operadic cotriple construction is studied more thoroughly in Rezk's thesis [54] to define a homology theory for operads.

The operadic bar construction of Getzler-Jones [17] and the Koszul construction of Ginzburg-Kapranov [18] are other constructions of the homology theory of operads. In [14], we prove that the operadic cotriple construction, the operadic bar construction and the Koszul construction are associated to free resolutions in categories of modules over operads, like the bar construction of algebras.

Classical theorems involving modules over algebras can be generalized to the context of operads: in [33], Kapranov and Manin use functors of the form $S_R(N) : {}_R\mathcal{E} \rightarrow {}_P\mathcal{E}$ to define Morita equivalences for categories of algebras over operads.

Our personal interest in modules over operads arose from the Lie theory of formal groups over operads. In summary, we use Lie algebras in right modules over operads to represent functors of the form $S_R(G) : {}_R\mathcal{E} \rightarrow {}_L\mathcal{E}$, where L refers to the operad of Lie algebras. Formal groups over an operad R are functors on nilpotent objects of the category of R-algebras. For a nilpotent R-algebra A, the object $S_R(G, A)$ forms a nilpotent Lie algebra and the Campbell-Hausdorff formula provides this object with a natural group structure. Thus the map $A \mapsto S_R(G, A)$ gives rise to a functor from nilpotent R-algebras to groups.

The Lie theory asserts that all formal groups over operads arise this way (see [11, 12, 13]). The operadic proof of this result relies on a generalization of the classical structure theorems of Hopf algebras to Hopf algebras in right modules over operads. Historically, the classification of formal groups was first obtained by Lazard in [37] in the formalism of "analyzers", an early precursor of the notion of an operad.

Recently, Patras-Schocker [49], Livernet-Patras [39] and Livernet [38] have observed that Hopf algebras in Σ_*-objects occur naturally to understand the structure of certain classical combinatorial Hopf algebras.

Lie algebras in Σ_*-objects were introduced before in homotopy theory by Barratt (see [2], see also [19, 61]) in order to model structures arising from Milnor's decomposition

$$\Omega\Sigma(X_1 \vee X_2) \sim \bigvee_w w(X_1, X_2),$$

where w runs over a Hall basis of the free Lie algebra in 2-generators x_1, x_2 and $w(X_1, X_2)$ refers to a smash product of copies of X_1, X_2 (one per occurrence of the variables x_1, x_2 in w).

In sequels [15, 16] we use modules over operads to define multiplicative structures on the bar complex of algebras. Recall that the bar complex $B(C^*(X))$ of a cochain algebra $A = C^*(X)$ is chain equivalent to the cochain complex of ΩX, the loop space of X (under standard completeness assumptions on X). We obtain that this cochain complex $B(C^*(X))$ comes naturally equipped with the structure of an E_∞-algebra so that $B(C^*(X))$ is equivalent to $C^*(\Omega X)$ as an E_∞-algebra.

Recall that an E_∞-operad refers to an operad E equipped with a weak-equivalence $\mathsf{E} \xrightarrow{\sim} \mathsf{C}$, where C is the operad of associative and commutative algebras. An E_∞-algebra is an algebra over some E_∞-operad. Roughly an E_∞-operad parameterizes operations that make an algebra structure commutative up to a whole set of coherent homotopies.

In the differential graded context, the bar construction $B(A)$ is defined naturally for algebras A over Stasheff's chain operad, the operad defined by the chain complexes of Stasheff's associahedra. We use the letter K to refer to this operad. We observe that the bar construction is identified with the functor $B(A) = S_{\mathsf{K}}(B_{\mathsf{K}}, A)$ associated to a right K-module B_{K}. We can restrict the bar construction to any category of algebras associated to an operad R equipped with a morphism $\eta : \mathsf{K} \to \mathsf{R}$. The functor obtained by this restriction process is also associated to a right R-module B_{R} obtained by an extension of structures from B_{K}. Homotopy equivalent operads $\mathsf{R} \xrightarrow{\sim} \mathsf{S}$ have homotopy equivalent modules $B_{\mathsf{R}} \xrightarrow{\sim} B_{\mathsf{S}}$.

The bar module B_{C} of the commutative operad C has a commutative algebra structure that reflects the classical structure of the bar construction of commutative algebras. The existence of an equivalence $B_{\mathsf{E}} \xrightarrow{\sim} B_{\mathsf{C}}$, where E is any E_∞-operad, allows us to transport the multiplicative structures of the bar module B_C to B_E and hence to obtain a multiplicative structure on the bar complex of E_∞-algebras. Constructions and theorems of this book are motivated by this application.

Note that modules over operads are applied differently in [25] in the study of structures on the bar construction: according to this article, modules over operads model morphisms between bar complexes of chain algebras.

In [15], we only deal with multiplicative structures on modules over operads and with multiplicative structures on the bar construction, but the bar complex forms naturally a coassociative coalgebra. In a subsequent paper [16], we address coalgebras and bialgebras in right modules over operads in order to extend constructions of [15] to the coalgebra setting and to obtain a bialgebra structure on the bar complex.

For a cochain algebra, the comultiplicative structure of the bar complex $B(C^*(X))$ models the multiplicative structure of the loop space ΩX. Bialgebras in right modules over operads give rise to Lie algebras, like the classical ones. One should expect that Lie algebras arising from the bar module B_R are related to Barratt's twisted Lie algebras.

Contents

The sections, paragraphs, and statements marked by the sign '¶' can be skipped in the course of a normal reading. These marks ¶ refer to refinement outlines.

Part I. Categorical and Operadic Background

The purpose of the first part of the book is to clarify the background of our constructions. This part does not contain any original result, but only changes of presentation in our approach of operads. Roughly, we make use of functors of symmetric monoidal categories in standard operadic constructions.

§1. Symmetric Monoidal Categories for Operads

First of all, we give the definition of a symmetric monoidal category $\mathcal{E}$ over a base symmetric monoidal category $\mathcal{C}$. Formally, a symmetric monoidal category over $\mathcal{C}$ is an object under $\mathcal{C}$ in the 2-category formed by symmetric monoidal categories and symmetric monoidal functors. In §1 we give equivalent axioms for this structure in a form suitable for our applications. Besides, we inspect properties of functors and adjunctions between symmetric monoidal categories.

§2. Symmetric Objects and Functors

We survey categorical properties of the functor $S : M \mapsto S(M)$ from Σ_*-objects to functors $F : \mathcal{E} \to \mathcal{E}$. More specifically, we recall the definition of the tensor product of Σ_*-objects, the operation that gives to Σ_*-objects the structure of a symmetric monoidal category over $\mathcal{C}$, and the definition of the composition product of Σ_*-objects, an operation that reflects the composition of functors. For the sake of completeness, we also recall that the functor $S : M \mapsto S(M)$ has a right adjoint $\Gamma : G \mapsto \Gamma(G)$. In the case $\mathcal{E} = \mathcal{C} = \Bbbk \operatorname{Mod}$, the category of modules over a ring $\Bbbk$, we use this construction to prove that the functor $S : M \mapsto S(M)$ is bijective on morphism sets for projective Σ_*-objects or if the ground ring is an infinite field.

§3. Operads and Algebras in Symmetric Monoidal Categories

We recall the definition of an algebra over an operad P. We review the basic examples of the commutative, associative, and Lie operads, which are associated to commutative and associative algebras, associative algebras and Lie algebras respectively.

We assume that the operad P belongs to the base category $\mathcal{C}$ and we define the category ${}_P\mathcal{E}$ of P-algebras in a symmetric monoidal category $\mathcal{E}$ over $\mathcal{C}$. We observe that any functor $\rho : \mathcal{D} \to \mathcal{E}$ of symmetric monoidal categories over $\mathcal{C}$ induces a functor on the categories of P-algebras $\rho : {}_P\mathcal{D} \to {}_P\mathcal{E}$, for any operad P in the base category $\mathcal{C}$. We review the classical constructions of free objects, extension and restriction functors, colimits in categories of algebras over operads, and we check that these constructions are invariant under changes of symmetric monoidal categories. We review the classical definition of endomorphism operads with similar functoriality questions in mind.

At this point, we study the example of algebras in Σ_*-objects and the example of algebras in functors. We make the observation that a left module over an operad P is equivalent to a P-algebra in Σ_*-objects. We also observe that a functor $F : \mathcal{X} \to {}_P\mathcal{E}$, where $\mathcal{X}$ is any source category, is equivalent to a P-algebra in the category of functors of the form $F : \mathcal{X} \to \mathcal{E}$. We use that $S : M \mapsto S(M)$ defines a symmetric monoidal functor to obtain the correspondence between left P-modules N and functors $S(N) : \mathcal{E} \to {}_P\mathcal{E}$.

§4. Miscellaneous Structures Associated to Algebras over Operads

We recall the definition of miscellaneous structures associated to algebras over operads: enveloping operads, which model comma categories of algebras over operads; enveloping algebras, which are associative algebras formed from the structure of enveloping operads; representations, which are nothing but modules over enveloping algebras; and modules of Kähler differentials, which represent the module of derivations of an algebra over an operad. We study applications of these notions to the usual operads: commutative, associative, and Lie. For each example, the operadic definition of an enveloping algebra is equivalent to a standard construction of algebra, and similarly as regards representations and Kähler differentials.

Part II. The Category of Right Modules over Operads and Functors

In the second part of the book, we study categorical structures of right modules over operads and functors. Roughly, we prove that the categories of right modules over operads are equipped with structures that reflect natural operations at the functor level. This part contains overlaps with the literature (with [13] and [54, Chapter 2] in particular). Nevertheless, we prefer to give a comprehensive account of definitions and categorical constructions on right modules over operads.

§5. Definitions and Basic Constructions

First of all, we recall the definition of a right module over an operad R and of the associated functors $S_R(M) : A \mapsto S_R(M, A)$, where the variable A ranges over R-algebras in any symmetric monoidal category $\mathcal{E}$ over the base category $\mathcal{C}$.

In the book, we use the notation $\mathcal{M}$ for the category of Σ_*-objects in the base category $\mathcal{C}$ and the notation $\mathcal{F}$ for the category of functors $F : \mathcal{E} \to \mathcal{E}$, where $\mathcal{E}$ is any given symmetric monoidal category over $\mathcal{C}$. The map $S : M \mapsto S(M)$ defines a functor $S : \mathcal{M} \to \mathcal{F}$. In a similar way, we use the notation $\mathcal{M}_R$ for the category of right R-modules and the notation $\mathcal{F}_R$ for the category of functors $F : {}_R\mathcal{E} \to \mathcal{E}$. The map $S_R : M \mapsto S_R(M)$ defines a functor $S_R : \mathcal{M}_R \to \mathcal{F}_R$.

§6. Tensor Products

For any source category $\mathcal{A}$, the category of functors $F : \mathcal{A} \to \mathcal{E}$ has a natural tensor product inherited pointwise from the symmetric monoidal category $\mathcal{E}$. In the first part of this introduction, we mention that the category of Σ_*-objects $\mathcal{M}$ comes also equipped with a natural tensor product, as well as the category of right R-modules $\mathcal{M}$. The definition of the tensor product of Σ_*-objects is recalled in §2. The tensor product of right R-modules is derived from the tensor product of Σ_*-objects and this definition is recalled in §6.

Classical theorems assert that the functor $S : M \mapsto S(M)$ satisfies $S(M \otimes N) \simeq S(M) \otimes S(N)$. In §6, we check that the functor $S_R : M \mapsto S_R(M)$ satisfies similarly $S_R(M \otimes N) \simeq S_R(M) \otimes S_R(N)$. Formally, we prove that the map $S_R : M \mapsto S_R(M)$ defines a functor of symmetric monoidal categories over $\mathcal{C}$:

$$(\mathcal{M}_R, \otimes, 1) \xrightarrow{S_R} (\mathcal{F}_R, \otimes, 1).$$

§7. Universal Constructions on Right Modules over Operads

An operad morphism $\psi : R \to S$ gives rise to extension and restriction functors $\psi_! : {}_R\mathcal{E} \rightleftarrows {}_S\mathcal{E} : \psi^!$. The composition of functors $F : {}_R\mathcal{E} \to \mathcal{E}$ with the restriction functor $\psi^! : {}_S\mathcal{E} \to {}_R\mathcal{E}$ defines an extension functor on functor categories: $\psi_! : \mathcal{F}_R \to \mathcal{F}_S$. In the converse direction, the composition of functors $G : {}_S\mathcal{E} \to \mathcal{E}$ with the extension functor $\psi_! : {}_R\mathcal{E} \to {}_S\mathcal{E}$ defines a restriction functor $\psi^! : \mathcal{F}_S \to \mathcal{F}_R$. These extension and restriction functors $\psi_! : \mathcal{F}_R \rightleftarrows \mathcal{F}_S : \psi^!$ form a pair of adjoint functors.

At the level of module categories, we also have extension and restriction functors $\psi_! : \mathcal{M}_R \rightleftarrows \mathcal{M}_S : \psi^!$ that generalize the classical extension and restriction functors of modules over associative algebras. We prove that these operations on modules correspond to the extension and restriction of functors. Explicitly, we have natural functor isomorphisms $\psi_! S_R(M) \simeq S_S(\psi_! M)$ and $\psi^! S_R(M) \simeq S_S(\psi^! M)$. Besides, we check the coherence of these isomorphisms with respect to adjunction relations and tensor structures.

In the particular case of the unit morphism of an operad, we obtain that the composite of a functor $S_R(M) : {}_R\mathcal{E} \to \mathcal{E}$ with the free R-algebra functor is identified with the functor $S(M) : \mathcal{E} \to \mathcal{E}$ associated to the underlying Σ_*-object of M. In the converse direction, for a Σ_*-object L, we obtain that the composite of the functor $S(L) : \mathcal{E} \to \mathcal{E}$ with the forgetful functor $U : {}_R\mathcal{E} \to \mathcal{E}$ is identified with the functor associated to a free right R-module associated to L.

§8. Adjunction and Embedding Properties

The functor $S_R : \mathcal{M}_R \to \mathcal{F}_R$ has a right adjoint $\Gamma_R : \mathcal{F}_R \to \mathcal{M}_R$, like the functor $S : \mathcal{M} \to \mathcal{F}$. In the case $\mathcal{E} = \mathcal{C} = \Bbbk \, \mathrm{Mod}$, the category of modules over a ring $\Bbbk$, we use this result to prove that the functor $S_R : M \mapsto S_R(M)$ is bijective on morphism sets in the same situations as the functor $S : M \mapsto S(M)$ on Σ_*-objects.

§9. Algebras in Right Modules over Operads

We study the structure of algebras in a category of right modules over an operad. We observe at this point that a bimodule over operads P, R is equivalent to a P-algebra in right R-modules. We use that $S_R : M \mapsto S_R(M)$ defines a symmetric monoidal functor to obtain the correspondence between P-R-bimodules N and functors $S(N) : {}_R\mathcal{E} \to {}_P\mathcal{E}$, as in the case of left P-modules. We review applications of the general theory of §3 to this correspondence between P-algebra structures.

§10. Miscellaneous Examples

To give an illustration of our constructions, we prove that enveloping operads, enveloping algebras and modules of Kähler differentials, whose definitions are recalled in §4, are instances of functors associated to modules over operads. Besides we examine the structure of these modules for classical operads, namely the operad of associative algebras A, the operad of Lie algebras L, and the operad of associative and commutative algebras C.

New examples of functors associated to right modules over operads can be derived by using the categorical operations of §6, §7 and §9.

Part III. Homotopical Background

The purpose of the third part of the book is to survey applications of homotopical algebra to the theory of operads. We review carefully axioms of model categories to define (semi-)model structures on categories of algebras over operads. We study with more details the (semi-)model categories of algebras in differential graded modules.

This part does not contain any original result, like our exposition of the background of operads, but only changes of presentations. More specifically, we observe that crucial verifications in the homotopy theory of algebras over operads are consequences of general results about the homotopy of functors associated to modules over operads. In this sense, this part gives first motivations to set a homotopy theory for modules over operads, the purpose of the fourth part of the book.

§11. Symmetric Monoidal Model Categories For Operads

First of all, we review the axioms of model categories and the construction of new model structures by adjunction from a given model category. The model categories introduced in this book are defined by this adjunction process.

The notion of a symmetric monoidal model category gives the background for the homotopy theory of operads. We give axioms for a symmetric monoidal model category $\mathcal{E}$ over a base symmetric monoidal model category $\mathcal{C}$.

We prove that the category of Σ_*-objects inherits a model structure and forms a symmetric monoidal model category over the base category. Then we study the homotopy invariance of the functor $S(M) : \mathcal{E} \to \mathcal{E}$ associated to a Σ_*-object M. We prove more globally that the bifunctor $(M, X) \mapsto S(M, X)$ satisfies an analogue of the pushout-product axiom of tensor products in symmetric monoidal model categories.

§12. The Homotopy of Algebras over Operads

To have a good homotopy theory for algebras over an operad P, we have to make assumptions on the operad P. In general, it is only reasonable to assume that the unit morphism of the operad $\eta : I \to P$ defines a cofibration in the underlying category of Σ_*-objects – we say that the operad P is Σ_*-cofibrant. In the differential graded context, this assumption implies that each component of the operad $P(n)$ forms a chain complex of projective Σ_n-modules. For certain good symmetric monoidal categories, we can simply assume that the unit morphism of the operad $\eta : I \to P$ defines a cofibration in the base category $\mathcal{C}$ – we say that the operad P is $\mathcal{C}$-cofibrant. In the differential graded context, this assumption implies that the chain complex $P(n)$ consists of projective $\Bbbk$-modules, but we do not assume that $P(n)$ forms a chain complex of projective Σ_n-modules.

The category of P-algebras $_P\mathcal{E}$ inherits natural classes of weak-equivalences, cofibrations, fibrations, which are defined by using the adjunction between $_P\mathcal{E}$ and the underlying category $\mathcal{E}$. But none of our assumptions, though reasonable, is sufficient to imply the full axioms of model categories: the lifting and cofibration axioms hold in the category of P-algebras for morphisms with a cofibrant domain only. In this situation, it is usual to say that P-algebras forms a semi-model category.

In fact, the structure of a semi-model category is sufficient to carry out every usual construction of homotopical algebra. For our purpose, we review the definition of model structures by adjunction and the notion of a Quillen adjunction in the context of semi-model categories.

We prove the existence of semi-model structures on the category of P-algebras $_P\mathcal{E}$ for every symmetric monoidal model category $\mathcal{E}$ over the base category $\mathcal{C}$ when the operad P is Σ_*-cofibrant. Since we prove that Σ_*-objects form a symmetric monoidal model category over $\mathcal{C}$, we can apply our result to obtain a semi-model structure on the category of P-algebras in Σ_*-objects, equivalently, on the category of left P-modules. Recall that a Σ_*-object is connected if $M(0) = 0$. We observe that the semi-model structure of the category of P-algebras in connected Σ_*-objects is well defined as long as the operad P is $\mathcal{C}$-cofibrant.

§13. The (Co)homology of Algebras over Operads

The category of algebras over an operad P has a natural cohomology theory $H_P^*(A, E)$ defined as a derived functor from the functor of derivations

$\mathrm{Der_P}(A, E)$. There is also a natural homology theory $H_*^P(A, E)$ defined from the functor of Kähler differentials $\Omega_P^1(A)$. We review the definition of these derived functors. The operadic (co)homology agrees with the usual Hochschild (co)homology for associative algebras, with the Chevalley-Eilenberg (co)homology for Lie algebras, and with the Harrison (co)homology for commutative algebras.

The cohomology of a P-algebra A has coefficients in representations of A, equivalently in left modules over the enveloping algebra of A. We prove the existence of universal coefficient spectral sequences which determine the cohomology $H_P^*(A, E)$ from Ext-functors $\mathrm{Ext}_U^*(H_P^*(A, U), E)$, where U refers to the enveloping algebra of A. We have similar results for the homology $H_*^P(A, E)$.

We observe that the universal coefficients spectral sequences degenerate for the associative and Lie operads, but not for the commutative operad. We retrieve from this observation that the cohomology (respectively, homology) of associative algebras is determined by an Ext-functor (respectively, by a Tor-functor), and similarly as regards the cohomology of Lie algebras.

An alternative construction of a (co)homology theory for algebras over an operad P comes from the cotriple construction of Beck. We recall the definition of the cotriple complex and we prove that the cotriple construction gives the same result as the abstract definition of the (co)homology by a derived functor.

Part IV. The Homotopy of Modules over Operads and Functors

In the fourth part of the book, we study the homotopical properties of the functor $\mathrm{S_R}(M) : A \mapsto \mathrm{S_R}(M, A)$ associated to a right R-module M, and more globally of the bifunctor $(M, A) \mapsto \mathrm{S_R}(M, A)$ on right R-modules M and R-algebras A. We aim to prove that modules over operads give a good homotopy model of functors.

§14. The Model Category of Right Modules over an Operad

First of all, we check that the category of right modules over an operad R inherits the structure of a cofibrantly generated symmetric monoidal model category from the base category $\mathcal{C}$, like the category of Σ_*-objects, as long as the operad R is $\mathcal{C}$-cofibrant.

According to §12, the category of P-algebras in right R-modules (equivalently, on the category of P-R-modules) inherits a semi-model structure provided that the operad P is Σ_*-cofibrant. If we restrict ourself to connected right R-modules, then this model structure is well defined as long as the operad P is $\mathcal{C}$-cofibrant.

§15. Modules and Homotopy Invariance of Functors

We study the homotopy invariance of the bifunctor $(M, A) \mapsto S_R(M, A)$. We prove that a weak-equivalence of right R-modules $f : M \xrightarrow{\sim} N$ induces a weak-equivalence $S_R(f, A) : S_R(M, A) \xrightarrow{\sim} S_R(N, A)$ provided that:

(1) the right R-module M, N are cofibrant as right R-modules and the R-algebra A defines a cofibrant object in the underlying category $\mathcal{E}$,

(2) or the operad R is Σ_*-cofibrant, the right R-module M, N are cofibrant as Σ_*-objects and the R-algebra A is cofibrant as an R-algebra.

We prove symmetrically that a weak-equivalence of R-algebras $f : A \xrightarrow{\sim} B$ induces a weak-equivalence $S_R(M, f) : S_R(M, A) \xrightarrow{\sim} S_R(M, B)$ provided that:

(1) the right R-module M is cofibrant as a right R-module and the R-algebras A, B define cofibrant objects in the underlying category $\mathcal{E}$,

(2) or the operad R is Σ_*-cofibrant, the right R-module M is cofibrant as a Σ_*-object and the R-algebras A, B are cofibrant as R-algebras.

Some instances of these assertions already occur in the literature, but the unifying statement is completely new. Assertions (1) are proved by standard homotopical arguments, but the technical verification of assertions (2), postponed to an appendix, takes a whole part of the book.

The assumption about the operad in assertions (2) ensures that the R-algebras form a semi-model category. But we note that, for certain good categories $\mathcal{E}$, the semi-model structure of R-algebras is well defined as long as the operad R is $\mathcal{C}$-cofibrant. In this situation, we can simply assume that the right R-modules are cofibrant as collections of objects $M(n) \in \mathcal{C}$ to have the homotopy invariance properties of assertions (2).

In §5, we observe that the relative composition product $M \circ_R N$ (a natural operation between a right R-module M and a left R-module N) is a particular instance of a construction of the form $S_R(M, N)$, where we identify left R-modules with R-algebras in Σ_*-objects. Accordingly, we can use our theorems to study the homotopy invariance of the relative composition product $M \circ_R N$. In §15.3, we study finer homotopy invariance properties of relative composition products $M \circ_R N$ which occur for a connected left R-module N.

§16. Extension and Restriction Functors and Model Structures

Let $\psi : R \to S$ an operad morphism. If the operads R, S are $\mathcal{C}$-cofibrant, then the extension and restriction functors $\psi_! : \mathcal{M}_R \rightleftarrows \mathcal{M}_S : \psi^!$ define a Quillen adjunction for modules. In the context of algebras over operads, we assume that the operads R, S are Σ_*-cofibrant or that the monoidal category $\mathcal{E}$ is good enough to ensure that R-algebras and S-algebras form model categories. In this situation, we obtain that the extension and restriction functors $\psi_! : {}_R\mathcal{E} \rightleftarrows {}_S\mathcal{E} : \psi^!$ define Quillen adjoint functors too.

In both contexts (modules and algebras), the extension and restriction functors define Quillen adjoint equivalences when the morphism ψ is an operad equivalence. In §16, we give a new general proof of this assertion, which has already been verified in various situations: in the context of simplicial sets and simplicial modules, [54, §3.6]; for certain operads in differential graded modules, [26]; in the context of spectra, [21, Theorem 1.2.4] and [23]; under the assumption that the underlying model category is left proper, [4].

For algebras, the crux is to check that the adjunction unit $\eta A : A \to \psi^! \psi_! A$ defines a weak-equivalence for every cofibrant R-algebra A. In §7 and §9, we observe that the functors $\psi_! : {}_R\mathcal{E} \rightleftarrows {}_S\mathcal{E} : \psi^!$ are associated to right modules over operads, as well as the functors $\psi_! : \mathcal{M}_R \rightleftarrows \mathcal{M}_S : \psi^!$. From this observation, the proof that $\eta A : A \to \psi^! \psi_! A$ defines a weak-equivalence occurs as an application of the results of §15 on the homotopy invariance of functors associated to modules over operads.

§17. Miscellaneous Applications

To conclude the book, we survey some applications of the homotopy theory of right modules of operads to the homotopy of algebras over operads. More specifically, we use results of §15 to study the homotopy invariance of usual operadic constructions, like enveloping operads, enveloping algebras and Kähler differentials, and to revisit the definition of the (co)homology of algebras over an operad. In particular, we prove in §17 that the (co)homology of algebras over an operad agrees with the cohomology of the cotriple construction.

In §13, we define the (co)homology of algebras over an operad R by using derived functors of Kähler differentials $\Omega_R^1(A)$. In §10, we prove that $\Omega_R^1(A)$ is the functor associated to a right R-module Ω_R^1. In §17, we use theorems of §15 to prove that the homology of R-algebras is defined by a Tor-functor (and the cohomology by an Ext-functor) if Ω_R^1 forms a free right R-module (when R is an operad in k-modules). As an example, we retrieve that the classical Hochschild homology of associative algebras, as well as the classical Chevalley-Eilenberg homology of Lie algebras, are given by Tor-functors (respectively, Ext-functors for the cohomology), unlike Harrison or André-Quillen homology of commutative algebras.

Part V. Appendix: Technical Verifications

The purpose of the appendix is to achieve technical verifications of §12 and §15. Namely we prove that the bifunctor $(M, A) \mapsto S_R(M, A)$ associated to an operad R satisfies an analogue of the pushout-product axiom of tensor products in symmetric monoidal model categories.

Part I
Categorical and Operadic Background

Foreword: Categorical Conventions

0.1 Functor Categories. In this book, we deal with categories of functors $F : \mathcal{A} \to \mathcal{X}$. Generally, the category $\mathcal{A}$ is not supposed to be small, but to avoid set-theoretic difficulties we assume tacitely that any category $\mathcal{A}$ considered in the book contains a small subcategory $\mathcal{A}_f$ such that every object $X \in \mathcal{A}$ is the filtered colimit of a diagram of $\mathcal{A}_f$*. Moreover, we consider tacitely only functors $F : \mathcal{A} \to \mathcal{X}$ that preserve filtered colimits and the notation $\mathcal{F}(\mathcal{A}, \mathcal{X})$ refers to the category formed by these functors. All functors which arise from our constructions satisfy this assumption. Under this convention, we obtain that the category $\mathcal{F}(\mathcal{A}, \mathcal{X})$ has small morphism sets and no actual set-theoretic difficulty occurs.

Besides, the existence of the small category $\mathcal{A}_f$ implies that a functor $\phi_! : \mathcal{A} \to \mathcal{X}$ admits a right adjoint $\phi^* : \mathcal{X} \to \mathcal{A}$ if and only if it preserves colimits, because the set-theoretic condition of the adjoint functor theorem is automatically fulfilled.

0.2 Notation for Colimits. Categories occur at two levels in our constructions: we use ground categories, which are usually symmetric monoidal categories, and categories of algebras over operads, which lie over an underlying ground category. To distinguish the role of these categories, we use two system of conventions to represent colimits: we adopt additive notation (0 for the initial object and $\oplus$ for the coproduct) for ground categories, and the base-set notation $\vee$ for the coproduct in categories of algebras over operads. Nevertheless, we return to base-set (or set) notation in particular instances of ground categories (sets, simplicial sets, topological spaces, ...) for which this convention is usual.

The base-set notation is applied to the category of operads in a base symmetric monoidal category (see §3.1.1). The base-set notation is also used for categories to which no role is assigned.

Note that ground categories are not assumed to be additive in general. The initial object 0 is not supposed to be a zero object, and the coproduct $\oplus$ is not supposed to be a bi-product.

---

* We only make an exception for the category of topological spaces.

0.3 Symmetric Monoidal Categories and Enriched Categories. The structure of a symmetric monoidal category gives the categorical background of the theory of operads. The definition of this notion is reviewed in the next chapter. For the moment, recall briefly that a symmetric monoidal category consists of a category $\mathcal{C}$ equipped with a tensor product $\otimes : \mathcal{C} \times \mathcal{C} \to \mathcal{C}$ and a unit object $1 \in \mathcal{C}$ which satisfies the usual unit, associativity and symmetry relations of the tensor product of modules over a ring.

The notation $\mathrm{Mor}_{\mathcal{E}}(X, Y)$ is used throughout the book to refer to the morphism-sets of any category $\mathcal{E}$. But many categories are assumed to be enriched over a base symmetric monoidal category and come equipped with a hom-bifunctor with value in $\mathcal{C}$, if $\mathcal{C}$ refers to the base category (see §1.1.12). The hom-objects of every enriched category $\mathcal{E}$ are denoted by $\mathrm{Hom}_{\mathcal{E}}(X, Y)$, to be distinguished from the morphism sets $\mathrm{Mor}_{\mathcal{E}}(X, Y)$.

0.4 Point-Set Symmetric Monoidal Categories. We illustrate our constructions by applications in categories of modules over a commutative ground ring, in categories of differential graded modules (dg-modules for short), in categories of Σ_*-objects (also called symmetric objects in English words), or in categories of right modules over an operad. In these instances of symmetric monoidal categories, the tensor product $X \otimes Y$ is spanned in a natural sense by certain tensors $x \otimes y$, where $(x, y) \in X \times Y$, and any morphism $\phi : X \otimes Y \to Z$ is equivalent to a kind of multilinear map $\phi : x \otimes y \mapsto \phi(x, y)$ on the set of generating tensors. The tensor product $\theta : (x, y) \mapsto x \otimes y$ represents itself a universal multilinear map $\theta : X \times Y \to X \otimes Y$. The representation of morphisms $\phi : X \otimes Y \to Z$ by actual multilinear maps can be extended to homomorphisms, elements of internal hom-objects of these categories. This pointwise representation, usual for modules over a ring, is formalized in §1.1.5 in the context of dg-modules, in §2.1.9 in the context of Σ_*-objects, and in §6.1.3 in the context of right modules over an operad.

In illustrations, we apply the pointwise representation of tensors to the categories, derived from the category of modules, which are mentioned in this paragraph. But, of course, pointwise representations of tensors hold in the category of sets, whose tensor product is defined by the cartesian product, and more generally in any point-set category derived from the cartesian category of sets, like the category of topological spaces or the category simplicial sets.

In the sequel, we speak abusively of a point-set context to refer to a symmetric monoidal category in which a pointwise representation of tensors holds.

0.5 The Principle of Generalized Point-Tensors. The pointwise representation, which makes many definitions more basic, is used in applications. To simplify, we may make explicit the example of modules over a ring only and we may omit the case of dg-modules (respectively, Σ_*-objects, right modules over operads) in illustrations. Usually, the generalization from modules to dg-modules (respectively, Σ_*-objects, right modules over operads) can be carried out automatically within the pointwise representation, without calling back the category formalism.

The only rule is to keep track of tensor permutations in mappings to replace a tensor $x \otimes y$ by a decorated tensor $\sigma \cdot x \otimes y$, where the decoration σ consists of a sign in the context of dg-modules (see §1.1.5), a block permutation in the context of Σ_*-objects and modules over operads (see §2.1.9), a sign combined with a block permutation in the context of symmetric objects in dg-modules ... The rule arises from the definition of the symmetry isomorphism $\tau : X \otimes Y \xrightarrow{\simeq} X \otimes Y$ in these symmetric monoidal categories.

To refer to these rules, we say that we apply the principle of generalized point-tensors.

Chapter 1
Symmetric Monoidal Categories for Operads

Introduction

The notion of a symmetric monoidal category is used to give a general background for the theory of operads. Standard examples of symmetric monoidal categories include categories of modules over a commutative ring, categories of differential graded modules, various categories of coalgebras, the category of sets together with the cartesian product, the category of simplicial sets, and the category of topological spaces. Other possible examples include the modern categories of spectra used to model stable homotopy, but in applications operads come often from the category of topological spaces or simplicial sets and categories of spectra are not used as the base category in our sense (see next).

The first purpose of this chapter is to survey definitions of symmetric monoidal categories.

To set our constructions, we fix a base symmetric monoidal category, usually denoted by $\mathcal{C}$, in which all small colimits, all small limits exist, and we assume that the tensor product of $\mathcal{C}$ preserves all colimits (see §1.1.1). In the sequel, operads are usually defined within this base category $\mathcal{C}$. But, in our constructions, we use naturally algebras over operads in extensions of the base category to which the operad belongs. For this reason, we review definitions of symmetric monoidal categories to have an appropriate axiomatic background for our applications. The axiomatic structure, needed to generalize the definition of an algebra over an operad, consists of a symmetric monoidal category over the base category $\mathcal{C}$, an object under $\mathcal{C}$ in the 2-category of symmetric monoidal categories (see §1.1.2).

Definitions of symmetric monoidal categories are reviewed in §1.1 with this aim in mind. The structure of a symmetric monoidal category over the base category is made explicit in that section.

In §1.2, we recall the definition and properties of particular colimits (namely, reflexive coequalizers and filtered colimits). We use repeatedly that these colimits are preserved by tensor powers in symmetric monoidal categories.

B. Fresse, *Modules over Operads and Functors*, Lecture Notes in Mathematics 1967, 21
DOI: 10.1007/978-3-540-89056-0_1, © Springer-Verlag Berlin Heidelberg 2009

The categories of spectra form natural examples of symmetric monoidal categories over the category of simplicial sets. In applications of the theory of operads, categories of spectra are often used as such and the actual base category in our sense is formed by the category of simplicial sets. The example of spectra is only examined in remarks of parts III-IV of the book.

1.1 Symmetric Monoidal Categories over a Base

In this section, we recall briefly the definition of a symmetric monoidal category, essentially to fix axiomatic requirements on base symmetric monoidal categories. Then we study the 2-category of symmetric monoidal categories and we give an axiomatic definition of the structure of a symmetric monoidal category over a base category. Besides we survey basic examples of symmetric monoidal categories, used throughout the book in examples of applications.

1.1.1 The Base Symmetric Monoidal Category. Roughly (we refer to [44, Chapter 11] for a detailed definition), a symmetric monoidal category consists of a category $\mathcal{C}$ equipped with a tensor product $\otimes : \mathcal{C} \times \mathcal{C} \to \mathcal{C}$ and a unit object $1 \in \mathcal{C}$ such that we have unit relations

$$1 \otimes C \simeq C \simeq C \otimes 1, \quad \forall C \in \mathcal{C},$$

an associativity relation

$$(A \otimes B) \otimes C \simeq A \otimes (B \otimes C), \quad \forall A, B, C \in \mathcal{C},$$

and a symmetry isomorphism

$$\tau(C, D) : C \otimes D \xrightarrow{\simeq} D \otimes C, \quad \forall C, D \in \mathcal{C}.$$

The natural isomorphisms that give these relations are part of the structure and, whenever relations are patched together, all composites of relation isomorphisms are supposed to return the same result. According to [44, Chapter 11], the verification of this coherence assumption reduces to the commutativity of a 4-term associativity pentagon, three 2-term unit triangles, and a 3-term symmetry hexagon.

Abusively, we omit to specify the unit, associativity, and symmetry isomorphisms in the notation of a symmetric monoidal category. Moreover, we use the notation $(\mathcal{C}, \otimes, 1)$ only to insist that we consider the symmetric monoidal structure of the category $\mathcal{C}$. Usually, we specify a symmetric monoidal category by the underlying category $\mathcal{C}$, assuming that $\mathcal{C}$ has a natural internal symmetric monoidal structure.

Throughout this book, we use the letter $\mathcal{C}$ to refer to a base symmetric monoidal category, fixed once and for all. Usual examples are surveyed next.

All small colimits, all small limits are supposed to exist in the base category $\mathcal{C}$ and the tensor product $\otimes : \mathcal{C} \times \mathcal{C} \to \mathcal{C}$ is assumed to preserve colimits in each variable separately. Explicitly, the natural morphism

$$\operatorname*{colim}_{i}(C_i \otimes D) \to (\operatorname*{colim}_{i} C_i) \otimes D,$$

is an isomorphism for all diagrams C_i, $i \in I$, and every fixed object $D \in \mathcal{C}$, and similarly for colimits on the right-hand side.

1.1.2 The Axioms of a Symmetric Monoidal Category over a Base.

In brief, a *symmetric monoidal category over* $\mathcal{C}$ consists of an object under $\mathcal{C}$ in the 2-category of symmetric monoidal categories. For our needs, we give a more explicit definition of the structure of a symmetric monoidal category over a base category and we prove afterwards that the explicit definition agrees with the abstract categorical definition.

For us, a symmetric monoidal category over $\mathcal{C}$ consists of a symmetric monoidal category $(\mathcal{E}, \otimes, 1)$ equipped with an external tensor product $\otimes : \mathcal{C} \times \mathcal{E} \to \mathcal{E}$ such that we have a unit relation

$$1 \otimes X \simeq X, \quad \forall X \in \mathcal{E},$$

an associativity relation

$$(C \otimes D) \otimes X \simeq C \otimes (D \otimes X), \quad \forall C, D \in \mathcal{C}, \forall X \in \mathcal{E},$$

and such that we have a distribution relation

$$C \otimes (X \otimes Y) \simeq (C \otimes X) \otimes Y \simeq X \otimes (C \otimes Y), \quad \forall C \in \mathcal{C}, \forall X, Y \in \mathcal{E},$$

between the external tensor product and the internal tensor product of $\mathcal{E}$. Again, the isomorphisms that give these relations are part of the structure and all composites of relation isomorphisms are supposed to return the same result whenever relations are patched together. The verification of this coherence assumption reduces to the commutativity of the associativity pentagons, unit triangles and symmetry hexagons of symmetric monoidal categories, but where internal and external tensor products are patched together in all possible ways.

Recall that a category $\mathcal{E}$ is *tensored over a base symmetric monoidal category* $\mathcal{C}$ if it is equipped with an external tensor product $\otimes : \mathcal{C} \times \mathcal{E} \to \mathcal{E}$ that satisfies the unit and associativity relations. The structure of a symmetric monoidal category over $\mathcal{C}$ put together an internal symmetric monoidal structure with the structure of a tensored category.

Again, we assume that all small colimits, all small limits exist in the category $\mathcal{E}$. To simplify (see next remark), we also assume that the internal tensor product of $\mathcal{E}$ preserves all colimits in each variable, and similarly as regards the external tensor product $\otimes : \mathcal{C} \times \mathcal{E} \to \mathcal{E}$.

The base category forms clearly a symmetric monoidal category over itself.

1.1.3 Remark. In constructions of parts I-II, (for instance, in the definition of the category of algebras in $\mathcal{E}$ associated to an operad in $\mathcal{C}$), it is only necessary to assume that:

- The internal tensor product of $\mathcal{E}$ preserves initial objects, filtered colimits and reflexive coequalizers on both sides (see §1.2 for recollections on these particular colimits);
- The external tensor product $\otimes : \mathcal{C} \otimes \mathcal{E} \to \mathcal{E}$ preserves all colimits on the left-hand side, but only initial objects, filtered colimits and reflexive coequalizers on the right-hand side.

The category of connected coalgebras used in [16] gives a motivating example of a symmetric monoidal category over a base for which these weakened axioms have to be taken.

In the next paragraphs, we recap our basic examples of base symmetric monoidal categories and we give simple examples of symmetric monoidal categories over the base. Usual tensor products are derived from two primitive constructions: the tensor product of modules over a commutative ground ring and the cartesian products of sets. Our exposition follows this subdivision.

1.1.4 Basic Examples: Symmetric Monoidal Categories of Modules and Differential Graded Modules. Our first instance of a base symmetric monoidal category is the category $\mathcal{C} = \Bbbk\,\mathrm{Mod}$ of modules over a fixed commutative ground ring $\Bbbk$ together with the tensor product over the ground ring $\otimes = \otimes_\Bbbk$. The ground ring is usually omitted in this notation and the tensor product of $\mathcal{C} = \Bbbk\,\mathrm{Mod}$ is denoted by $\otimes$, like every tensor product of symmetric monoidal category.

The usual categories of differential graded modules (non-negatively graded, lower graded, ...) form other classical instances of symmetric monoidal categories. In the sequel, we use the notation $\mathrm{dg}\,\Bbbk\,\mathrm{Mod}$ for the category of differential lower $\mathbb{Z}$-graded $\Bbbk$-modules, whose objects (called dg-modules for short) consists of a $\Bbbk$-module C equipped with a grading $C = \oplus_{*\in\mathbb{Z}} C_*$ and with an internal differential, usually denoted by $\delta : C \to C$, that decreases degrees by 1.

The tensor product of dg-modules $C, D \in \mathrm{dg}\,\Bbbk\,\mathrm{Mod}$ is the dg-module $C \otimes D$ defined by the usual formula

$$(C \otimes D)_n = \bigoplus_{p+q=n} C_p \otimes D_q$$

together with the differential such that $\delta(x \otimes y) = \delta(x) \otimes y + \pm x \otimes \delta(y)$, where $\pm = (-1)^p$ for homogeneous elements $x \in C_p$, $y \in C_q$. The tensor product of dg-modules is obviously associative, has a unit given by the ground ring $\Bbbk$ (put in degree 0 to form a dg-module), and has a symmetry isomorphism $\tau(C, D) : C \otimes D \xrightarrow{\simeq} D \otimes C$ defined by the usual formula $\tau(C, D)(x \otimes y) = \pm y \otimes x$, where $\pm = (-1)^{pq}$ for homogeneous elements $x \in C_p$, $y \in C_q$. Hence we obtain that the category of dg-modules comes equipped with the structure of a symmetric monoidal category.

The symmetric monoidal category of dg-modules forms obviously a symmetric monoidal category over $\Bbbk$-modules.

The symmetry isomorphism of dg-modules follows the standard sign convention of differential graded algebra, according to which a permutation of homogeneous objects of degree p and q produces a sign $\pm = (-1)^{pq}$. In the sequel, we do not make explicit the signs which are determined by this rule.

1.1.5 The Pointwise Representation of dg-Tensors. The pointwise representation of tensors of §§0.4-0.5 can be applied to the category of dg-modules. In this representation, the tensor product $C \otimes D$ is considered as a dg-module spanned by tensor products $x \otimes y$ of homogeneous elements $(x, y) \in C_p \times D_q$. To justify this rule, observe that a morphism of dg-modules $\phi : C \otimes D \to E$ is equivalent to a collection of homogeneous maps $\phi : C_p \times D_q \to E_n$, which are multilinear in the usual sense of linear algebra, together with a commutation relation with respect to differentials. Hence a morphism $\phi : C \otimes D \to E$ is well determined by a mapping $x \otimes y \mapsto \phi(x, y)$ on the set of generating tensors $x \otimes y$, where $(x, y) \in C_p \times D_q$.

The definition of the symmetry isomorphism $\tau(C, D) : X \otimes Y \xrightarrow{\simeq} Y \otimes X$ gives the commutation rule $x \otimes y \mapsto \pm y \otimes x$, with a sign added.

1.1.6 Remark: Symmetric Monoidal Categories of Graded Objects. Let $\mathrm{gr}\,\Bbbk\,\mathrm{Mod}$ denote the category of graded $\Bbbk$-modules, whose objects consists simply of $\Bbbk$-modules $C \in \Bbbk\,\mathrm{Mod}$ equipped with a splitting $C = \bigoplus_{* \in \mathbb{Z}} C_*$. Any object in that category can be identified with a dg-module equipped with a trivial differential. The category $\mathrm{gr}\,\Bbbk\,\mathrm{Mod}$ inherits a symmetric monoidal structure from this identification, since a tensor product of dg-modules equipped with a trivial differential has still a trivial differential. The symmetry isomorphism of this symmetric monoidal structure includes the sign of differential graded algebra.

In the context of graded modules, we can drop this sign to obtain another natural symmetry isomorphism $\tau(C, D) : C \otimes D \xrightarrow{\simeq} D \otimes C$. This convention gives another symmetric monoidal structure on graded modules, which is not equivalent to the symmetric monoidal structure of differential graded algebra.

These symmetric monoidal categories of graded modules form both symmetric monoidal categories over $\Bbbk$-modules.

1.1.7 Basic Examples: Symmetric Monoidal Structures Based on the Cartesian Product. The category of sets Set together with the cartesian product $\times$ forms an obvious instance of a symmetric monoidal category. The cartesian product is also used to define a symmetric monoidal structure in the usual category of topological spaces Top, and in the usual category of simplicial sets $\mathcal{S}$. The one-point set, which gives the unit of these symmetric monoidal categories, is represented by the notation $*$.

Suppose C is any category with colimits. For a set $K \in$ Set and an object $C \in C$, we have a tensor product $K \otimes C \in C$ formed by the coproduct over the set K of copies of the object C :

$$K \otimes C := \bigoplus_{k \in K} C.$$

This construction provides any category with colimits with a tensor product over the category of sets. If we assume further that C is a symmetric monoidal category so that the tensor product $\otimes : C \times C \to C$ preserves colimits, then our external tensor product $\otimes :$ Set $\times C \to C$ satisfies the distribution relation of §1.1.2. Hence every symmetric monoidal category C whose tensor product $\otimes : C \times C \to C$ preserves colimits forms naturally a symmetric monoidal category over the category of sets.

Recall that a simplicial object in a category C consists of a collection of objects $X_n \in \mathcal{E}$, $n \in \mathbb{N}$, equipped with faces $d_i : X_n \to X_{n-1}$, $i = 0, \ldots, n$, and degeneracies $s_j : X_n \to X_{n+1}$, $j = 0, \ldots, n$, that satisfy the universal relations of faces and degeneracies of simplicial sets. The category of simplicial objects in C is denoted by C_Δ.

If C is a symmetric monoidal category, then C_Δ inherits a tensor product and forms still a symmetric monoidal category. The construction is standard: the tensor product of simplicial objects $C, D \in C_\Delta$ is defined dimensionwise by $(C \otimes D)_n = C_n \otimes D_n$ and comes equipped with the diagonal action of faces and degeneracies. If $\mathcal{E}$ is a symmetric monoidal category over a base symmetric monoidal category C, then the category of simplicial objects $\mathcal{E}_\Delta$ forms similarly a symmetric monoidal category over C_Δ.

If we apply this general construction to the base category of sets and to any symmetric monoidal category C such that the tensor product $\otimes : C \times C \to C$ preserves colimits, then we obtain that the category of simplicial objects C_Δ forms a symmetric monoidal category over simplicial sets $\mathcal{S}$. Next, in §1.1.11, we shall observe that the category of topological spaces Top forms another instance of a symmetric monoidal category over the category of simplicial sets $\mathcal{S}$.

The categories of spectra, alluded to in the introduction of this chapter, form other instances of symmetric monoidal categories over simplicial sets, but we defer applications of this idea to brief remarks in parts III-IV of the book.

1.1.8 Functors and Natural Transformations. A functor $\rho : \mathcal{D} \to \mathcal{E}$ is a *functor of symmetric monoidal categories over* C if we have a relation

$$\rho(X \otimes Y) \simeq \rho(X) \otimes \rho(Y)$$

for the internal tensor product of objects $X, Y \in \mathcal{D}$, a relation

$$\rho(C \otimes Y) \simeq C \otimes \rho(Y),$$

for the external tensor product of $Y \in \mathcal{D}$ with $C \in \mathcal{C}$, and an isomorphism

$$\rho(1) \simeq 1$$

for the unit object $1 \in \mathcal{E}$. Again, the isomorphisms that give these relations are part of the structure and are supposed to satisfy coherence relations with respect to the isomorphisms that give the internal relations of symmetric monoidal categories.

A natural transformation $\theta(X) : \rho(X) \to \sigma(X)$, where $\rho, \sigma : \mathcal{D} \to \mathcal{E}$ are functors of symmetric monoidal categories over $\mathcal{C}$, is a *natural transformation of symmetric monoidal categories over* $\mathcal{C}$ if we have $\theta(1) = \text{id}$, for unit objects, $\theta(X \otimes Y) = \theta(X) \otimes \theta(Y)$, for all $X, Y \in \mathcal{D}$, and $\theta(C \otimes Y) = C \otimes \theta(Y)$, for all $C \in \mathcal{C}$, $Y \in \mathcal{D}$. In these equations, we identify abusively the relations of symmetric monoidal functors with identities. The relations are defined properly by commutative diagrams obtained by patching together the natural transformations with the isomorphisms that give these relations.

Adjoint functors $\rho_! : \mathcal{D} \rightleftarrows \mathcal{E} : \rho^*$ define an *adjunction of symmetric monoidal categories over* $\mathcal{C}$ if $\rho_!$ and ρ^* are functors of symmetric monoidal categories over $\mathcal{C}$, and the adjunction unit $\eta X : X \to \rho^* \rho_! X$, as well as the adjunction augmentation $\epsilon Y : \rho_! \rho^* Y \to Y$, are natural transformations of symmetric monoidal categories over $\mathcal{C}$.

The next assertion is straightforward:

1.1.9 Proposition. *For any symmetric monoidal category $\mathcal{E}$ over $\mathcal{C}$, we have a symmetric monoidal functor $\eta : (\mathcal{C}, \otimes, 1) \to (\mathcal{E}, \otimes, 1)$ defined by $\eta(C) = C \otimes 1$, where $C \otimes 1$ refers to the external tensor product of $C \in \mathcal{C}$ with the unit object of $\mathcal{E}$.*

Conversely, any symmetric monoidal category $(\mathcal{E}, \otimes, 1)$ equipped with a symmetric monoidal functor $\eta : (\mathcal{C}, \otimes, 1) \to (\mathcal{E}, \otimes, 1)$ forms a symmetric monoidal category over $\mathcal{C}$ so that $C \otimes X = \eta(C) \otimes X$, for all $C \in \mathcal{C}$, $X \in \mathcal{E}$. In addition, the functor $\eta : \mathcal{C} \to \mathcal{E}$ preserves colimits if and only if the associated external tensor product $C \otimes X = \eta(C) \otimes X$ preserves colimits on the left.

These constructions define inverse equivalences of 2-categories between the category of symmetric monoidal categories over $\mathcal{C}$ and the comma category of objects under $(\mathcal{C}, \otimes, 1)$ in the 2-category of symmetric monoidal categories.

$\square$

The functor $\eta : \mathcal{C} \to \mathcal{E}$ is usually omitted in the notation of the object $\eta(C) \in \mathcal{E}$ associated to $C \in \mathcal{C}$. Note however that the identification $\eta(C) = C$ may be excessively abusive, though the functor $\eta : \mathcal{C} \to \mathcal{E}$ is canonically determined by the structure of $\mathcal{E}$, because $\eta : \mathcal{C} \to \mathcal{E}$ is not supposed to be faithful.

Proposition 1.1.9 makes clear the following transitivity relation:

1.1.10 Observation. *Let $\mathcal{D}$ be a symmetric monoidal category over the base category $\mathcal{C}$. Suppose $\mathcal{E}$ is a symmetric monoidal category over $\mathcal{D}$. Then $\mathcal{E}$ forms itself a symmetric monoidal category over the base category $\mathcal{C}$.*

1.1.11 Examples.

(a) In §1.1.7, we observe that any symmetric monoidal category $\mathcal{C}$ forms naturally a symmetric monoidal category over the cartesian category of sets provided that the tensor product of $\mathcal{C}$ preserves colimits. The canonical functor $\eta : \mathrm{Set} \to \mathcal{C}$ determined by this structure is given by the explicit formula $\eta(K) = K \otimes 1 = \bigoplus_{k \in K} 1$. Throughout the book, we also adopt the notation $1[K] = K \otimes 1$ for this construction.

(b) The standard geometric realization of simplicial sets defines a functor of symmetric monoidal categories $|-| : (\mathcal{S}, \times, *) \to (\mathrm{Top}, \times, *)$ since the classical Eilenberg-Zilber subdivision of prisms gives rise to a natural homeomorphism

$$\nabla : |K| \times |L| \xrightarrow{\simeq} |K \times L|$$

which satisfies the coherence relations alluded to in §1.1.8. Thus the category of topological spaces forms a symmetric monoidal category over simplicial sets.

(c) The normalized chain complex of simplicial sets defines a functor $N_*(-) : \mathcal{S} \to \mathrm{dg}\,\Bbbk\,\mathrm{Mod}$, but in this context we only have a weak-equivalence

$$\nabla : N_*(K) \times N_*(L) \xrightarrow{\sim} N_*(K \times L),$$

for $K, L \in \mathcal{S}$, and the category of dg-modules does not form a symmetric monoidal category over simplicial sets in our sense.

1.1.12 Enriched Symmetric Monoidal Categories over the Base Category. Throughout this book, we assume that all categories $\mathcal{E}$ which are tensored over the base category $\mathcal{C}$ satisfy the assumption that the external tensor product $\otimes : \mathcal{C} \times \mathcal{E} \to \mathcal{E}$ preserves colimits on the left-hand side, as asserted in §1.1.2 for symmetric monoidal categories over $\mathcal{C}$. According to our convention (see §0.1 in the foreword), this assumption implies the existence of an external hom-bifunctor

$$Hom_\mathcal{E}(-,-) : \mathcal{E}^{op} \times \mathcal{E} \to \mathcal{C}$$

such that
$$\mathrm{Mor}_\mathcal{E}(C \otimes X, Y) = \mathrm{Mor}_\mathcal{C}(C, Hom_\mathcal{E}(X, Y)),$$

for all $C \in \mathcal{C}$, $X, Y \in \mathcal{E}$. Thus we obtain that the category $\mathcal{E}$ is naturally enriched over the base category $\mathcal{C}^*$.

---

* In the context of topological spaces, the set-theoretic convention of §0.1 is not satisfied, but the existence of a good homomorphism object is a classical issue, which can be solved if we take a good category of topological spaces (see the survey and the bibliographical references of [28, §§2.4.21–26]).

In a point-set context, where the objects of $\mathcal{C}$ have an actual underlying set, we call *homomorphisms* the elements of $Hom_{\mathcal{E}}(X, Y)$, which have to be distinguished from the actual *morphisms* $f \in Mor_{\mathcal{E}}(X, Y)$.

1.1.13 Examples.

(a) For $\mathcal{E} = \mathcal{C} = \Bbbk\,\mathrm{Mod}$, the category of $\Bbbk$-modules, we have simply $Hom_{\Bbbk\,\mathrm{Mod}}(X, Y) = Mor_{\Bbbk\,\mathrm{Mod}}(X, Y)$ since the set of $\Bbbk$-module morphisms forms itself a $\Bbbk$-module and this $\Bbbk$-module satisfies the adjunction relation of hom-objects.

(b) For $\mathcal{E} = \mathcal{C} = \mathrm{dg}\,\Bbbk\,\mathrm{Mod}$, the category of dg-modules, the dg-object $Hom_{\mathrm{dg}\,\Bbbk\,\mathrm{Mod}}(C, D)$ is spanned in degree d by homogeneous maps $f : C \to D$ which raise degrees by d. The differential of a map $f \in Hom_{\mathrm{dg}\,\Bbbk\,\mathrm{Mod}}(C, D)$ is given by the usual commutator formula $\delta(f) = \delta \cdot f - \pm f \cdot \delta$ in which we use the internal differential of the dg-modules C and D.

Observe that a homogeneous element $f \in Hom_{\mathrm{dg}\,\Bbbk\,\mathrm{Mod}}(C \otimes D, E)$ of degree d is also equivalent to a collection of actual multilinear maps $f : C_p \times D_q \to E_n$, where $n = p + q + d$. Thus the pointwise representation of §1.1.4 can be extended to homomorphisms of the category of dg-modules.

(c) For $\mathcal{E} = \mathcal{C} = \mathrm{Set}$, the category of sets, we have simply $Hom_{\mathrm{Set}}(X, Y) = Mor_{\mathrm{Set}}(X, Y)$, the set of maps $f : X \to Y$.

(d) The n-simplex Δ^n in the category of simplicial sets $\mathcal{S}$ is characterized by the relation $Mor_{\mathcal{S}}(\Delta^n, K) = K_n$, for all $K \in \mathcal{S}$. The collection Δ^n, $n \in \mathbb{N}$, form a cosimplicial object in $\mathcal{S}$. For a category $\mathcal{E}$ tensored over the category of simplicial sets $\mathcal{S}$, we have a hom-object $Hom_{\mathcal{E}}(C, D) \in \mathcal{S}$ defined by $Hom_{\mathcal{E}}(C, D)_n = Mor_{\mathcal{E}}(\Delta^n \otimes C, D)$, together with faces $d_i : Mor_{\mathcal{E}}(\Delta^n \otimes C, D) \to Mor_{\mathcal{E}}(\Delta^{n-1} \otimes C, D)$ and degeneracies $s_j : Mor_{\mathcal{E}}(\Delta^n \otimes C, D) \to Mor_{\mathcal{E}}(\Delta^{n+1} \otimes C, D)$ induced by the faces $d^i : \Delta^{n-1} \to \Delta^n$ and the degeneracies $s^j : \Delta^{n+1} \to \Delta^n$ of the n-simplex Δ^n. This definition is forced by the relations

$$Hom_{\mathcal{E}}(X, Y)_n = Mor_{\mathcal{S}}(\Delta^n, Hom_{\mathcal{E}}(X, Y)) = Mor_{\mathcal{E}}(\Delta^n \otimes X, Y).$$

Use that the simplices Δ^n, $n \in \mathbb{N}$, generate the category of simplicial sets to extend the adjunction relation $Mor_{\mathcal{S}}(\Delta^n, Hom_{\mathcal{E}}(X, Y)) = Mor_{\mathcal{E}}(\Delta^n \otimes X, Y)$ to all simplicial sets.

1.1.14 Functors on Enriched Categories. Recall that $1[-] : \mathrm{Set} \to \mathcal{C}$ refers to the canonical functor of symmetric monoidal categories such that $1[K] = \bigoplus_{k \in K} 1$. For $K = Mor_{\mathcal{E}}(X, Y)$, we have a morphism

$$1[Mor_{\mathcal{E}}(X, Y)] \otimes X \simeq \bigoplus_{f \in Mor_{\mathcal{E}}(X, Y)} X \xrightarrow{(f)_*} Y$$

that determines a natural morphism

$$1[Mor_{\mathcal{E}}(X, Y)] \to Hom_{\mathcal{E}}(X, Y).$$

A functor $\rho : \mathcal{D} \to \mathcal{E}$, where the categories $\mathcal{D}$ and $\mathcal{E}$ are tensored over $\mathcal{C}$, is a *functor in the enriched sense* if we have a natural transformation $\rho : Hom_{\mathcal{D}}(X,Y) \to Hom_{\mathcal{E}}(\rho X, \rho Y)$ which preserves compositions in hom-objects and so that the diagram

$$
\begin{array}{ccc}
1[\mathrm{Mor}_{\mathcal{D}}(X,Y)] & \xrightarrow{\quad 1[\rho] \quad} & 1[\mathrm{Mor}_{\mathcal{E}}(\rho X, \rho Y)] \\
\downarrow & & \downarrow \\
Hom_{\mathcal{D}}(X,Y) & \xdashrightarrow[\rho]{\quad\quad} & Hom_{\mathcal{E}}(\rho X, \rho Y)
\end{array}
$$

commutes, for all $X, Y \in \mathcal{D}$.

The next assertion is formal:

1.1.15 Proposition. *Let $\rho : \mathcal{D} \to \mathcal{E}$ be a functor, where the categories $\mathcal{D}$ and $\mathcal{E}$ are tensored over $\mathcal{C}$. If ρ preserves external tensor products, then the map $f \mapsto \rho(f)$, defined for morphisms in $\mathcal{E}$, extends to a morphism*

$$ Hom_{\mathcal{D}}(X,Y) \xrightarrow{\rho} Hom_{\mathcal{E}}(\rho X, \rho Y), $$

so that we have a commutative diagram

$$
\begin{array}{ccc}
\mathrm{Mor}_{\mathcal{C}}(C, Hom_{\mathcal{D}}(X,Y)) & \xdashrightarrow{\quad \rho_* \quad} & \mathrm{Mor}_{\mathcal{C}}(C, Hom_{\mathcal{E}}(\rho X, \rho Y)) \\
& & \downarrow \simeq \\
\simeq \downarrow & & \mathrm{Mor}_{\mathcal{E}}(C \otimes \rho X, \rho Y) \\
& & \uparrow \simeq \\
\mathrm{Mor}_{\mathcal{D}}(C \otimes X, Y) & \xrightarrow{\quad \rho \quad} & \mathrm{Mor}_{\mathcal{E}}(\rho(C \otimes X), \rho Y)
\end{array}
$$

for all $C \in \mathcal{C}$, and ρ defines a functor in the enriched sense.　　□

For our needs we check further:

1.1.16 Proposition. *Let $\rho_! : \mathcal{D} \rightleftarrows \mathcal{E} : \rho^*$ be adjoint functors, where the categories $\mathcal{D}$ and $\mathcal{E}$ are tensored over $\mathcal{C}$. If $\rho_!$ preserves external tensor products, then the functors $\rho_! : \mathcal{D} \rightleftarrows \mathcal{E} : \rho^*$ satisfy an enriched adjunction relation*

$$ Hom_{\mathcal{E}}(\rho_! X, Y) \simeq Hom_{\mathcal{D}}(X, \rho^* Y), $$

where morphism sets are replaced by hom-objects over $\mathcal{C}$. Moreover, the morphism on hom-objects induced by $\rho_! : \mathcal{D} \to \mathcal{E}$ fits a commutative diagram

$$
\begin{array}{ccc}
Hom_{\mathcal{D}}(X,Y) & \xrightarrow{\quad\quad\quad \rho_! \quad\quad\quad} & Hom_{\mathcal{E}}(\rho_! X, \rho_! Y) \\
& \searrow{\scriptstyle \eta Y_*} \qquad \nearrow{\scriptstyle \simeq} & \\
& Hom_{\mathcal{D}}(X, \rho^* \rho_! Y) &
\end{array}
$$
,

where ηY_* denotes the morphism induced by the adjunction unit $\eta Y : Y \to \rho^* \rho_! Y$.

Proof. Apply the adjunction relation on morphism sets to a tensor product $C \otimes Y \in \mathcal{D}$, where $C \in \mathcal{C}$ and $X \in \mathcal{D}$. Since we assume $\rho_!(C \otimes X) \simeq C \otimes \rho_!(X)$, we obtain natural isomorphisms

$$\begin{array}{ccc}
\mathrm{Mor}_{\mathcal{E}}(C \otimes \rho_! X, Y) \xrightarrow{\simeq} \mathrm{Mor}_{\mathcal{E}}(\rho_!(C \otimes X), Y) \xrightarrow{\simeq} \mathrm{Mor}_{\mathcal{D}}(C \otimes X, \rho^* Y) \\
\Big\downarrow{\simeq} \qquad\qquad\qquad\qquad\qquad\qquad\qquad\qquad \Big\downarrow{\simeq} \\
\mathrm{Mor}_{\mathcal{C}}(C, \mathrm{Hom}_{\mathcal{E}}(\rho_! X, Y)) \cdots\cdots\cdots\cdots\xrightarrow{\simeq}\cdots\cdots\cdots\cdots \mathrm{Mor}_{\mathcal{C}}(C, \mathrm{Hom}_{\mathcal{D}}(X, \rho^* Y))
\end{array}$$

from which we deduce the existence of an isomorphism

$$\mathrm{Hom}_{\mathcal{E}}(\rho_! X, Y) \simeq \mathrm{Hom}_{\mathcal{D}}(X, \rho^* Y),$$

for all $X \in \mathcal{D}$, $Y \in \mathcal{E}$.

For $X, Y \in \mathcal{D}$, we have a commutative diagram on morphism sets

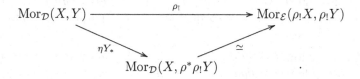

This assertion is a formal consequence of the definition of an adjunction unit. Again, we apply this diagram to a tensor product $X := C \otimes X \in \mathcal{D}$, where $C \in \mathcal{C}$ and $X \in \mathcal{D}$, and we use the relation $\rho_!(C \otimes X) \simeq C \otimes \rho_!(X)$ to check that the diagram commutes at the level of hom-objects. $\qquad\qquad\square$

1.1.17 Reduced Symmetric Monoidal Categories over a Base. In the sequel, we also consider *reduced symmetric monoidal categories* $\mathcal{E}^0$ which come equipped with a symmetric internal tensor product and with an external tensor product over $\mathcal{C}$, but which have no internal unit object. In this case, we assume all axioms of a symmetric monoidal category over a base, except the axioms that involve the unit of $\mathcal{E}$.

Any reduced symmetric monoidal category $\mathcal{E}^0$ is equivalent to a symmetric monoidal category over $\mathcal{C}$ of the form $\mathcal{E} = \mathcal{C} \times \mathcal{E}^0$, together with the functor $\eta : \mathcal{C} \to \mathcal{E}$ such that $\eta C = (C, 0)$ and the internal tensor product so that:

$$(C, X^0) \otimes Y = \eta C \otimes Y \oplus (0, X^0) \otimes Y, \quad \forall (C, X^0) \in \mathcal{C} \times \mathcal{E}^0, \forall Y \in \mathcal{E},$$
$$X \otimes (D, Y^0) = X \otimes \eta D \oplus X \otimes (0, Y^0), \quad \forall X \in \mathcal{E}, \forall (D, Y^0) \in \mathcal{C} \times \mathcal{E}^0,$$
$$\eta C \otimes (0, X^0) = (0, X^0) \otimes \eta C = (0, C \otimes X^0), \quad \forall (C, X^0) \in \mathcal{C} \times \mathcal{E}^0,$$
$$\text{and} \quad (0, X^0) \otimes (0, Y^0) = (0, X^0 \otimes Y^0), \quad \forall (X^0, Y^0) \in \mathcal{E}^0 \times \mathcal{E}^0.$$

The coproduct in $\mathcal{E} = \mathcal{C} \times \mathcal{E}^0$ is given by the obvious formula $(C, X^0) \oplus (D, Y^0) = (C \oplus D, X^0 \oplus Y^0)$.

This construction gives readily:

1.1.18 Proposition. *The category of reduced symmetric monoidal categories over $\mathcal{C}$ is equivalent to a subcategory of the 2-category of symmetric monoidal categories over $\mathcal{C}$.*   □

1.2 Reflexive Coequalizers and Filtered Colimits

Recall that a coequalizer

$$X_1 \underset{d_1}{\overset{d_0}{\rightrightarrows}} X_0 \longrightarrow \mathrm{coker}(X_1 \rightrightarrows X_0)$$

is *reflexive* if there exists a morphism $s_0 : X_0 \to X_1$ such that $d_0 s_0 = \mathrm{id} = d_1 s_0$. In this context, we say that d_0, d_1 forms a *reflexive pair of morphisms*.

The importance of reflexive coequalizers for our purpose comes from the following assertions:

1.2.1 Proposition (see [20, p. 46] and [54, Lemma 2.3.2]). *Suppose $T : \mathcal{A} \times \mathcal{B} \to \mathcal{E}$ is a bifunctor that preserves reflexive coequalizers in each variable separately. Explicitly: the natural morphism*

$$\mathrm{coker}(T(X_1, Y) \rightrightarrows T(X_0, Y)) \to T(\mathrm{coker}(X_1 \rightrightarrows X_0), Y),$$

is an isomorphism, for all reflexive pairs of morphisms $d_0, d_1 : X_1 \rightrightarrows X_0$ and all objects Y, and similarly for coequalizers in the second variable. Then:

(a) The functor T preserves reflexive coequalizers in two variables. Explicitly, the natural morphism

$$\mathrm{coker}(T(X_1, Y_1) \rightrightarrows T(X_0, Y_0)) \to T(\mathrm{coker}(X_1 \rightrightarrows X_0), \mathrm{coker}(Y_1 \rightrightarrows Y_0))$$

is an isomorphism for all reflexive pairs of morphisms $d_0, d_1 : X_1 \rightrightarrows X_0$ and $d_0, d_1 : Y_1 \rightrightarrows Y_0$.

(b) In the case $\mathcal{A} = \mathcal{B}$, the composite of $(X, Y) \mapsto T(X, Y)$ with the diagonal functor $X \mapsto (X, X)$ preserves reflexive coequalizers as well.

Proof. Check that the obvious implication

$$\epsilon T(d_0, \mathrm{id}) = \epsilon T(d_1, \mathrm{id}) \quad \text{and} \quad \epsilon T(\mathrm{id}, d_0) = \epsilon T(\mathrm{id}, d_1)$$
$$\Rightarrow \quad \epsilon T(d_0, d_0) = \epsilon T(d_1, d_1)$$

is an equivalence if the parallel morphisms $d_0, d_1 : X_1 \to X_0$ and $d_0, d_1 : Y_1 \to Y_0$ have a reflection. Deduce from this observation:

$$\mathrm{coker}(T(X_1, Y_1) \rightrightarrows T(X_0, Y_0)) \simeq \mathop{\mathrm{colim}}_{(i,j)} T(X_i, Y_j),$$

where $\mathrm{colim}_{(i,j)} T(X_i, Y_j)$ refers to the colimit over the diagram

$$\begin{array}{ccc} T(X_1, Y_1) & \rightrightarrows & T(X_0, Y_1) \\ \Downarrow & & \Downarrow \\ T(X_1, Y_0) & \rightrightarrows & T(X_0, X_0), \end{array}$$

and use the assumption to conclude:

$$\mathrm{coker}(T(X_1, Y_1) \rightrightarrows T(X_0, Y_0)) \simeq \mathop{\mathrm{colim}}_{(i,j)} T(X_i, Y_j)$$
$$\simeq \mathop{\mathrm{colim}}_{i} \mathop{\mathrm{colim}}_{j} T(X_i, Y_j) \simeq T(\mathrm{coker}(X_1 \rightrightarrows X_0), \mathrm{coker}(Y_1 \rightrightarrows Y_0)).$$

Use that diagonal functors preserve all colimits to obtain assertion (b).   □

The same assertions hold for filtered colimits. Recall that a filtered colimit refers to a colimit over a filtered category, and a category I is filtered if (see [44, Chapter 9]):

(1) Any pair of objects $i, j \in I$ can be joined together by morphisms

in I;

(2) Any pair of parallel morphisms $u, v : i \to j$ can be equalized by a morphism

$$i \underset{v}{\overset{u}{\rightrightarrows}} j \overset{w}{\dashrightarrow} k$$

in I.

As an example, any ordinal (see [27, 28]) forms a filtered category.

1.2.2 Proposition (see [54, Lemma 2.3.2]). *Suppose $T : \mathcal{A} \times \mathcal{B} \to \mathcal{E}$ is a bifunctor that preserves filtered colimits in each variable separately. Explicitly: the natural morphism*

$$\mathop{\mathrm{colim}}_{i} T(X_i, Y) \to T(\mathop{\mathrm{colim}}_{i} X_i, Y),$$

is an isomorphism, for all I-diagrams $i \mapsto X_i$, where I is a filtered category, and similarly with respect to the second variable. Then:

(a) *The functor T preserves filtered colimits in two variables. Explicitly, the natural morphism*

$$\operatorname*{colim}_i T(X_i, Y_i) \to T(\operatorname*{colim}_i X_i, \operatorname*{colim}_i Y_i)$$

is an isomorphism for all I-diagrams $i \mapsto X_i$ and $i \mapsto Y_i$, where I is a filtered category.

(b) *In the case $\mathcal{A} = \mathcal{B}$, the composite of $(X, Y) \mapsto T(X, Y)$ with the diagonal functor $X \mapsto (X, X)$ preserves filtered colimits as well.*

Proof. Observe that the diagonal $\Delta : I \to I \times I$ is a final functor when I is filtered (see [44, Chapter 9] for the notion of a final functor) to conclude that

$$\operatorname*{colim}_{i \in I} T(X_i, Y_i) \simeq \operatorname*{colim}_{(i,j) \in I \times I} T(X_i, Y_j)$$

$$\simeq \operatorname*{colim}_{i \in I} \operatorname*{colim}_{j \in I} T(X_i, Y_j) \simeq T(\operatorname*{colim}_i X_i, \operatorname*{colim}_j Y_j).$$

Use again that diagonal functors preserve all colimits to obtain assertion (b).
□

These assertions imply:

1.2.3 Proposition. *The tensor power functors $\mathrm{Id}^{\otimes r} : X \mapsto X^{\otimes r}$ in a symmetric monoidal category $\mathcal{E}$ preserve reflexive coequalizers and filtered colimits.*   □

Chapter 2
Symmetric Objects and Functors

Introduction

In this chapter, we recall the definition of the category of Σ_*-objects and we review the relationship between Σ_*-objects and functors. In short, a Σ_*-*object* (in English words, a *symmetric sequence of objects*, or simply a *symmetric object*) is the coefficient sequence of a generalized symmetric functor $S(M)$: $X \mapsto S(M, X)$, defined by a formula of the form

$$ S(M, X) = \bigoplus_{r=0}^{\infty} (M(r) \otimes X^{\otimes r})_{\Sigma_r}. $$

In §2.1, we recall the definition of the tensor product of Σ_*-objects, the operation which reflects the pointwise tensor product of functors and which provides the category of Σ_*-objects with the structure of a symmetric monoidal category over the base category.

Beside the tensor product, the category of Σ_*-objects comes equipped with a composition product that reflects the composition of functors. The definition of this composition structure is recalled in §2.2.

The map $S : M \mapsto S(M)$ defines a functor $S : \mathcal{M} \to \mathcal{F}$, where $\mathcal{M}$ denotes the category of Σ_*-objects and $\mathcal{F}$ denotes the category of functors $F : \mathcal{E} \to \mathcal{E}$ on any symmetric monoidal category $\mathcal{E}$ over the base category $\mathcal{C}$. The adjoint functor theorem implies that this functor has a right adjoint $\Gamma : \mathcal{F} \to \mathcal{M}$. In §2.3 we give an explicit construction of this adjoint functor by using that the symmetric monoidal category $\mathcal{E}$ is enriched over the base category $\mathcal{C}$. In addition, we prove that the map $S : M \mapsto S(M)$ defines a faithful functor in the enriched sense as long as the category $\mathcal{E}$ is equipped with a faithful functor $\eta : \mathcal{C} \to \mathcal{E}$. In the case $\mathcal{E} = \mathcal{C} = \Bbbk \operatorname{Mod}$, the category of modules over a ring $\Bbbk$, we use the explicit construction of the adjoint functor $\Gamma : G \mapsto \Gamma(G)$ to prove that the functor $S : M \mapsto S(M)$ is bijective on object sets under mild conditions on Σ_*-objects or on the ground ring $\Bbbk$.

B. Fresse, *Modules over Operads and Functors*, Lecture Notes in Mathematics 1967, 35
DOI: 10.1007/978-3-540-89056-0_2, © Springer-Verlag Berlin Heidelberg 2009

In §§2.1-2.3, we deal with global structures of the category of Σ_*-objects. In §2.4, we study the image of colimits under the functor $S(M) : \mathcal{E} \to \mathcal{E}$ associated to a Σ_*-object M. Explicitly, we record that the functor $S(M) : \mathcal{E} \to \mathcal{E}$ preserves filtered colimits and reflexive coequalizers (but not all colimits). This verification is required by our conventions on functors (see §0.1) and is also used in §3.3, where we address the construction of colimits in categories of algebras over operads.

2.1 The Symmetric Monoidal Category of Σ_*-Objects and Functors

Formally, a Σ_*-object in a category $\mathcal{C}$ consists of a sequence $M(n)$, $n \in \mathbb{N}$, where $M(n)$ is an object of $\mathcal{C}$ equipped with an action of the symmetric group Σ_n. A morphism of Σ_*-objects $f : M \to N$ consists of a sequence of morphisms $f : M(n) \to N(n)$ that commute with the action of symmetric groups.

Usually, we have a base category $\mathcal{C}$, fixed once and for all, and we deal tacitly with Σ_*-objects in that category $\mathcal{C}$. Otherwise we specify explicitly the category in which we define our Σ_*-object. We may use the notation $\mathcal{E}^{\Sigma_*}$ to refer to the category of Σ_*-objects in a given category $\mathcal{E}$, but we usually adopt the short notation $\mathcal{M}$ for the category of Σ_*-objects in the base category $\mathcal{E} = \mathcal{C}$.

In the introduction of the chapter, we recall that $\mathcal{M}$ forms a symmetric monoidal category over $\mathcal{C}$. In this section, we address the definition and applications of this categorical structure. More specifically, we use the formalism of symmetric monoidal categories over a base category to express the relationship between the tensor product of Σ_*-objects and the pointwise tensor product of functors on a symmetric monoidal category $\mathcal{E}$ over $\mathcal{C}$. Formally, the category $\mathcal{F}$ of functors $F : \mathcal{E} \to \mathcal{E}$ inherits the structure of a symmetric monoidal category over $\mathcal{C}$ and the map $S : M \mapsto S(M)$ defines a functor of symmetric monoidal categories over $\mathcal{C}$:

$$(\mathcal{M}, \otimes, 1) \xrightarrow{S} (\mathcal{F}, \otimes, 1).$$

2.1.1 The Functor Associated to a Σ_*-Object. First of all, we recall the definition of the functor $S(M) : \mathcal{E} \to \mathcal{E}$ associated to a Σ_*-object M, for $\mathcal{E}$ a symmetric monoidal category over $\mathcal{C}$. The image of an object $X \in \mathcal{E}$ under this functor, denoted by $S(M, X) \in \mathcal{E}$, is defined by the formula

$$S(M, X) = \bigoplus_{r=0}^{\infty} (M(r) \otimes X^{\otimes r})_{\Sigma_r},$$

where we consider the coinvariants of the tensor products $M(r) \otimes X^{\otimes r}$ under the action of the symmetric groups Σ_r. We use the internal tensor product of $\mathcal{E}$ to form the tensor power $X^{\otimes r}$, the external tensor product to form the object $M(r) \otimes X^{\otimes r}$ in $\mathcal{E}$, and the existence of colimits in $\mathcal{E}$ to form the coinvariant object $(M(r) \otimes X^{\otimes r})_{\Sigma_r}$ and $S(M, X)$.

In §2.1.4, we introduce pointwise operations on functors $F : \mathcal{E} \to \mathcal{E}$ that correspond to tensor operations on the target. In light of these structures on functors, we have a functor identity

$$S(M) = \bigoplus_{r=0}^{\infty} (M(r) \otimes \mathrm{Id}^{\otimes r})_{\Sigma_r},$$

where $\mathrm{Id} : \mathcal{E} \to \mathcal{E}$ denotes the identity functor on $\mathcal{E}$.

The construction $S : M \mapsto S(M)$ is clearly functorial in $\mathcal{E}$. Explicitly, for a functor $\rho : \mathcal{D} \to \mathcal{E}$ of symmetric monoidal categories over $\mathcal{C}$, the diagram of functors

$$
\begin{array}{ccc}
\mathcal{D} & \xrightarrow{\ \rho\ } & \mathcal{E} \\
{\scriptstyle S(M)}\big\downarrow & & \big\downarrow{\scriptstyle S(M)} \\
\mathcal{D} & \xrightarrow[\ \rho\ ]{} & \mathcal{E}
\end{array}
$$

commutes up to natural isomorphisms. Equivalently, we have a natural functor isomorphism $S(M) \circ \rho \simeq \rho \circ S(M)$, for every $M \in \mathcal{M}$.

In the point-set context, the element of $S(M, V)$ represented by the tensor $\xi \otimes (x_1 \otimes \cdots \otimes x_r) \in M(r) \otimes V^{\otimes r}$ is denoted by $\xi(x_1, \ldots, x_r) \in S(M, V)$. The coinvariant relations read $\sigma\xi(x_1, \ldots, x_r) = \xi(x_{\sigma(1)}, \ldots, x_{\sigma(r)})$, for $\sigma \in \Sigma_r$.

Clearly, the map $S : M \mapsto S(M)$ defines a functor $S : \mathcal{M} \to \mathcal{F}$, where $\mathcal{F} = \mathcal{F}(\mathcal{E}, \mathcal{E})$ denotes the category of functors $F : \mathcal{E} \to \mathcal{E}$. (Because of our conventions on functor categories, we should check that $S(M) : \mathcal{E} \to \mathcal{E}$ preserves filtered colimits, but we postpone the simple verification of this assertion to §2.4.)

The category $\mathcal{M}$ is equipped with colimits and limits created termwise in $\mathcal{C}$. The category of functors $\mathcal{F} = \mathcal{F}(\mathcal{E}, \mathcal{E})$ is equipped with colimits as well, inherited pointwise from the category $\mathcal{E}$. By interchange of colimits, we obtain immediately that the functor $S : \mathcal{M} \to \mathcal{F}(\mathcal{E}, \mathcal{E})$ preserves colimits.

2.1.2 Constant Σ_*-Objects and Constant Functors. Recall that a Σ_*-object M is *constant* if we have $M(r) = 0$ for all $r > 0$. The base category $\mathcal{C}$ is isomorphic to the full subcategory of $\mathcal{M}$ formed by constant objects. Explicitly, to any object $C \in \mathcal{C}$, we associate the constant Σ_*-object $\eta(C)$ such that $\eta(C)(0) = C$. This constant Σ_*-object is associated to the constant functor $S(C, X) \equiv C$.

2.1.3 Connected Σ_*-Objects and Functors. The category embedding $\eta : \mathcal{C} \to \mathcal{M}$ has an obvious left-inverse $\epsilon : \mathcal{M} \to \mathcal{C}$ defined by $\epsilon(M) = M(0)$. The category of *connected Σ_*-objects* $\mathcal{M}^0$ is the full subcategory of $\mathcal{M}$ formed by Σ_*-objects M such that $\epsilon(M) = M(0) = 0$, the initial object of $\mathcal{C}$. Clearly, connected Σ_*-objects are associated to functors $\mathrm{S}(M) : \mathcal{E} \to \mathcal{E}$ such that $\mathrm{S}(M, 0) = 0$.

In the case of a connected Σ_*-object $M \in \mathcal{M}^0$, we can extend the construction of §2.1.1 to reduced symmetric monoidal categories. To be explicit, for objects X in a reduced symmetric monoidal category $\mathcal{E}^0$ over $\mathcal{C}$, we set

$$\mathrm{S}^0(M, X) = \bigoplus_{n=1}^{\infty} (M(n) \otimes X^{\otimes n})_{\Sigma_n}$$

to obtain a functor $\mathrm{S}^0(M) : \mathcal{E}^0 \to \mathcal{E}^0$.

2.1.4 The Symmetric Monoidal Category of Functors. Let $\mathcal{F} = \mathcal{F}(\mathcal{A}, \mathcal{C})$ denote the category of functors $F : \mathcal{A} \to \mathcal{C}$, where $\mathcal{A}$ is any category (see §0.1). Recall that $\mathcal{F} = \mathcal{F}(\mathcal{A}, \mathcal{C})$ has all small colimits and limits, inherited pointwise from the base category $\mathcal{C}$.

Observe that the category $\mathcal{F}$ is equipped with an internal tensor product $\otimes : \mathcal{F} \otimes \mathcal{F} \to \mathcal{F}$ and with an external tensor product $\otimes : \mathcal{C} \otimes \mathcal{F} \to \mathcal{F}$, inherited from the base symmetric monoidal category, so that $\mathcal{F}$ forms a symmetric monoidal category over $\mathcal{C}$. Explicitly: the internal tensor product of functors $F, G : \mathcal{A} \to \mathcal{C}$ is defined pointwise by $(F \otimes G)(X) = F(X) \otimes G(X)$; for all $X \in \mathcal{A}$, the tensor product of a functor $G : \mathcal{A} \to \mathcal{C}$ with an object $C \in \mathcal{C}$ is defined by $(C \otimes F)(X) = C \otimes F(X)$; the constant functor $\mathbf{1}(X) \equiv \mathbf{1}$, where $\mathbf{1}$ is the unit object of $\mathcal{C}$, represents the unit object in the category of functors.

The functor of symmetric monoidal categories

$$\eta : (\mathcal{C}, \otimes, \mathbf{1}) \to (\mathcal{F}, \otimes, \mathbf{1})$$

determined by this structure identifies an object $C \in \mathcal{C}$ with the constant functor $\eta(C)(X) \equiv C$. If $\mathcal{A}$ is equipped with a base object $0 \in \mathcal{A}$, then we have a natural splitting $\mathcal{F} = \mathcal{C} \times \mathcal{F}^0$, where $\mathcal{F}^0$ is the reduced symmetric monoidal category over $\mathcal{C}$ formed by functors F such that $F(0) = 0$, the initial object of $\mathcal{C}$.

Obviously, we can extend the observations of this paragraph to a category of functors $\mathcal{F} = \mathcal{F}(\mathcal{A}, \mathcal{E})$, where $\mathcal{E}$ is a symmetric monoidal category over the base category $\mathcal{C}$. In this case, the category $\mathcal{F} = \mathcal{F}(\mathcal{A}, \mathcal{E})$ forms a symmetric monoidal category over $\mathcal{E}$, and hence over the base category by transitivity.

We have:

2.1.5 Proposition (*cf.* [12, §1.1.3] or [14, §1.2] or [54, Lemma 2.2.4]). *The category $\mathcal{M}$ is equipped with the structure of a symmetric monoidal category over $\mathcal{C}$ so that the map $\mathrm{S} : M \mapsto \mathrm{S}(M)$ defines a functor of symmetric monoidal categories over $\mathcal{C}$*

$$\mathrm{S} : (\mathcal{M}, \otimes, \mathbf{1}) \to (\mathcal{F}(\mathcal{E}, \mathcal{E}), \otimes, \mathbf{1}),$$

functorially in $\mathcal{E}$, for every symmetric monoidal category $\mathcal{E}$ over $\mathcal{C}$. □

The functoriality claim asserts explicitly that, for any functor $\rho : \mathcal{D} \to \mathcal{E}$ of symmetric monoidal categories over $\mathcal{C}$, the tensor isomorphisms $S(M \otimes N) \simeq S(M) \otimes S(N)$ and the functoriality isomorphisms $S(M) \circ \rho \simeq \rho \circ S(M)$ fit a commutative hexagon

$$
\begin{array}{ccc}
S(M \otimes N) \circ \rho & \xrightarrow{\;\simeq\;} & \rho \circ S(M \otimes N) \\
{\scriptstyle \simeq} \downarrow & & \downarrow {\scriptstyle \simeq} \\
(S(M) \otimes S(N)) \circ \rho & & \rho \circ (S(M) \otimes S(N)) \\
\qquad {}_{=} \searrow & & \swarrow {}_{\simeq} \qquad \\
S(M) \circ \rho \otimes S(N) \circ \rho & \xrightarrow[\;\simeq\;]{} & \rho \circ S(M) \otimes \rho \circ S(N)
\end{array}
$$

and similarly for the isomorphism $S(1) \simeq 1$.

We have further:

2.1.6 Proposition. *The category $\mathcal{M}^0$ of connected Σ_*-objects forms a reduced symmetric monoidal category over $\mathcal{C}$.*

The category $\mathcal{M}$ admits a splitting $\mathcal{M} = \mathcal{C} \times \mathcal{M}^0$ and is isomorphic to the symmetric monoidal category over $\mathcal{C}$ associated to the reduced category $\mathcal{M}^0$. The functor $S : M \mapsto S(M)$ fits a diagram of symmetric monoidal categories over $\mathcal{C}$

$$
\begin{array}{ccc}
\mathcal{M} & \xrightarrow{\;\;S\;\;} & \mathcal{F}(\mathcal{E}, \mathcal{E}) \\
{\scriptstyle \simeq} \uparrow & & \uparrow {\scriptstyle \simeq} \\
\mathcal{C} \times \mathcal{M}^0 & \xrightarrow[\mathrm{Id} \times S]{} & \mathcal{C} \times \mathcal{F}(\mathcal{E}, \mathcal{E})^0
\end{array} \qquad \square
$$

We refer to the literature for the proof of the assertions of propositions 2.1.5-2.1.6. For our needs, we recall simply the explicit construction of the tensor product $M \otimes N$. This construction also occurs in the definition of the category of symmetric spectra in stable homotopy (see [30, §2.1]).

2.1.7 The Tensor Product of Σ_*-Objects. The terms of the tensor product of Σ_*-objects are defined explicitly by a formula of the form

$$
(M \otimes N)(n) = \bigoplus_{p+q=n} \Sigma_n \otimes_{\Sigma_p \times \Sigma_q} M(p) \otimes N(q),
$$

where we use the tensor product over the category of sets, defined explicitly in §1.1.7. In the construction, we use the canonical group embedding $\Sigma_p \times \Sigma_q \subset \Sigma_{p+q}$ which identifies a permutation $\sigma \in \Sigma_p$ (respectively, $\tau \in \Sigma_q$) to a permutation of the subset $\{1, \ldots, p\} \subset \{1, \ldots, p, p+1, \ldots, p+q\}$ (respectively, $\{p+1, \ldots, p+q\} \subset \{1, \ldots, p, p+1, \ldots, p+q\}$). The tensor product $M(p) \otimes N(q)$ forms a $\Sigma_p \times \Sigma_q$-object in $\mathcal{C}$. The group $\Sigma_p \times \Sigma_q$ acts on Σ_n by translations on the right. The quotient in the tensor product makes this right $\Sigma_p \times \Sigma_q$-action agree with the left $\Sigma_p \times \Sigma_q$-action on $M(p) \otimes N(q)$.

The group Σ_n also acts on Σ_n by translation on the left. This left Σ_n-action induces a left Σ_n-action on $(M \otimes N)(n)$ and determines the Σ_*-object structure of the collection $M \otimes N = \{(M \otimes N)(n)\}_{n \in \mathbb{N}}$.

The constant Σ_*-object $\mathbf{1}$ such that

$$\mathbf{1}(n) = \begin{cases} \mathbf{1} \ (\text{the unit object of } \mathcal{C}), & \text{if } n = 0, \\ 0, & \text{otherwise,} \end{cases}$$

defines a unit for this tensor product. The associativity of the tensor product of Σ_*-objects is inherited from the base category. Let $\tau(p, q) \in \Sigma_n$ be the permutation such that:

$$\tau(p, q)(i) = p + i, \text{ for } i = 1, \ldots, q,$$
$$\tau(p, q)(q + i) = i, \text{ for } i = 1, \ldots, p.$$

The symmetry isomorphism $\tau(M, N) : M \otimes N \to N \otimes M$ is induced componentwise by morphisms of the form

$$\Sigma_n \otimes M(p) \otimes N(q) \xrightarrow{\tau(p,q)^* \otimes \tau} \Sigma_n \otimes N(q) \otimes M(p)$$

where we use the symmetry isomorphism $\tau : M(p) \otimes N(q) \to N(q) \otimes M(p)$ of the category $\mathcal{C}$ and a translation of the right by the block transposition $\tau(p, q)$ on the symmetric group Σ_n.

The functor $\eta : \mathcal{C} \to \mathcal{M}$ which identifies the objects of $\mathcal{C}$ to constant Σ_*-objects defines a functor of symmetric monoidal categories

$$\eta : (\mathcal{C}, \otimes, \mathbf{1}) \to (\mathcal{M}, \otimes, \mathbf{1})$$

and makes $(\mathcal{M}, \otimes, \mathbf{1})$ into a symmetric monoidal category over $\mathcal{C}$. By an immediate inspection of definitions, we obtain that the external tensor product of a Σ_*-object M with an object $C \in \mathcal{C}$ is given by the obvious formula $(C \otimes M)(r) = C \otimes M(r)$.

2.1.8 Tensor Powers. For the needs of §3.2, we make explicit the structure of tensor powers $M^{\otimes r}$ in the category of Σ_*-objects.

For all $n \in \mathbb{N}$, we have obviously:

$$M^{\otimes r}(n) = \bigoplus_{n_1 + \cdots + n_r = n} \Sigma_n \otimes_{\Sigma_{n_1} \times \cdots \times \Sigma_{n_r}} (M(n_1) \otimes \cdots \otimes M(n_r)).$$

In this formula, we use the canonical group embedding $\Sigma_{n_1} \times \cdots \times \Sigma_{n_r} \hookrightarrow \Sigma_n$ which identifies a permutation of Σ_{n_i} to a permutation of the subset $\{n_1 + \cdots + n_{i-1} + 1, \ldots, n_1 + \cdots + n_{i-1} + n_i\} \subset \{1, \ldots, n\}$. Again the quotient in the tensor product makes agree the internal Σ_{n_i}-action on $M(n_i)$ with the action of Σ_{n_i} by right translations on Σ_n.

The tensor power $M^{\otimes r}$ is equipped with a Σ_r-action, deduced from the symmetric structure of the tensor product of Σ_*-objects. Let $w \in \Sigma_r$ be any permutation. For any partition $n = n_1 + \cdots + n_r$, we form the block permutation $w(n_1, \ldots, n_r) \in \Sigma_n$ such that:

$$w(n_1, \ldots, n_r)(n_{w(1)} + \cdots + n_{w(i-1)} + k) = n_1 + \cdots + n_{w(i)-1} + k,$$
$$\text{for } k = 1, \ldots, n_{w(i)}, \ i = 1, \ldots, r.$$

The tensor permutation $w^* : M^{\otimes r} \to M^{\otimes r}$ is induced componentwise by morphisms of the form

$$\Sigma_n \otimes M(n_1) \otimes \cdots \otimes M(n_r) \xrightarrow{w(n_1, \ldots, n_r) \otimes w^*} \Sigma_n \otimes M(n_{w(1)}) \otimes \cdots \otimes M(n_{w(r)})$$

where we use the tensor permutation $w^* : M(n_1) \otimes \cdots \otimes M(n_r) \to M(n_{w(1)}) \otimes \cdots \otimes M(n_{w(r)})$ within the category $\mathcal{C}$ and a left translation by the block permutation $w(n_1, \ldots, n_r)$ on the symmetric group Σ_n. This formula extends obviously the definition of §2.1.7 in the case $r = 2$. To prove the general formula, check the definition of associativity isomorphisms for the tensor product of Σ_*-objects and observe that composites of block permutations are still block permutations to determine composites of symmetry isomorphisms.

2.1.9 The Pointwise Representation of Tensors in Σ_*-Objects. In the point-set context, we use the notation $w \cdot x \otimes y \in M \otimes N$ to represent the element defined by $w \otimes x \otimes y \in \Sigma_n \otimes M(p) \otimes N(q)$ in the tensor product of Σ_*-objects

$$M \otimes N(n) = \bigoplus_{p+q=n} \Sigma_n \otimes_{\Sigma_p \times \Sigma_q} M(p) \otimes N(q),$$

and the notation $x \otimes y \in M \otimes N$ in the case where $w = \mathrm{id}$ is the identity permutation.

By definition, the action of a permutation w on $M \otimes N$ maps the tensor $x \otimes y$ to $w \cdot x \otimes y$. Accordingly, the tensor product $M \otimes N$ is spanned, as a Σ_*-object, by the tensors $x \otimes y \in M(p) \otimes N(q)$, where $(x, y) \in M(p) \times N(q)$.

In our sense (see §§0.4-0.5), the tensor product of Σ_*-objects inherits a pointwise representation from the base category. To justify our pointwise representation, we also use the next assertion which identifies morphisms $f : M \otimes N \to T$ with actual multilinear maps on the set of generating tensors.

The abstract definition of §2.1.7 implies that the symmetry isomorphism $\tau(M, N) : M \otimes N \xrightarrow{\simeq} N \otimes M$ maps the tensor $x \otimes y \in M \otimes N$ to a tensor of the form $\tau(p, q) \cdot y \otimes x \in N \otimes M$, where $\tau(p, q)$ is a block permutation. Thus the permutation rule of tensors in Σ_*-objects is determined by the mapping $x \otimes y \mapsto \tau(p, q) \cdot y \otimes x$.

2.1.10 Fact. *For any Σ_*-object T, a morphism $f : M \otimes N \to T$ is equivalent to a collection of morphisms*

$$f : M(p) \otimes N(q) \to T(p+q)$$

which commute with the action of the subgroup $\Sigma_p \times \Sigma_q \subset \Sigma_{p+q}$.

This assertion is an obvious consequence of the definition of the tensor product in §2.1.7.

2.1.11 Enriched Category Structures. In §1.1.12, we observe that any symmetric monoidal category over $\mathcal{C}$ that satisfies the convention of §0.1 is naturally enriched over $\mathcal{C}$. An explicit construction of external hom-objects for categories of functors $\mathcal{F} = \mathcal{F}(\mathcal{A}, \mathcal{E})$ and the category of Σ_*-objects $\mathcal{M}$ can be derived from the existence of hom-objects in $\mathcal{E}$ (respectively, $\mathcal{C}$)*.

The external hom of the functor category $\mathcal{F} = \mathcal{F}(\mathcal{A}, \mathcal{E})$ is given by the end

$$Hom_{\mathcal{F}}(F, G) = \int_{X \in \mathcal{A}} Hom_{\mathcal{E}}(F(X), G(X)).$$

The adjunction relation

$$Mor_{\mathcal{F}}(C \otimes F, G) = Mor_{\mathcal{C}}(C, Hom_{\mathcal{F}}(F, G)),$$

for $C \in \mathcal{C}$, $F, G \in \mathcal{F}$, is equivalent to the definition of an end.

The external hom of the category of Σ_*-objects is defined by a product of the form

$$Hom_{\mathcal{M}}(M, N) = \prod_{n=0}^{\infty} Hom_{\mathcal{C}}(M(n), N(n))^{\Sigma_n}.$$

The hom-object $Hom_{\mathcal{C}}(M(n), N(n))$ inherits a conjugate action of the symmetric group from the Σ_n-objects $M(n)$ and $N(n)$. The expression $Hom_{\mathcal{C}}(M(n), N(n))^{\Sigma_n}$ refers to the invariant object with respect to this action of Σ_n. The adjunction relation of hom-objects

$$Mor_{\mathcal{M}}(C \otimes M, N) = Mor_{\mathcal{C}}(C, Hom_{\mathcal{M}}(F, G))$$

for $C \in \mathcal{C}$, $M, N \in \mathcal{M}$, is immediate.

2.1.12 Generating Σ_*-Objects. The identity functor $\text{Id} : \mathcal{E} \to \mathcal{E}$ is identified with the functor $S(I) = \text{Id}$ associated to a Σ_*-object I defined by:

$$I(n) = \begin{cases} 1, & \text{if } n = 1, \\ 0, & \text{otherwise.} \end{cases}$$

This object I represents the unit of the composition product of Σ_*-objects defined next.

---

* But serious set-theoretic difficulties occur for the category of functors $\mathcal{F} = \mathcal{F}(\mathcal{A}, \mathcal{E})$ if $\mathcal{A}$ does not satisfy the condition of §0.1, for instance when we take $\mathcal{A} = \mathcal{E} = \text{Top}$, the category of topological spaces.

For $r \in \mathbb{N}$, let $F_r = I^{\otimes r}$ be the rth tensor power of I in $\mathcal{M}$. Since $S(F_r) = S(I)^{\otimes r} = \mathrm{Id}^{\otimes r}$, we obtain that $S(F_r) : \mathcal{E} \to \mathcal{E}$ represents the rth tensor power functor $\mathrm{Id}^{\otimes r} : X \mapsto X^{\otimes r}$.

The definition of the tensor product of Σ_*-objects (see §2.1.7) implies that $F_r = I^{\otimes r}$ satisfies

$$F_r(n) = \begin{cases} 1[\Sigma_r], & \text{if } n = r, \\ 0, & \text{otherwise.} \end{cases}$$

Recall that $1[\Sigma_r]$ denotes the Σ_r-object in $\mathcal{C}$ formed by the sum over Σ_r of copies of the tensor unit $1 \in \mathcal{C}$.

The symmetric group Σ_r acts on $F_r(r) = 1[\Sigma_r]$ equivariantly by translations on the right, and hence acts on F_r on the right by automorphisms of Σ_*-objects. This symmetric group action corresponds to the action by tensor permutations on tensor powers $I^{\otimes r}$.

The Σ_*-objects F_r, $r \in \mathbb{N}$, are characterized by the following property:

2.1.13 Proposition. *We have a natural Σ_r-equivariant isomorphism*

$$\omega_r(M) : M(r) \xrightarrow{\simeq} Hom_{\mathcal{M}}(F_r, M),$$

for all $M \in \mathcal{M}$.

Proof. Immediate: we have

$$Hom_{\mathcal{M}}(F_r, M) \simeq Hom_{\mathcal{C}}(1[\Sigma_r], M(r))^{\Sigma_r}$$

$$\text{and} \quad Hom_{\mathcal{C}}(1[\Sigma_r], M(r))^{\Sigma_r} \simeq Hom_{\mathcal{C}}(1, M(r)) \simeq M(r).$$

One checks readily that the Σ_r-action by right translations on $1[\Sigma_r]$ corresponds to the internal Σ_r-action of $M(r)$ under the latter isomorphisms. Hence we obtain a Σ_r-equivariant isomorphism

$$\omega_r(M) : M(r) \xrightarrow{\simeq} Hom_{\mathcal{M}}(F_r, M),$$

as stated. $\qquad\qquad\qquad\qquad\qquad\qquad\qquad\qquad\qquad\qquad\qquad\qquad\qquad\qquad$ □

2.1.14 Canonical Generating Morphisms. Observe that

$$(M(r) \otimes F_r(n))_{\Sigma_r} \simeq \begin{cases} M(r), & \text{if } n = r, \\ 0, & \text{otherwise.} \end{cases}$$

Accordingly, for a Σ_*-object M, we have obvious morphisms

$$\iota_r(M) : (M(r) \otimes F_r)_{\Sigma_r} \to M$$

that sum up to an isomorphism

$$\iota(M) : \bigoplus_{r=0}^{\infty}(M(r) \otimes F_r)_{\Sigma_r} \xrightarrow{\simeq} M.$$

At the functor level, we have $S((M(r) \otimes F_r)_{\Sigma_r}) \simeq (M(r) \otimes \mathrm{Id}^{\otimes r})_{\Sigma_r}$ and $S(\iota_r(M))$ represents the canonical morphism

$$(M(r) \otimes \mathrm{Id}^{\otimes r})_{\Sigma_r} \to \bigoplus_{r=0}^{\infty}(M(r) \otimes \mathrm{Id}^{\otimes r})_{\Sigma_r} = S(M).$$

The morphism $Hom(F_r, M) \otimes F_r \to M$ induces a natural morphism $(Hom(F_r, M) \otimes F_r)_{\Sigma_r} \to M$. We check readily that the isomorphism of proposition 2.1.13 fits a commutative diagram

$$
\begin{array}{ccc}
(M(r) \otimes F_r)_{\Sigma_r} & \xdashrightarrow{\ \simeq\ } & (Hom(F_r, M) \otimes F_r)_{\Sigma_r} \\
& {\scriptstyle \iota_r(M)} \searrow & \downarrow {\scriptstyle \epsilon} \\
& & M.
\end{array}
$$

Equivalently, the isomorphism $\omega_r(M)$ corresponds to the morphism $\iota_r(M)$ under the adjunction relation

$$\mathrm{Mor}_{\mathcal{M}}((M(r) \otimes F_r)_{\Sigma_r}, M) \simeq \mathrm{Mor}_{\mathcal{C}}(M(r), Hom_{\mathcal{M}}(F_r, M))^{\Sigma_r}.$$

To conclude, proposition 2.1.13 and the discussion of §2.1.14 imply:

2.1.15 Proposition. *The objects F_r, $r \in \mathbb{N}$, define small projective generators of $\mathcal{M}$ in the sense of enriched categories. Explicitly, the functors*

$$Hom_{\mathcal{M}}(F_r, -) : M \mapsto Hom_{\mathcal{M}}(F_r, M)$$

preserve filtered colimits and coequalizers and the canonical morphism

$$\bigoplus_{r=0}^{\infty} Hom_{\mathcal{M}}(F_r, M) \otimes F_r \to M$$

is a regular epi, for all $M \in \mathcal{M}$. □

Note that the functors $S(F_r) = \mathrm{Id}^{\otimes r}$ do not generate $\mathcal{F}$ and do not form projective objects in $\mathcal{F}$ in general.

2.1.16 Remark. Since $F_r = I^{\otimes r}$, the isomorphism of §2.1.14 can be identified with an isomorphism

$$S(M, I) = \bigoplus_{r=0}^{\infty}(M(r) \otimes I^{\otimes r})_{\Sigma_r} \simeq M$$

between M and the Σ_*-object $S(M, I) \in \mathcal{M}$ associated to $I \in \mathcal{M}$ by the functor $S(M) : \mathcal{E} \to \mathcal{E}$ for $\mathcal{E} = \mathcal{M}$. This observation can be used to recover a Σ_*-object M from the associated collection of functors $S(M) : \mathcal{E} \to \mathcal{E}$, where $\mathcal{E}$ runs over all monoidal symmetric categories over $\mathcal{C}$.

2.2 Composition of Σ_*-Objects and Functors

The category of functors $\mathcal{F} = \mathcal{F}(\mathcal{E}, \mathcal{E})$ is equipped with another (non-symmetric) monoidal structure $(\mathcal{F}, \circ, \mathrm{Id})$ defined by the composition of functors $F, G \mapsto F \circ G$, together with the identity functor Id as a unit object. The category of Σ_*-objects has a (non-symmetric) monoidal structure that reflects the composition structure of functors. Formally, we have:

2.2.1 Proposition (see [17, 56]). *The category of Σ_*-objects $\mathcal{M}$ is equipped with a monoidal structure $(\mathcal{M}, \circ, I)$ so that the map $S : M \mapsto S(M)$ defines a functor of monoidal categories*

$$S : (\mathcal{M}, \circ, I) \to (\mathcal{F}(\mathcal{E}, \mathcal{E}), \circ, \mathrm{Id}),$$

for all symmetric monoidal categories $\mathcal{E}$ over $\mathcal{C}$.   □

The composition product of Σ_*-objects refers to the operation $M, N \mapsto M \circ N$ that yields this monoidal structure. For our purposes, we recall the construction of [14, §1.3] which uses the symmetric monoidal structure of the category of Σ_*-objects in the definition of the composition product $M, N \mapsto M \circ N$.

2.2.2 The Monoidal Composition Structure of the Category of Σ_*-Objects. In fact, the composite $M \circ N$ is defined by a generalized symmetric tensor construction formed in the category $\mathcal{E} = \mathcal{M}$:

$$M \circ N = S(M, N) = \bigoplus_{r=0}^{\infty} (M(r) \otimes N^{\otimes r})_{\Sigma_r}.$$

Since the functor $S : M \mapsto S(M)$ preserves colimits and tensor products, we have identities

$$S(M \circ N) = \bigoplus_{r=0}^{\infty} S(M(r) \otimes N^{\otimes r})_{\Sigma_r} = \bigoplus_{r=0}^{\infty} (M(r) \otimes S(N)^{\otimes r})_{\Sigma_r}.$$

Hence, we obtain immediately that this composition product $M \circ N$ satisfies the relation $S(M \circ N) \simeq S(M) \circ S(N)$, asserted by proposition 2.2.1.

The unit of the composition product is the object I, defined in §2.1.12, which corresponds to the identity functor $S(I) = \mathrm{Id}$. The isomorphism of §2.1.14, identified with

$$S(M, I) = \bigoplus_{r=0}^{\infty} (M(r) \otimes I^{\otimes r})_{\Sigma_r} \simeq M$$

(see §2.1.16), is equivalent to the right unit relation $M \circ I \simeq M$.

2.2.3 The Distribution Relation Between Tensor and Composition Products.
In the category of functors, the tensor product and the composition product satisfy the distribution relation $(F \otimes G) \circ S = (F \circ S) \otimes (G \circ S)$. In the category of Σ_*-modules, we have a natural distribution isomorphism

$$\theta(M, N, P) : (M \otimes N) \circ P \xrightarrow{\simeq} (M \circ P) \otimes (N \circ P)$$

which arises from the relation $S(M \otimes N, P) \simeq S(M, P) \otimes S(N, P)$ yielded by proposition 2.1.5. This distribution isomorphism reflects the distribution relation at the functor level. Formally, we have a commutative hexagon

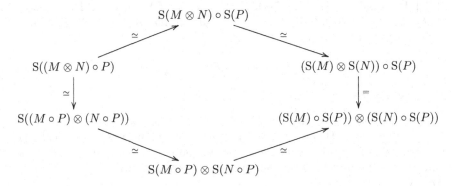

that connects the distribution isomorphism $\theta(M, N, P)$ to the functor identity $(S(M) \otimes S(N)) \circ S(P) = (S(M) \circ S(P)) \otimes (S(N) \circ S(P))$.

To summarize, we obtain:

2.2.4 Observation. *Let $\mathcal{F} = \mathcal{F}(\mathcal{E}, \mathcal{E})$. For any functor $S \in \mathcal{F}$, the composition product $F \mapsto F \circ S$ defines a functor of symmetric monoidal categories over $\mathcal{C}$*

$$- \circ S : (\mathcal{F}, \otimes, 1) \to (\mathcal{F}, \otimes, 1).$$

For any $N \in \mathcal{M}$, the composition product $M \mapsto M \circ N$ defines a functor of symmetric monoidal categories over $\mathcal{C}$

$$- \circ N : (\mathcal{M}, \otimes, 1) \to (\mathcal{M}, \otimes, 1)$$

and the diagram of functors

$$
\begin{array}{ccc}
\mathcal{M} & \xrightarrow{\ S\ } & \mathcal{F} \\
{\scriptstyle -\circ N}\downarrow & & \downarrow{\scriptstyle -\circ S(N)} \\
\mathcal{M} & \xrightarrow[\ S\ ]{} & \mathcal{F}
\end{array}
$$

commutes up to a natural equivalence of symmetric monoidal categories over C.

Besides, we check readily:

2.2.5 Observation. *The distribution isomorphisms $\theta(M, N, P)$ satisfy*

$$\theta(M, N, I) = \mathrm{id}$$

for the unit object $C = I$ and make commute the triangles

$$
\begin{array}{ccc}
(M \otimes N) \circ P \circ Q & \xrightarrow{\quad \theta(M,N,P \circ Q) \quad} & (M \circ P \circ Q) \otimes (N \circ P \circ Q) \\
& \theta(M,N,P) \circ Q \searrow \qquad \nearrow \theta(M \circ P, N \circ P, Q) & \\
& ((M \circ P) \otimes (N \circ P)) \circ Q &
\end{array}
\quad ,
$$

for all $M, N, P, Q \in \mathcal{M}$.

These coherence relations are obvious at the functor level since all isomorphisms are identities in this case.

2.3 Adjunction and Embedding Properties

In the context of a module category $\mathcal{E} = \mathcal{C} = \Bbbk\,\mathrm{Mod}$, where $\Bbbk$ is an infinite field, we recall in [14, §1.2] that the functor $S : M \mapsto S(M)$ is full and faithful. To prove this assertion, one can observe that the functor $S : M \mapsto S(M)$ has a right adjoint $\Gamma : G \mapsto \Gamma(G)$ so that the adjunction unit $\eta(M) : M \to \Gamma(S(M))$ forms an isomorphism (see proposition 1.2.5 in *loc. cit.*). In the general case of a module category $\mathcal{E} = \mathcal{C} = \Bbbk\,\mathrm{Mod}$, where $\Bbbk$ is any ground ring, we obtain further that $\eta(M) : M \to \Gamma(S(M))$ forms an isomorphism if M is a projective Σ_*-module (see proposition 2.3.12).

The aim of this section is to review these properties in the context of a symmetric monoidal category $\mathcal{E}$ over $\mathcal{C}$. For short, we set $\mathcal{F} = \mathcal{F}(\mathcal{E}, \mathcal{E})$.

Since we observe that the functor $S : \mathcal{M} \to \mathcal{F}$ preserves colimits, we obtain that this functor has a right adjoint $\Gamma : \mathcal{F} \to \mathcal{M}$. In a first part, we give an explicit construction of this adjoint functor $\Gamma : G \mapsto \Gamma(G)$. For this purpose, we assume that $\mathcal{C}$ has an internal hom, $\mathcal{E}$ is enriched over $\mathcal{C}$, and we generalize a construction of [14, §1.2]. In a second part, we observe that $S : M \mapsto S(M)$ extends to a functor of enriched categories and we prove that this functor $S : \mathcal{M} \mapsto \mathcal{F}$ is faithful in an enriched sense, at least if the category $\mathcal{E}$ is equipped with a faithful functor $\eta : \mathcal{C} \to \mathcal{E}$. Equivalently, we obtain that the adjunction unit $\eta(M) : M \to \Gamma(S(M))$ defines a monomorphism.

This account is motivated by the subsequent generalization of §8 in the context of right modules over operad. The results and constructions of this section are not used anywhere else in the book.

2.3.1 The Endomorphism Module of a Pair. Observe first that the functor $M \mapsto S(M, Y)$, for a fixed object $Y \in \mathcal{E}$, has a right adjoint. For this aim, form, for $X, Y \in \mathcal{E}$, the Σ_*-object $End_{X,Y}$ such that

$$End_{X,Y}(r) = Hom_{\mathcal{E}}(X^{\otimes r}, Y).$$

In §8.1.1, we observe that this Σ_*-object defines naturally a right module over End_X, the endomorphism operad of X, and we call this structure the *endomorphism module of the pair* (X, Y).

For the moment, observe simply:

2.3.2 Proposition (*cf.* [54, Proposition 2.2.7]). *We have a natural isomorphism*

$$Mor_{\mathcal{E}}(S(M, X), Y) \simeq Mor_{\mathcal{M}}(M, End_{X,Y})$$

for all $M \in \mathcal{M}$ and $X, Y \in \mathcal{E}$.

Proof. This adjunction relation arises from the canonical isomorphisms:

$$Mor_{\mathcal{E}}\left(\bigoplus_{r=0}^{\infty}(M(r) \otimes X^{\otimes r})_{\Sigma_r}, Y\right) \simeq \prod_{r=0}^{\infty} Mor_{\mathcal{E}}((M(r) \otimes X^{\otimes r})_{\Sigma_r}, Y)$$

$$\simeq \prod_{r=0}^{\infty} Mor_{\mathcal{C}}(M(r), Mor_{\mathcal{E}}(X^{\otimes r}, Y))^{\Sigma_r}$$

$$= Mor_{\mathcal{M}}(M, End_{X,Y}).$$

$\square$

2.3.3 Observation. Next (see observation 3.2.15) we observe that the map $S(N) : X \mapsto S(N, X)$ defines a functor $S(N) : \mathcal{E} \to {}_P\mathcal{E}$ to the category ${}_P\mathcal{E}$ of algebras over an operad P when N is equipped with the structure of a left P-module. One can observe that the endomorphism module $End_{X,Y}$ forms a left module over the endomorphism operad of Y. As a corollary, if $Y = B$ is a P-algebra, then we obtain that $End_{X,B}$ forms a left module over P by restriction of structures. In the context of P-algebras, we have an adjunction relation

$$Mor_{P\mathcal{E}}(S(N, X), B) \simeq Mor_{P\mathcal{M}}(N, End_{X,B})$$

for all $N \in {}_P\mathcal{M}$, $X \in \mathcal{E}$ and $B \in {}_P\mathcal{E}$, where ${}_P\mathcal{M}$ refers to the category of left P-modules (see §§3.2.9-3.2.10).

2.3.4 Definition of the Adjoint Functor $\Gamma : \mathcal{F} \to \mathcal{M}$. We apply the pointwise adjunction relation of proposition 2.3.2 to the category of functors $\mathcal{F}$.

In §2.1.1, we notice that the functor $S(M)$ satisfies

$$S(M) = \bigoplus_{r=0}^{\infty}(M(r) \otimes \mathrm{Id}^{\otimes r})_{\Sigma_r} = S(M, \mathrm{Id}),$$

where Id is the identity functor on $\mathcal{E}$. According to this relation, if we set $\Gamma(G) = End_{\mathrm{Id},G}$ for $G \in \mathcal{F}$, then proposition 2.3.2 returns:

2.3.5 Proposition. *The functor* $\Gamma : \mathcal{F} \rightarrow \mathcal{M}$ *defined by the map* $G \mapsto End_{\mathrm{Id},G}$ *is right adjoint to* $\mathrm{S} : \mathcal{M} \rightarrow \mathcal{F}$. $\qquad\square$

By proposition 1.1.16, we have as well:

2.3.6 Proposition. *The functors* $\mathrm{S} : \mathcal{M} \rightleftarrows \mathcal{F} : \Gamma$ *satisfy an enriched adjunction relation*

$$Hom_{\mathcal{M}}(\mathrm{S}(M), G) \simeq Hom_{\mathcal{F}}(M, \Gamma(G)),$$

where morphism sets are replaced by hom-objects over $\mathcal{C}$. $\qquad\square$

Proposition 1.1.15 implies that any functor of symmetric monoidal categories over $\mathcal{C}$, like $\mathrm{S} : \mathcal{M} \rightarrow \mathcal{F}$, defines a functor in the enriched sense. Accordingly, the map $f \mapsto \mathrm{S}(f)$, defined for morphisms of Σ_*-objects, extends to a morphism on hom-objects:

$$Hom_{\mathcal{M}}(M, N) \xrightarrow{\mathrm{S}} Hom_{\mathcal{F}}(\mathrm{S}(M), \mathrm{S}(N)).$$

By proposition 1.1.16, we obtain further:

2.3.7 Proposition. *The diagram*

commutes. $\qquad\square$

According to this assertion, we can use the adjunction unit $\eta(N) : N \rightarrow \Gamma(\mathrm{S}(N))$ and the adjunction relation between $\mathrm{S} : \mathcal{M} \rightarrow \mathcal{F}$ and $\Gamma : \mathcal{F} \rightarrow \mathcal{M}$ to determine $\mathrm{S} : Hom_{\mathcal{M}}(M, N) \rightarrow Hom_{\mathcal{F}}(\mathrm{S}(M), \mathrm{S}(N))$. In the converse direction, we can apply the morphism $\mathrm{S} : Hom_{\mathcal{M}}(M, N) \rightarrow Hom_{\mathcal{F}}(\mathrm{S}(M), \mathrm{S}(N))$ to the generating Σ_*-objects $M = F_r = I^{\otimes r}$, $r \in \mathbb{N}$, in order to determine the adjunction unit:

2.3.8 Proposition. *The component* $\eta(N) : N(r) \rightarrow Hom_{\mathcal{F}}(\mathrm{Id}^{\otimes r}, \mathrm{S}(N))$ *of the adjunction unit* $\eta(N) : N \rightarrow \Gamma(\mathrm{S}(N))$ *coincides with the morphism*

$$N(r) \xrightarrow{\simeq} Hom_{\mathcal{M}}(F_r, N) \xrightarrow{\mathrm{S}} Hom_{\mathcal{F}}(\mathrm{S}(F_r), \mathrm{S}(N)) \xrightarrow{\simeq} Hom_{\mathcal{F}}(\mathrm{Id}^{\otimes r}, \mathrm{S}(N))$$

formed by the composite of the isomorphism $\omega_r(N) : N(r) \xrightarrow{\simeq} Hom_{\mathcal{M}}(F_r, N)$ *of proposition 2.1.13, the morphism induced by the functor* $\mathrm{S} : \mathcal{M} \rightarrow \mathcal{F}$ *on hom-objects, and the isomorphism induced by the relation* $\mathrm{S}(F_r) \simeq \mathrm{Id}^{\otimes r}$.

Proof. This proposition is a consequence of proposition 2.3.7. In the case $M = F_r$, we obtain a commutative diagram:

$$
\begin{array}{ccc}
N(r) & \xrightarrow{\;\simeq\;} & Hom_{\mathcal{M}}(F_r, N) \\
{\scriptstyle \eta(N)}\downarrow & & {\scriptstyle \eta(N)_*}\downarrow \qquad \searrow^{\;S} \\
\Gamma(S(N))(r) & \xrightarrow[\simeq]{} Hom_{\mathcal{M}}(F_r, \Gamma(S(N))) & \xrightarrow[\simeq]{} Hom_{\mathcal{M}}(S(F_r), S(N)).
\end{array}
$$

One proves by a straightforward verification that the composite

$$
\Gamma(S(N))(r) \xrightarrow{\;\simeq\;} Hom_{\mathcal{M}}(F_r, \Gamma(S(N))) \xrightarrow{\;\simeq\;} Hom_{\mathcal{M}}(S(F_r), S(N))
$$
$$
\downarrow{\scriptstyle \simeq}
$$
$$
Hom_{\mathcal{M}}(\mathrm{Id}^{\otimes r}, N)
$$

is the identity morphism of $\Gamma(S(N))(r) = Hom_{\mathcal{M}}(\mathrm{Id}^{\otimes r}, S(N))$ and the proposition follows. $\square$

In the remainder of this section, we check that the morphism $S :$ $Hom_{\mathcal{M}}(M, N) \to Hom_{\mathcal{F}}(S(M), S(N))$ is mono under the assumption that the symmetric monoidal category $\mathcal{E}$ is equipped with a faithful functor $\eta : \mathcal{C} \to \mathcal{E}$. The proof of this observation is based on the next lemma:

2.3.9 Lemma. *Let $1^{\oplus r} = 1_1 \oplus \cdots \oplus 1_r$ be the sum of r copies of the unit object $1 \in \mathcal{C}$. For $M \in \mathcal{M}$, we have a canonical isomorphism*

$$
S(M, 1^{\oplus r}) \simeq \bigoplus_{n_1 + \cdots + n_r = n} M(n_1 + \cdots + n_r)_{\Sigma_{n_1} \times \cdots \times \Sigma_{n_r}}.
$$

Proof. We have Σ_n-equivariant isomorphisms

$$
(1_1 \oplus \cdots \oplus 1_r)^{\otimes n} \simeq \bigoplus_{(i_1, \ldots, i_n)} 1_{i_1} \otimes \cdots \otimes 1_{i_n}
$$
$$
\simeq \bigoplus_{(i_1, \ldots, i_n)} 1
$$

where the symmetric group Σ_n acts on n-tuples $(i_1, \ldots, i_n)$ by permutations of terms. We have an identification

$$
\bigoplus_{(i_1, \ldots, i_n)} 1 = \bigoplus_{n_1 + \cdots + n_r = n} 1[\Sigma_{n_1} \times \cdots \times \Sigma_{n_r} \backslash \Sigma_n],
$$

from which we deduce the splitting

$$(M(n) \otimes (1_1 \oplus \cdots \oplus 1_r)^{\otimes n})_{\Sigma_n} \simeq \bigoplus_{n_1 + \cdots + n_r = n} (M(n) \otimes 1[\Sigma_{n_1} \times \cdots \times \Sigma_{n_r} \backslash \Sigma_n])_{\Sigma_n}$$

$$\simeq \bigoplus_{n_1 + \cdots + n_r = n} M(n_1 + \cdots + n_r)_{\Sigma_{n_1} \times \cdots \times \Sigma_{n_r}}$$

and the lemma follows.   □

We deduce from lemma 2.3.9:

2.3.10 Proposition. *The functor* $S : \mathcal{M} \to \mathcal{F}(\mathcal{E}, \mathcal{E})$ *is faithful for all symmetric monoidal categories over* $\mathcal{C}$ *equipped with a faithful functor* $\eta : \mathcal{C} \to \mathcal{E}$.
   Moreover, the functor $S : \mathcal{M} \to \mathcal{F}(\mathcal{E}, \mathcal{E})$ *is faithful in an enriched sense. Explicitly, the morphism induced by* S *on hom-objects*

$$Hom_{\mathcal{M}}(M, N) \xrightarrow{S} Hom_{\mathcal{F}}(S(M), S(N))$$

is mono in $\mathcal{C}$, *for all* $M, N \in \mathcal{M}$.

Proof. The object $M(r)$ is isomorphic to the component $n_1 = \cdots = n_r = 1$ in the decomposition of lemma 2.3.9. As a byproduct, lemma 2.3.9 implies the existence of a natural monomorphism $\sigma(M) : M(r) \to S(M, 1^{\oplus r})$, for all $M \in \mathcal{M}$. From this assertion we deduce that S induces an injection on hom-sets

$$\mathrm{Mor}_{\mathcal{M}}(M, N) \xrightarrow{S} \mathrm{Mor}_{\mathcal{C}}(S(M, 1^{\oplus r}), S(N, 1^{\oplus r})),$$

for all $M, N \in \mathcal{M}$. If $\mathcal{E}$ is a symmetric monoidal category equipped with a faithful functor $\eta : \mathcal{C} \to \mathcal{E}$, then the map

$$\mathrm{Mor}_{\mathcal{C}}(S(M, 1^{\oplus r}), S(N, 1^{\oplus r})) \to \mathrm{Mor}_{\mathcal{E}}(S(M, 1^{\oplus r}), S(N, 1^{\oplus r}))$$

is injective as well. Hence we conclude readily that S induces an injection on hom-sets

$$\mathrm{Mor}_{\mathcal{M}}(M, N) \xrightarrow{S} \int_{X \in \mathcal{E}} \mathrm{Mor}_{\mathcal{C}}(S(M, X), S(N, X)) = \mathrm{Mor}_{\mathcal{F}}(S(M), S(N)),$$

for all $M, N \in \mathcal{M}$, and defines a faithful functor $S : \mathcal{M} \to \mathcal{F}$.
   In the context of enriched categories, we obtain that the map on hom-sets

$$\mathrm{Mor}_{\mathcal{M}}(C \otimes M, N) \xrightarrow{S} \mathrm{Mor}_{\mathcal{F}}(S(C \otimes M), S(N)) \simeq \mathrm{Mor}_{\mathcal{F}}(C \otimes S(M), S(N))$$

is injective for all $C \in \mathcal{C}$, $M, N \in \mathcal{M}$. By adjunction, we conclude immediately that

$$Hom_{\mathcal{M}}(M, N) \xrightarrow{S} Hom_{\mathcal{F}}(S(M), S(N))$$

is mono.   □

By proposition 2.3.7 and proposition 2.3.8, we have equivalently:

2.3.11 Proposition. *The adjunction unit*

$$\eta(N) : N \to \Gamma(\mathrm{S}(N))$$

is mono in $\mathcal{M}$, *for all* $N \in \mathcal{M}$. $\qquad\qquad\qquad\qquad\qquad\qquad\square$

We record stronger results in the case $\mathcal{E} = \mathcal{C} = \Bbbk\,\mathrm{Mod}$:

2.3.12 Proposition. *In the case* $\mathcal{E} = \mathcal{C} = \Bbbk\,\mathrm{Mod}$, *the category of modules over a ring* $\Bbbk$, *the adjunction unit* $\eta(M) : M \to \Gamma(\mathrm{S}(M))$ *is an isomorphism as long as* M *is a projective* Σ_*-*module or the ground ring is an infinite field.*

Proof. The case of an infinite ground field, recalled in the introduction of this section, is stated explicitly in [14, Proposition 1.2.5]. In the other case, one can check directly that the adjunction unit $\eta(M) : M \to \Gamma(\mathrm{S}(M))$ forms an isomorphism for the generating projective Σ_*-modules $M = F_r$, $r \in \mathbb{N}$. This implies that $\eta(M) : M \to \Gamma(\mathrm{S}(M))$ forms an isomorphism if M is a projective Σ_*-module. $\qquad\qquad\qquad\qquad\qquad\square$

2.4 Colimits

In §2.1.1, we observe that the functor $\mathrm{S} : M \mapsto \mathrm{S}(M)$ preserves colimits. Since colimits in functor categories are obtained pointwise, we obtain equivalently that the bifunctor $(M, X) \mapsto \mathrm{S}(M, X)$ preserves colimits in M, for any fixed object $X \in \mathcal{E}$.

In contrast, one can observe that the functor $\mathrm{S}(M) : \mathcal{E} \to \mathcal{E}$ associated to a fixed Σ_*-object does not preserve all colimits. Equivalently, the bifunctor $(M, X) \mapsto \mathrm{S}(M, X)$ does not preserve colimits in X in general.

Nevertheless:

2.4.1 Proposition (*cf.* [54, Lemma 2.3.3]). *The functor* $\mathrm{S}(M) : \mathcal{E} \to \mathcal{E}$ *associated to a* Σ_*-*object* $M \in \mathcal{M}$ *preserves filtered colimits and reflexive coequalizers.*

Proof. In proposition 1.2.1 we observe that the tensor power functors $\mathrm{Id}^{\otimes r} : X \mapsto X^{\otimes r}$ preserves filtered colimits and reflexive coequalizers. By assumption, the external tensor products $Y \mapsto M(r) \otimes Y$ preserves these colimits. By interchange of colimits, we deduce readily from these assertions that the functor $\mathrm{S}(M, X) = \bigoplus_{r=0}^{\infty} (M(r) \otimes X^{\otimes r})_{\Sigma_r}$ preserves filtered colimits and reflexive coequalizers as well. $\qquad\qquad\qquad\qquad\qquad\square$

As regards reflexive coequalizers, a first occurrence of proposition 2.4.1 appears in [51, §B.3] in the particular case of the symmetric algebra $V \mapsto \mathrm{S}(V)$ on dg-modules.

Chapter 3
Operads and Algebras in Symmetric Monoidal Categories

Introduction

Operads are used to define categories of algebras, usually within a fixed underlying symmetric monoidal category $\mathcal{C}$. But in the sequel, we form algebras in extensions of the category in which the operad is defined. Formally, we define algebras in symmetric monoidal categories $\mathcal{E}$ over the base category $\mathcal{C}$.

In principle, we can use the functor $\eta : \mathcal{C} \to \mathcal{E}$ associated to the structure of a symmetric monoidal category over $\mathcal{C}$ to transport all objects in $\mathcal{E}$. This functor maps an operad in $\mathcal{C}$ to an operad in $\mathcal{E}$. Thus we could just change the underlying category to retrieve standard definitions.

But this point of view is not natural in applications and supposes to forget the natural category in which the operad is defined. Besides, universal structures, like operads, are better determined by a collection rather than by a single category of algebras. For this reason, we fix the underlying category of operads, but we allow the underlying category of algebras to vary. The purpose of this section is to review basic constructions of the theory of operads in this setting.

In §3.1 we recall the definition of an operad and the usual examples of the commutative, associative and Lie operads, which are used throughout the book to illustrate our constructions. Though we study algebras over operads rather than operad themselves, we can not avoid this survey of basic definitions, at least to fix conventions.

In §3.2 we review the definition of an algebra over an operad in the context of symmetric monoidal categories over a base and we study the structure of algebras in functors and in Σ_*-objects. In §3.3 we review the construction of categorical operations (colimits, extension and restriction functors) in categories of algebras associated to operads. In §3.4 we address the definition of endomorphism operads in the context of enriched symmetric monoidal categories.

B. Fresse, *Modules over Operads and Functors*, Lecture Notes in Mathematics 1967,   53
DOI: 10.1007/978-3-540-89056-0_3, © Springer-Verlag Berlin Heidelberg 2009

3.1 Recollections: Operads

The purpose of this section is to recall the definition of an operad, at least to fix conventions. Though we have to extend the setting in which algebras over operads are defined, we still use the standard notion of an operad in a symmetric monoidal category.

For the sake of completeness, we also recall the definition of usual operad constructions, like free operads, and the classical examples of operads mentioned in the introduction of this chapter.

3.1.1 The Category of Operads. By definition, an *operad* in a symmetric monoidal category $\mathcal{C}$ consists of a Σ_*-object $P \in \mathcal{M}$ equipped with a composition product $\mu : P \circ P \to P$ and a unit morphism $\eta : I \to P$ that make commute the diagrams

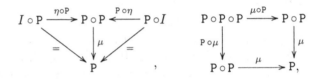

equivalent to usual unit and associativity relations of monoids. In short, an operad can be defined abstractly as a *monoid object* with respect to the composition structure of the category of Σ_*-objects (see [17] or [46, §1.8] for this definition). An operad morphism consists obviously of a morphism of Σ_*-objects $\phi : P \to Q$ that preserves operad structures. In principle, we use the notation $\mathcal{O}_\mathcal{C}$ for the category of operads in $\mathcal{C}$.

Most often, we consider operads within a base symmetric monoidal category $\mathcal{C}$, which is supposed to be fixed. In this context, we can omit to refer to the category in the definition of an operad and we adopt the short notation $\mathcal{O} = \mathcal{O}_\mathcal{C}$ for the category of operads in $\mathcal{C}$.

The initial object of the category of operads is obviously given by the composition unit I of the category of Σ_*-objects. For the category of operads, we use the base-set notation $\vee$ to denote the coproduct.

3.1.2 The Composition Morphisms of an Operad. The composition morphism of an operad $\mu : P \circ P \to P$ is equivalent to a collection of morphisms

$$P(r) \otimes P(n_1) \otimes \cdots \otimes P(n_r) \xrightarrow{\mu} P(n_1 + \cdots + n_r)$$

that satisfy natural equivariance properties. This assertion follows from the explicit definition of the composition product of Σ_*-objects. The unit morphism $\eta : I \to P$ is equivalent to a morphism $\eta : 1 \to P(1)$ in $\mathcal{C}$. The unit and associativity relations of operads have natural expressions in terms of these componentwise composition and unit morphisms (May's axioms, see [47]).

In the point-set context, we use the notation $p(q_1, \ldots, q_r) \in P(n_1 + \cdots + n_r)$ to refer to the composite of the elements $p \in P(r)$ and $q_1 \in P(n_1), \ldots, q_r \in P(n_r)$ in the operad. The unit morphism is determined by a unit element $1 \in P(1)$.

3.1.3 Partial Composites. We also use partial composites

$$P(r) \otimes P(s) \xrightarrow{\circ_e} P(r + s - 1),$$

defined by composites

$$P(r) \otimes P(s) \simeq P(r) \otimes (1 \otimes \cdots \otimes P(s) \otimes \cdots \otimes 1)$$
$$\to P(r) \otimes (P(1) \otimes \cdots \otimes P(s) \otimes \cdots \otimes P(1)) \xrightarrow{\mu} P(r + s - 1),$$

where operad units $\eta : I \to P(1)$ are applied at positions $k \neq e$ of the tensor product. In the point-set context, we have $p \circ_e q = p(1, \ldots, q, \ldots, 1)$, where $q \in P(s)$ is set at the eth entry of $p \in P(r)$.

The unit and associativity relations imply that the composition morphism of an operad is determined by its partial composition products. The unit and associativity relations of the composition morphism have an equivalent formulation in terms of partial composition products (see [46]).

3.1.4 The Intuitive Interpretation of Operads. The elements of an operad $p \in P(n)$ have to be interpreted as operations of n-variables

$$p = p(x_1, \ldots, x_n).$$

The composition morphism of an operad models composites of such operations. The partial composites represent composites of the form

$$p \circ_e q = p(x_1, \ldots, x_{e-1}, q(x_e, \ldots, x_{e+n-1}), x_{i+n}, \ldots, x_{m+n-1}),$$

for $p \in P(m)$, $q \in P(n)$. The action of permutations $w \in \Sigma_n$ on $P(n)$ models permutations of variables:

$$wp = p(x_{w(1)}, \ldots, x_{w(n)}).$$

The unit element $1 \in P(1)$ has to be interpreted as an identity operation $1(x_1) = x_1$.

3.1.5 Free Operads. The obvious forgetful functor $U : \mathcal{O} \to \mathcal{M}$ from operads to Σ_*-objects has a left adjoint $F : \mathcal{M} \to \mathcal{O}$ which maps any Σ_*-object M to an associated free operad $F(M)$.

In the point-set context, the free operad $F(M)$ on a Σ_*-object M consists roughly of all formal composites

$$(\cdots ((\xi_1 \circ_{e_2} \xi_2) \circ_{e_3} \cdots) \circ_{e_r} \xi_r$$

of generating elements $\xi_1 \in M(m_1), \ldots, \xi_r \in M(m_r)$. The definition of the partial composition product of the free operad $\circ_i : F(M) \otimes F(M) \to F(M)$ is deduced from associativity relations of partial composites.

The proper explicit construction of the free operad, for which we refer to [18], [14, §1.1.9], [46], uses the language of trees. This construction comes from Boardman-Vogt' original monograph [6] where free objects are introduced in the category of algebraic theories, the structures, more general than operads, which model operations of the form $p : X^{\times m} \to X^{\times n}$.

3.1.6 Limits and Colimits of Operads. The forgetful functor $U : \mathcal{O} \to \mathcal{M}$ which maps an operad to its underlying Σ_*-object creates all limits, creates reflexive coequalizers and filtered colimits as well, but does not preserve all colimits.

In fact, for a limit of operads, we can use the canonical morphism $(\lim_i P_i) \circ (\lim_i P_i) \to \lim_i(P_i \circ P_i)$ to define a composition product $\mu : (\lim_i P_i) \circ (\lim_i P_i) \to (\lim_i P_i)$. For colimits, the natural morphism $\mathrm{colim}_i(P_i \circ P_i) \to (\mathrm{colim}_i P_i) \circ (\mathrm{colim}_i P_i)$ goes in the wrong direction to define a composition product on $\mathrm{colim}_i P_i$. But proposition 2.4.1 and proposition 1.2 imply that this natural morphism $\mathrm{colim}_i(P_i \circ P_i) \to (\mathrm{colim}_i P_i) \circ (\mathrm{colim}_i P_i)$ is iso in the case of a reflexive coequalizer and in the case of a filtered colimit. Therefore, for these particular colimits, we can form a composition product

$$(\mathrm{colim}_i P_i) \circ (\mathrm{colim}_i P_i) \xleftarrow{\simeq} \mathrm{colim}_i(P_i \circ P_i) \to \mathrm{colim}_i P_i$$

to provide $\mathrm{colim}_i P_i$ with an operad structure. The conclusion follows readily.

The existence of reflexive coequalizers and filtered colimits in the category of operads together with the existence of free objects suffices to prove the existence of all colimits in the category of operads. But we insist that the forgetful functor $U : \mathcal{O} \to \mathcal{M}$ does not preserves all colimits.

3.1.7 Operads Defined by Generators and Relations. The classical basic examples of operads, the commutative operad C, the associative operad A, and the Lie operad L, are associated to the classical structure of an associative and commutative algebra, of an associative algebras, and of a Lie algebra respectively. These operads can be defined naturally by generators and relations. This definition reflects the usual definition of a commutative algebra, of an associative algebra, of a Lie algebra, as an object equipped with generating operations that satisfy a set of relations.

In this paragraph we recall briefly the general construction of operads by generators and relations. For more details on this construction we refer to [18] and [14, §1.1.9]. The examples of the commutative operad C, the associative operad A, and the Lie operad L are addressed in the next paragraph.

In the usual context of $\Bbbk$-modules, a presentation of an operad P by generators and relations consists of a Σ_*-object M, whose elements represent the generating operations of P, together with a collection $R \subset F(M)$ whose elements represent generating relations between composites of the generating operations $\xi \in M$. The operad is defined as the quotient $\mathsf{P} = \mathsf{F}(M)/(R)$, where (R) refers to the ideal generated by (R) in the operadic sense. This ideal (R) is spanned by composites which include a factor of the form $p = w\rho$, where $w \in \Sigma_n$ and $\rho \in R(n)$. The definition of an operad ideal implies essentially that a morphism on the free operad $\nabla_f : \mathsf{F}(M) \to \mathsf{Q}$ induces a morphism on the quotient operad $\overline{\nabla}_f : \mathsf{F}(M)/(R) \to \mathsf{Q}$ if and only if ∇_f cancels the generating relations $\rho \in R$.

In the general setting of symmetric monoidal categories, a presentation of an operad P by generators and relations is defined by a coequalizer of the form

$$\mathsf{F}(R) \underset{d_1}{\overset{d_0}{\rightrightarrows}} \mathsf{F}(M) \longrightarrow \mathsf{P} \ .$$

The generating relations are represented by identities $\rho^0 \equiv \rho^1$ in the free operad $\mathsf{F}(M)$, for pairs $\rho^0 = d^0(\rho)$, $\rho^1 = d^1(\rho)$, where $\rho \in R$. The quotient of an operad by an ideal $\mathsf{P} = \mathsf{F}(M)/(R)$ is equivalent to a presentation of this form where $d_1 = 0$. By definition of free operads, a morphism $\nabla_f : \mathsf{F}(M) \to \mathsf{Q}$ induces a morphism on the coequalizer $\overline{\nabla}_f : \mathsf{P} \to \mathsf{Q}$ if and only if the identity $\nabla_f d^0 = \nabla_f d^1$ holds on $R \subset \mathsf{F}(R)$.

3.1.8 Basic Examples in $\Bbbk$-Modules. The classical examples of operads, the commutative operad C, the associative operad A, and the Lie operad L, are all defined by a presentation $\mathsf{P} = \mathsf{F}(M)/(R)$, such that the Σ_*-object M is spanned by operations of 2-variables $\xi = \xi(x_1, x_2)$. In all examples, we use the notation τ to refer to the transposition $\tau = (1\ 2) \in \Sigma_2$.

(a) The *commutative operad* C is defined by a presentation of the form

$$\mathsf{C} = \mathsf{F}(\Bbbk\,\mu)/(\text{associativity}),$$

where the generating Σ_*-object $M = \Bbbk\,\mu$ is spanned by a single operation $\mu = \mu(x_1, x_2)$ so that $\tau\mu = \mu$. Equivalently, the operation $\mu = \mu(x_1, x_2)$ satisfies the symmetry relation

$$\mu(x_1, x_2) = \mu(x_2, x_1).$$

The ideal (associativity) is generated by the element $\mu \circ_1 \mu - \mu \circ_2 \mu \in \mathsf{F}(\Bbbk\,\mu)(3)$ which represents the associativity relation

$$\mu(\mu(x_1, x_2), x_3) \equiv \mu(x_1, \mu(x_2, x_3)).$$

Thus the operation $\mu(x_1, x_2)$ represents a commutative and associative product $\mu(x_1, x_2) = x_1 \cdot x_2$.

Any element $p \in \mathsf{C}(n)$ has a unique representative of the form

$$p = \mu(\cdots \mu(\mu(x_1, x_2), x_3), \ldots), x_n).$$

Equivalently, the commutative operad is spanned in arity n by the single multilinear monomial of n commutative variables:

$$C(n) = \Bbbk\, x_1 \cdot \ldots \cdot x_n.$$

The symmetric group acts trivially on $C(n)$.

This construction gives an operad without unitary operations, an operad such that $C(0) = 0$.

(b) The *associative operad* A is defined by a presentation of the same form as the commutative operad

$$A = F(\Bbbk\, \mu \oplus \Bbbk\, \tau\mu)/(\text{associativity}),$$

but where the generating Σ_*-object $M = \Bbbk\, \mu \oplus \Bbbk\, \tau\mu$ is spanned by an operation $\mu = \mu(x_1, x_2)$ and its transposite $\tau\mu = \mu(x_2, x_1)$. The ideal (associativity) is generated again by the element $\mu \circ_1 \mu - \mu \circ_2 \mu \in F(\Bbbk\, \mu \oplus \Bbbk\, \tau\mu)(3)$ which gives an associativity relation in the quotient. Thus the operation $\mu(x_1, x_2)$ represents an associative but non-commutative product $\mu(x_1, x_2) = x_1 \cdot x_2$.

The elements of the associative operad can be represented by multilinear monomials of n non-commutative variables. Thus we have an identity

$$A(n) = \bigoplus_{(i_1, \ldots, i_n)} \Bbbk\, x_{i_1} \cdot \ldots \cdot x_{i_n}$$

where the indices $(i_1, \ldots, i_n)$ range over permutations of $(1, \ldots, n)$. The symmetric group acts by translation on $A(n)$. Thus we can also identify $A(n)$ with the regular representation $A(n) = \Bbbk[\Sigma_n]$ for $n > 0$. This construction gives an operad without unitary operations, an operad such that $A(0) = 0$.

(c) The *Lie operad* L is defined by a presentation of the form

$$L = F(\Bbbk\, \gamma)/(\text{Jacobi}),$$

where the generating operation $\gamma = \mu(x_1, x_2)$ satisfies $\tau\gamma = -\gamma$. Equivalently, the operation $\gamma = \gamma(x_1, x_2)$ satisfies the antisymmetry relation $\gamma(x_1, x_2) = -\gamma(x_2, x_1)$. The ideal (Jacobi) is generated by the element $(1 + c + c^2) \cdot \gamma \circ_1 \gamma \in F(\Bbbk\, \gamma)(3)$, where c represents the 3-cycle $c = (1\ 2\ 3) \in \Sigma_3$. This element gives the Jacobi relation

$$\gamma(\gamma(x_1, x_2), x_3) + \gamma(\gamma(x_1, x_2), x_3) + \gamma(\gamma(x_1, x_2), x_3) \equiv 0$$

of a Lie bracket $\gamma(x_1, x_2) = [x_1, x_2]$.

The elements of the Lie operad can be represented by multilinear Lie monomials of n variables. According to [53, §5.6], we have an identity

$$L(n) = \bigoplus_{(i_1,\dots,i_n)} \Bbbk[\cdots[[x_{i_1}, x_{i_2}],\dots], x_{i_n}],$$

where the indices $(i_1, \dots, i_n)$ range over the permutations of $(1, \dots, n)$ such that $i_1 = 1$, but the action of the symmetric group on this monomial basis is non-trivial.

3.1.9 The Commutative and Associative Operads in Sets.

In fact, the commutative and the associative operads can be defined within the category of sets. The commutative and associative operads in $\Bbbk$-modules consist of free $\Bbbk$-modules on the set versions of these operads. This assertion can be deduced from the explicit description of the previous paragraph.

For the set commutative operad, we have $C(n) = *$, the base point equipped with a trivial Σ_n-action. For the set associative operad, we have $A(n) = \Sigma_n$, the set of permutations equipped with the action of the symmetric group by left translations. In both cases, we assume $C(0) = A(0) = \emptyset$ to have operads without unitary operations.

In the case of the commutative operad, the partial composition products $\circ_e : C(m) \times C(n) \to C(m + n - 1)$ are trivially given by identity maps. In the case of the commutative operad, the partial composition products $\circ_e : A(m) \times A(n) \to A(m + n - 1)$ have a natural combinatorial expression in terms of permutations: identify any permutation $w \in \Sigma_n$ with the sequence of its values $w = (w(1), \dots, w(r))$; to form the composite $s \circ_e t$ of $s = (i_1, \dots, i_m)$ with $t = (j_1, \dots, j_n)$, we replace the term $i_k = e$ in the sequence $s = (i_1, \dots, i_m)$ by the sequence $t = (j_1, \dots, j_n)$ and we perform standard shifts $i_* \mapsto i_* + n$ on $i_* = e + 1, \dots, m$, respectively $j_* \mapsto j_* + e$ on $j_* = 1, \dots, n$. As an example, we have $(1, 3, \underline{2}, 4) \circ_2 (2, 1, 3) = (1, 5, \underline{3}, 2, \underline{4}, 6)$. In the sequel, we also refer to this structure as the *permutation operad*.

The definition of the commutative and associative operads in sets can also be deduced from the general construction of operads by generators and relations in symmetric monoidal categories.

One can use the natural functor $1[-] : \text{Set} \to \mathcal{C}$, associated to any symmetric monoidal category $\mathcal{C}$, to transport the commutative and associative operads to any such category $\mathcal{C}$. In the context of $\Bbbk$-modules, we get the commutative and associative operads defined in §3.1.8 since we have identifications $C(n) = \Bbbk$ and $A(n) = \Bbbk[\Sigma_n]$ for all $n > 0$.

3.1.10 Non-unitary Operads and the Unitary Commutative and Associative Operads.

In the sequel, we say that an operad P in a symmetric monoidal category $\mathcal{C}$ is *non-unitary* if we have $P(0) = 0$, the initial object of $\mathcal{C}$. In the literature, non-unitary operads are also called *connected* [14], or *reduced* [4]. The term *unitary* is used to replace the term *unital* used [47] to refer to an operad P such that $P(0) = 1$, the monoidal unit of $\mathcal{C}$. The term unital is now used to refer to an operad equipped with a unit element

$1 \in P(1)$ (see [46]). Therefore we prefer to introduce another terminology to avoid confusions.

The operads C and A, defined in §§3.1.8-3.1.9, are non-unitary versions of the commutative and associative operads.

The commutative and associative operads have unitary versions, C_+ and A_+, such that $C_+(n) = \Bbbk$, for all $n \in \mathbb{N}$, respectively $A_+(n) = \Bbbk[\Sigma_n]$, for all $n \in \mathbb{N}$. In the context of sets, we have similarly $C_+(n) = *$, for all $n \in \mathbb{N}$, respectively $A_+(n) = \Sigma_n$, for all $n \in \mathbb{N}$.

The partial composition products of C_+, respectively A_+, are defined by the formula of §3.1.9 which has a natural and obvious extension for unitary operations.

3.1.11 Initial Unitary Operads. In contrast to non-unitary operads, we have operads $*_C$, associated to objects $C \in \mathcal{C}$, so that $*_C(0) = C$, $*_C(1) = 1$, and $*_C(n) = 0$ for $n > 1$. The composition morphisms of such operads are reduced to a component $\mu : *_C(1) \otimes *_C(0) \to *_C(1)$ given, according to the unit relation of an operad, by the natural isomorphism $1 \otimes C \simeq C$.

We call this operad $*_C$ the *initial unitary operad associated to the object* $C \in \mathcal{C}$. We shorten the notation to $* = *_1$ for the unit object of the monoidal category.

3.2 Basic Definitions: Algebras over Operads

In the classical theory of operads, one uses that the functor $S(P)$ associated to an operad P forms a monad to define the category of algebras associated to P. The first purpose of this section is to review this definition in the relative context of a symmetric monoidal category over the base. To extend the usual definition, we use simply that a functor $S(P) : \mathcal{E} \to \mathcal{E}$ is defined on every symmetric monoidal category $\mathcal{E}$ over the base category $\mathcal{C}$. By definition, a P-algebra in $\mathcal{E}$ is an algebra over the monad $S(P) : \mathcal{E} \to \mathcal{E}$.

In the previous chapter, we recall that the category of Σ_*-objects $\mathcal{M}$, as well as categories of functors $F : \mathcal{A} \to \mathcal{E}$, are instances of symmetric monoidal categories over $\mathcal{C}$. We examine the structure of a P-algebra in the category of Σ_*-objects $\mathcal{M}$ and of a P-algebra in a category of functors $F : \mathcal{A} \to \mathcal{E}$. We obtain that the former are equivalent to left P-modules, the latter are equivalent to functors $F : \mathcal{A} \to {}_P\mathcal{E}$ to the category of P-algebras in $\mathcal{E}$.

3.2.1 Algebras over Operads. Let P be any operad in the base symmetric monoidal category $\mathcal{C}$. Let $\mathcal{E}$ be any symmetric monoidal category over $\mathcal{C}$. The structure of a P-algebra in $\mathcal{E}$ consists explicitly of an object $A \in \mathcal{E}$ equipped with an evaluation morphism $\lambda : S(P, A) \to A$ that makes commute the diagrams

$$S(I,A) \xrightarrow{S(\eta,A)} S(P,A) \qquad\qquad S(P \circ P, A) \xrightarrow{\;\simeq\;} S(P, S(P,A)) \xrightarrow{S(P,\lambda)} S(P,A)$$

$$S(I,A) \xrightarrow{\;\simeq\;} A, \qquad S(\mu,A)\big\downarrow \qquad\qquad\qquad\qquad\qquad\qquad \big\downarrow \lambda$$

$$\lambda \big\downarrow \qquad\qquad S(P,A) \xrightarrow{\qquad\qquad \lambda \qquad\qquad} A,$$

equivalent to usual unit and associativity relations of monoid actions.

The category of P-algebras in $\mathcal{E}$ is denoted by $_P\mathcal{E}$, a morphism of P-algebras $f : A \to B$ consists obviously of a morphism in $\mathcal{E}$ which commutes with evaluation morphisms.

The definition of $S(P,A)$ implies immediately that the evaluation morphism $\lambda : S(P,A) \to A$ is also equivalent to a collection of morphisms

$$\lambda : P(n) \otimes A^{\otimes n} \to A,$$

which are equivariant with respect to the action of symmetric groups, and for which we have natural unit and associativity relations equivalent to the unit and associativity relations of operad actions. In the standard setting, the tensor product $P(n) \otimes A^{\otimes n}$ is formed within the base category $\mathcal{C}$. In our context, we use the internal and external tensor products of $\mathcal{E}$ to form this tensor product. In a point-set context, the evaluation of the operation $p \in P(n)$ on elements $a_1, \ldots, a_n \in A$ is usually denoted by $p(a_1, \ldots, a_n) \in A$. The unit relation reads $1(a) = a$, $\forall a \in A$, and the associativity relation can be written in terms of partial composites:

$$p \circ_e q(a_1, \ldots, a_{m+n-1}) = p(a_1, \ldots, a_{e-1}, q(a_e, \ldots, a_{e+n-1}), a_{e+n}, \ldots, a_{m+n-1}),$$

for $p \in P(m)$, $q \in P(n)$, $a_1, \ldots, a_{m+n-1} \in A$.

3.2.2 Non-unitary Operads and Algebras in Reduced Symmetric Monoidal Categories. Recall that an operad P is *non-unitary* if we have $P(0) = 0$. One checks readily that the image of $P(0) \in \mathcal{C}$ under the canonical functor $\eta : \mathcal{C} \to \mathcal{E}$ defines the initial object of the category of P-algebras in $\mathcal{E}$. Accordingly, an operad P is non-unitary if and only if the initial object of the category of P-algebras in $\mathcal{E}$ is defined by the initial object of $\mathcal{E}$, for every symmetric monoidal category over $\mathcal{C}$.

For a non-unitary operad P, we can generalize the definition of a P-algebra to every reduced symmetric monoidal category over $\mathcal{C}$. Use simply the functor $S^0(P) : \mathcal{E}^0 \to \mathcal{E}^0$, defined on any such category $\mathcal{E}^0$ by the formula of §2.1.3:

$$S^0(P, X) = \bigoplus_{n=1}^{\infty} (P(n) \otimes X^{\otimes n})_{\Sigma_n},$$

to form the evaluation morphism of P-algebras in $\mathcal{E}^0$.

The next constructions of this chapter (free objects, colimits, extension and restriction functors, endomorphism operads) have a straightforward generalization in the context of non-unitary operads and reduced symmetric monoidal categories.

3.2.3 Algebras over Free Operads and over Operads Defined by Generators and Relation.

If $P = F(M)$ is a free operad, then the structure of a P-algebra on an object $A \in \mathcal{E}$ is equivalent to a collection of morphisms $\lambda : M(n) \otimes A^{\otimes n} \to A$, $n \in \mathbb{N}$, with no relation required outside Σ_n-equivariance relations. Informally, the associativity of operad actions implies that the action of formal composites $(\cdots ((\xi_1 \circ_{e_2} \xi_2) \circ_{e_3} \cdots) \circ_{e_r} \xi_r$, which span the free operad $F(M)$, is determined by the action of the generating operations $\xi_i \in M(n_i)$ on A. To prove the assertion properly, one has to use the endomorphism operad End_A, whose definition is recalled later on (in §3.4) and the equivalence of proposition 3.4.3 between operad actions $\lambda : S(P, A) \to A$ and operad morphisms $\nabla : P \to End_A$. In the case of a free operad, an operad morphism $\nabla : F(M) \to End_A$ is equivalent to a morphism of Σ_*-objects $\nabla : M \to End_A$ and this equivalence imply our assertion.

If $P = F(M)/(R)$ is an operad given by generators and relations in $\Bbbk$-modules, then the structure of a P-algebra on an object $A \in \mathcal{E}$ is determined by a collection of morphisms $\lambda : M(n) \otimes A^{\otimes n} \to A$ such that the generating relations $\rho \in R$ are canceled by the induced action of the free operad $F(M)$ on A. This assertion is also an immediate consequence of the equivalence of proposition 3.4.3 applied to a quotient operad $P = F(M)/(R)$.

3.2.4 Examples: Algebras over Classical Operads in $\Bbbk$-Modules.

The usual structures of commutative algebras, respectively associative algebras, have natural generalizations in the setting of a symmetric monoidal category. The generalized commutative algebra structures, respectively associative algebras, are also called commutative monoids, respectively associative monoids, because these structures are equivalent to usual commutative monoids, respectively associative monoids, in the context of sets. The structure of a Lie algebra has as well a natural generalization in the setting of a symmetric monoidal category over $\mathbb{Z}$-modules.

The equivalence of §3.2.3 implies that the structures of algebras over the usual operads $P = C, L, A$, agree with such generalizations in the context of symmetric monoidal categories over $\Bbbk$-modules. Indeed, for these operads $P = C, L, A$, the assertions of §3.2.3 return:

(a) The structure of a C-algebra in $\mathcal{E}$ is determined by a product $\mu : A \otimes A \to A$ formed in the category $\mathcal{E}$, which satisfies the symmetry relation $\mu \cdot \tau^* = \mu$, where $\tau^* : X \otimes Y \to Y \otimes X$ refers to the symmetry isomorphism of $\mathcal{E}$, and the associativity relation $\mu(\mu, \mathrm{id}) = \mu(\mathrm{id}, \mu)$.

(b) The structure of an A-algebra in $\mathcal{E}$ is determined by a product $\mu : A \otimes A \to A$ which satisfies the associativity relation $\mu(\mu, \mathrm{id}) = \mu(\mathrm{id}, \mu)$, but without any commutativity requirement.

(c) The structure of an L-algebra in $\mathcal{E}$ is determined by a morphism $\gamma : A \otimes A \to A$ which satisfies the antisymmetry relation $\gamma \cdot \tau^* + \gamma = 0$ and a Jacobi relation

$$\gamma(\gamma, \mathrm{id}) + \gamma(\gamma, \mathrm{id}) \cdot c^* + \gamma(\gamma, \mathrm{id}) \cdot c^* \cdot c^* = 0,$$

where $c^* : A \otimes A \otimes A \to A \otimes A \otimes A$ refers to the cyclic permutation of 3-tensors in $\mathcal{E}$.

To summarize, say that C-algebras (respectively, A-algebras) in $\mathcal{E}$ are equivalent to commutative (respectively, associative) algebras (without unit) in $\mathcal{E}$, and L-algebras in $\mathcal{E}$ are equivalent to Lie algebras in $\mathcal{E}$. Note that L-algebras in $\Bbbk$-modules do not necessarily satisfy the relation $\gamma(a, a) = 0$, where the element a is repeated. In our sense, a Lie algebra is only assumed to satisfy an antisymmetry relation $\gamma(a, b) + \gamma(b, a) = 0$.

If we take the set versions of the commutative (respectively associative operads), then the identity between C-algebras (respectively, A-algebras) and usual commutative (respectively, associative) algebras (without unit) holds in any symmetric monoidal category, and not only in symmetric monoidal categories over $\Bbbk$-modules. This result can still be deduced from the presentation of these operads in sets, by an obvious generalization of the equivalence of §3.2.3. In the next paragraph, we reprove this equivalence by another approach to illustrate the abstract definition of an algebra over an operad.

3.2.5 More on Commutative and Associative Algebras. In this paragraph, we consider the set-theoretic version of the commutative operad C (respectively, of the associative operad A) and we adopt the set-monoid language to refer to commutative (respectively, associative) algebras in symmetric monoidal categories.

Since the operads $P = C, A$ have a simple underlying Σ_*-object, we can easily make explicit the monad $S(P) : \mathcal{E} \to \mathcal{E}$ associated to these operads and the evaluation morphism $S(P, A) \to A$ for algebras over these operads. Indeed, the identity $C(n) = *$ gives immediately

$$S(C, X) = \bigoplus_{n=1}^{\infty} (X^{\otimes n})_{\Sigma_n}.$$

The identity $A(n) = \Sigma_n$ gives the relation

$$S(A, X) = \bigoplus_{n=1}^{\infty} (\Sigma_n \otimes X^{\otimes n})_{\Sigma_n} \simeq \bigoplus_{n=1}^{\infty} X^{\otimes n}.$$

The components $\lambda : (A^{\otimes n})_{\Sigma_n} \to A$ of the evaluation morphism of a C-algebra are equivalent to the n-fold products of a commutative and associative monoid. The product is yielded by the 2-ary component of the operad action: $\lambda : (A^{\otimes 2})_{\Sigma_2} \to A$. The identity between the n-ary component of the operad action $\lambda : (A^{\otimes n})_{\Sigma_n} \to A$ and a composite of the product $\lambda : (A^{\otimes 2})_{\Sigma_2} \to A$ comes from the associativity relation for operad actions. In this manner we still obtain that C-algebras in $\mathcal{E}$ are equivalent to commutative and associative monoids (without unit) in $\mathcal{E}$.

Similarly, the components $\lambda : A^{\otimes n} \to A$ of the evaluation morphism of an A-algebra are equivalent to the n-fold products of an associative monoid, and we can still obtain that A-algebras in $\mathcal{E}$ are equivalent to associative monoids (without unit) in $\mathcal{E}$.

For the unitary versions of the commutative and associative operads, the evaluation morphisms have an additional component $\lambda : * \to A$ since $C_+(0) = A_+(0) = *$. This morphism provides C_+-algebras (respectively, A_+-algebras) with a unit. Hence we obtain that algebras over the unitary commutative operad C_+ (respectively over the unitary associative operad C_+) are equivalent to commutative and associative monoids with unit in $\mathcal{E}$ (respectively, to associative monoids with unit).

3.2.6 Algebras over Initial Unitary Operads. We can also easily determine the category of algebras associated to the initial unitary operads such that $*_C(0) = C$, for an object $C \in \mathcal{C}$. Since we have $*_C(n) = 0$ for $n > 0$, and $\lambda : *_C(1) \otimes A = 1 \otimes A \to A$ is necessarily the natural isomorphism $1 \otimes A \simeq A$, we obtain that the structure of a $*_C$-algebra is completely determined by a morphism $\lambda : C \to A$ in the category $\mathcal{E}$. Hence the category of $*_C$-algebras in $\mathcal{E}$ is isomorphic to the comma category of morphisms $\lambda : C \to X$, where $X \in \mathcal{E}$.

3.2.7 Algebras in Functor Categories and Functors to Algebras over Operads. In §2.1.4, we observe that a category of functors $\mathcal{F} = \mathcal{F}(\mathcal{A}, \mathcal{E})$, where $\mathcal{E}$ is a symmetric monoidal category over $\mathcal{C}$, forms naturally a symmetric monoidal category over $\mathcal{C}$. Hence, we can use our formalism to define the notion of a P-algebra in $\mathcal{F}$, for P an operad in $\mathcal{C}$.

By definition, the action of an operad P on a functor $F : \mathcal{A} \to \mathcal{E}$ is determined by collections of morphisms $\lambda(X) : P(r) \otimes F(X)^{\otimes r} \to F(X)$ that define a natural transformation in $X \in \mathcal{A}$. Accordingly, we obtain readily that a functor $F : \mathcal{A} \to \mathcal{E}$ forms a P-algebra in $\mathcal{F}$ if and only if all objects $F(X)$, $X \in \mathcal{A}$, are equipped with a P-algebra structure so that the map $X \mapsto F(X)$ defines a functor from $\mathcal{A}$ to the category of P-algebras in $\mathcal{E}$.

Thus, we obtain:

3.2.8 Observation. *The category of* P-*algebras in* $\mathcal{F}(\mathcal{A}, \mathcal{E})$ *is isomorphic to* $\mathcal{F}(\mathcal{A}, {}_P\mathcal{E})$, *the category of functors* $F : \mathcal{A} \to {}_P\mathcal{E}$ *from* $\mathcal{A}$ *to the category* P-*algebras in* $\mathcal{E}$.

Equivalently, we have a category identity ${}_P\mathcal{F}(\mathcal{A}, \mathcal{E}) = \mathcal{F}(\mathcal{A}, {}_P\mathcal{E})$.

For a fixed symmetric monoidal category over $\mathcal{C}$, we use the short notation $\mathcal{F}$ to refer to the category of functors $\mathcal{F} = \mathcal{F}(\mathcal{E}, \mathcal{E})$. According to observation 3.2.8, the notation ${}_P\mathcal{F}$ is coherent with our conventions to refer to the category of functors ${}_P\mathcal{F} = \mathcal{F}(\mathcal{E}, {}_P\mathcal{E})$ and we shall use this short notation in the sequel.

3.2.9 Algebras in Σ_*-Objects and Left Modules over Operads. In parallel with P-algebras in functors, we study P-algebras in the category of Σ_*-objects $\mathcal{M}$ for P an operad in $\mathcal{C}$. In §2.2.2, we observe that the generalized symmetric power functor $S(M) : \mathcal{E} \to \mathcal{E}$ for $\mathcal{E} = \mathcal{M}$ represents the composition product $N \mapsto M \circ N$ in the category of Σ_*-objects. As a consequence, the structure of a P-*algebra in* Σ_*-*objects* is determined by a morphism

$$S(\mathsf{P}, N) = \mathsf{P} \circ N \xrightarrow{\lambda} N$$

which provides the object $N \in \mathcal{M}$ with a left action of the monoid object defined by the operad P in the monoidal category $(\mathcal{M}, \circ, I)$. In [14, §2.1.6], we call this structure a *left* P-*module*.

To summarize, we have obtained:

3.2.10 Observation. *The structure of a P-algebra in the category of Σ_*-objects $\mathcal{M}$ is identified with the structure of a left P-module.*

We adopt the notation $_\mathsf{P}\mathcal{M}$ for the category of left P-modules. This notation $_\mathsf{P}\mathcal{M}$ is coherent with our conventions for categories of algebras over operads.

To make explicit the structure of a left module over the usual commutative C, associative A, and Lie operads L, we use the next fact, which extends the assertion of fact 2.1.10 and, like this statement, is an immediate consequence of the explicit definition of the tensor product of Σ_*-objects (§§2.1.7-2.1.8):

3.2.11 Fact. *For any Σ_*-object M, a morphism $\phi : M^{\otimes r} \to M$ is equivalent to a collection of morphisms*

$$\phi : M(n_1) \otimes \cdots \otimes M(n_r) \to M(n_1 + \cdots + n_r), \quad n_1, \ldots, n_r \in \mathbb{N},$$

which commute with the action of the subgroup $\Sigma_{n_1} \times \cdots \times \Sigma_{n_r} \subset \Sigma_{n_1 + \cdots + n_r}$.

The composite of $\phi : M^{\otimes r} \to M$ with a tensor permutation $w^* : M^{\otimes r} \to M^{\otimes r}$, where $w \in \Sigma_r$, is determined by composites of the form:

$$M(n_1) \otimes \cdots \otimes M(n_r) \xrightarrow{w^*} M(n_{w(1)}) \otimes \cdots \otimes M(n_{w(r)})$$
$$\xrightarrow{\phi} M(n_{w(1)} + \cdots + n_{w(r)}) \xrightarrow{w(n_1, \ldots, n_r)} M(n_1 + \cdots + n_r),$$

where we perform a permutation of tensors on the source together with an action of the bloc permutation $w(n_1, \ldots, n_r) \in \Sigma_{n_1 + \cdots + n_r}$ on the target $M(n_1 + \cdots + n_r)$.

From this fact and the assertions of §3.2.4, we obtain:

3.2.12 Proposition. *Suppose $\mathcal{E} = C = \Bbbk \operatorname{Mod}$, or $\mathcal{E}$ is any category with a pointwise representation of tensor products (see §0.4).*

(a) *The structure of a C-algebra in Σ_*-objects consists of a Σ_*-object N equipped with a collection of $\Sigma_p \times \Sigma_q$-equivariant morphisms*

$$\mu : N(p) \otimes N(q) \to N(p + q), \quad p, q \in \mathbb{N},$$

that satisfy the twisted symmetry relation

$$\mu(a, b) = \tau(p, q) \cdot \mu(b, a), \quad \forall a \in N(p), \forall b \in N(q),$$

and the associativity relation

$$\mu(\mu(a, b), c) = \mu(a, \mu(b, c)), \quad \forall a \in N(p), \forall b \in N(q), \forall c \in N(r).$$

(b) *The structure of an* A-*algebra in* Σ_*-*objects consists of a* Σ_*-*object* N *equipped with a collection of* $\Sigma_p \times \Sigma_q$-*equivariant morphisms*

$$\mu : N(p) \otimes N(q) \to N(p + q), \quad p, q \in \mathbb{N},$$

that satisfy the associativity relation

$$\mu(\mu(a, b), c) = \mu(a, \mu(b, c)), \quad \forall a \in N(p), \forall b \in N(q), \forall c \in N(r).$$

(c) *The structure of an* L-*algebra in* Σ_*-*objects consists of a* Σ_*-*object* N *equipped with a collection of* $\Sigma_p \times \Sigma_q$-*equivariant morphisms*

$$\gamma : N(p) \otimes N(q) \to N(p + q), \quad p, q \in \mathbb{N},$$

that satisfy the twisted antisymmetry relation

$$\gamma(a, b) + \tau(p, q) \cdot \gamma(b, a) = 0, \quad \forall a \in N(p), \forall b \in N(q),$$

and the twisted Jacobi relation

$$\gamma(\gamma(a, b), c) + c(p, q, r) \cdot \gamma(\gamma(b, c), a) + c^2(p, q, r) \cdot \gamma(\gamma(c, a), b) = 0,$$
$$\forall a \in N(p), \forall b \in N(q), \forall c \in N(r),$$

where c *denotes the 3-cycle* $c = (1\ 2\ 3) \in \Sigma_3$.

This proposition illustrates the principle (formalized in §0.5) of generalized point-tensors: to obtain the definition of a commutative (respectively, associative, Lie) algebra in Σ_*-objects from the usual pointwise definition in the context of $\Bbbk$-modules, we take simply point-tensors in Σ_*-objects and we add a permutation, determined by the symmetry isomorphisms of the category of Σ_*-objects, to every tensor commutation.

3.2.13 Free Objects. Since the category of P-algebras in $\mathcal{E}$ is defined by a monad, we obtain that this category has a free object functor $P(-) : \mathcal{E} \to {}_P\mathcal{E}$, left adjoint to the forgetful functor $U : {}_P\mathcal{E} \to \mathcal{E}$. Explicitly, the *free* P-*algebra* in $\mathcal{E}$ generated by $X \in \mathcal{E}$ is defined by the object $P(X) = S(P, X) \in \mathcal{E}$ together with the evaluation product $\lambda : S(P, P(X)) \to P(X)$ defined by the morphism

$$S(P, P(X)) = S(P \circ P, X) \xrightarrow{S(\mu, X)} S(P, X)$$

induced by the composition product of the operad.

In this book, we use the notation $P(X)$ to refer to the free P-algebra generated by X and the notation $S(P, X)$ to refer to the underlying object in $\mathcal{E}$.

In the context of a category of functors $\mathcal{F} = \mathcal{F}(\mathcal{A}, \mathcal{E})$, we obtain readily

$$P(F)(X) = P(F(X)),$$

for all $X \in \mathcal{A}$, where on the right-hand side we consider the free P-algebra in $\mathcal{E}$ generated by the object $F(X) \in \mathcal{E}$ associated to $X \in \mathcal{A}$ by the functor $F : \mathcal{E} \to \mathcal{E}$.

In the context of Σ_*-modules, we have an identity

$$P(M) = P \circ M.$$

In the formalism of left P-modules, the operad P acts on the composite $P \circ M$ by the natural morphism

$$P \circ P \circ M \xrightarrow{\mu \circ M} P \circ M,$$

where $\mu : P \circ P \to P$ refers to the composition product of P.

The next assertion is an easy consequence of definitions:

3.2.14 Observation (compare with [54, Corollary 2.4.5]). *Let* P *be an operad in* C. *Any functor of symmetric monoidal categories over* C

$$\rho : \mathcal{D} \to \mathcal{E}$$

restricts to a functor on P*-algebras so that we have a commutative diagram*

$$
\begin{array}{ccc}
\mathcal{D} & \xrightarrow{\;\rho\;} & \mathcal{E} \\
\uparrow{\scriptstyle U} & & \uparrow{\scriptstyle U} \\
{}_P\mathcal{D} & \cdots\!\!\xrightarrow{\;\rho\;}\!\!\cdots & {}_P\mathcal{E},
\end{array}
$$

where $U : A \mapsto U(A)$ *denotes the forgetful functors on the category of* P*-algebras in* $\mathcal{D}$ *(respectively, in* $\mathcal{E}$*).*

If ρ *preserves colimits, then the diagram*

$$
\begin{array}{ccc}
\mathcal{D} & \xrightarrow{\;\rho\;} & \mathcal{E} \\
\downarrow{\scriptstyle P(-)} & & \downarrow{\scriptstyle P(-)} \\
{}_P\mathcal{D} & \xrightarrow{\;\rho\;} & {}_P\mathcal{E},
\end{array}
$$

where $P(-) : X \mapsto P(X)$ *denotes the free* P*-algebra functor, commutes as well.*

As a corollary, in the case of the functor $S : \mathcal{M} \to \mathcal{F}$, where $\mathcal{F} = \mathcal{F}(\mathcal{E}, \mathcal{E})$, we obtain:

3.2.15 Observation. *Let* P *be an operad in* $\mathcal{C}$. *Set* $\mathcal{F} = \mathcal{F}(\mathcal{E}, \mathcal{E})$ *and* $_P\mathcal{F} = \mathcal{F}(\mathcal{E}, _P\mathcal{E})$.

The functor S : $\mathcal{M} \to \mathcal{F}$ *restricts to a functor*

$$S : {}_P\mathcal{M} \to {}_P\mathcal{F}$$

so that for the free P-*algebra* $N = P(M)$ *generated by* $M \in \mathcal{M}$ *we have*

$$S(P(M), X) = P(S(M, X)),$$

for all $X \in \mathcal{E}$, *where on the right-hand side we consider the free* P-*algebra in* $\mathcal{E}$ *generated by the object* $S(M, X) \in \mathcal{E}$ *associated to* $X \in \mathcal{E}$ *by the functor* $S(M) : \mathcal{E} \to \mathcal{E}$.

This construction is functorial in $\mathcal{E}$. *Explicitly, for any functor* $\rho : \mathcal{D} \to \mathcal{E}$ *of symmetric monoidal categories over* $\mathcal{C}$, *the diagram*

$$
\begin{array}{ccc}
\mathcal{D} & \xrightarrow{\rho} & \mathcal{E} \\
{\scriptstyle S(N)}\downarrow & & \downarrow{\scriptstyle S(N)} \\
{}_P\mathcal{D} & \xrightarrow{\rho} & {}_P\mathcal{E}
\end{array}
$$

commutes up to natural functor isomorphisms, for all $N \in {}_P\mathcal{M}$.

3.2.16 Restriction of Functors. Let $\alpha : \mathcal{A} \to \mathcal{B}$ be a functor. For any target category $\mathcal{X}$, we have a functor $\alpha^* : \mathcal{F}(\mathcal{B}, \mathcal{X}) \to \mathcal{F}(\mathcal{A}, \mathcal{X})$ induced by α, defined by $\alpha^* G(A) = G(\alpha(A))$, for all $G : \mathcal{B} \to \mathcal{X}$

In the case $\mathcal{X} = \mathcal{E}$, the map $G \mapsto \alpha^*(G)$ defines clearly a functor of symmetric monoidal categories over $\mathcal{C}$

$$\alpha^* : (\mathcal{F}(\mathcal{B}, \mathcal{E}), \otimes, 1) \to (\mathcal{F}(\mathcal{A}, \mathcal{E}), \otimes, 1).$$

By observations of §§3.2.7-3.2.8, the induced functor on P-algebras, obtained by the construction of observation 3.2.14, is identified with the natural functor

$$\alpha^* : (\mathcal{F}(\mathcal{B}, _P\mathcal{E}), \otimes, 1) \to (\mathcal{F}(\mathcal{A}, _P\mathcal{E}), \otimes, 1),$$

which is induced by α for the target category $\mathcal{X} = {}_P\mathcal{E}$.

3.3 Universal Constructions for Algebras over Operads

The forgetful functor $U : {}_P\mathcal{E} \to \mathcal{E}$, from a category of algebras over an operad P to the underlying category $\mathcal{E}$, creates all limits. This assertion is proved by a straightforward inspection. On the other hand, the example of commutative algebras shows that the forgetful functor $U : {}_P\mathcal{E} \to \mathcal{E}$ does not preserves colimits in general. This difference relies on the direction of natural morphisms

$$S(P, \lim_{i \in I} X_i) \rightarrow \lim_{i \in I} S(P, X_i),$$

respectively $\quad \text{colim}_{i \in I} S(P, X_i) \rightarrow S(P, \text{colim}_{i \in I} X_i),$

when the functor $S(P) : \mathcal{E} \rightarrow \mathcal{E}$ is applied to a limit, respectively to a colimit. For a limit, this direction is the good one to form an evaluation product on $\lim_{i \in I} A_i$ from evaluation products on each A_i. For a colimit, this direction is the wrong one.

But we prove in §2.4 that the functor $S(M) : \mathcal{E} \rightarrow \mathcal{E}$ associated to any Σ_*-object M preserves filtered colimits and reflexive coequalizers. If we apply this assertion, then we obtain as an easy corollary:

3.3.1 Proposition (*cf.* [12, Lemma 1.1.9], [54, Proposition 2.3.5]). *Let P be an operad in $\mathcal{C}$. The forgetful functor $U : {}_P\mathcal{E} \rightarrow \mathcal{E}$, from the category of P-algebras in $\mathcal{E}$ to the underlying category $\mathcal{E}$, creates filtered colimits and the coequalizers which are reflexive in $\mathcal{E}$.* ☐

Proposition 3.3.1 can be applied to prove the existence of other universal constructions in categories of algebras other operads: colimits over every small category and extension functors $\phi_! : {}_P\mathcal{E} \rightarrow {}_Q\mathcal{E}$. The purpose of this section is to review these constructions.

In the context of symmetric monoidal categories over a base, the universality of the constructions is used to prove the functoriality of colimits and extensions with respect to functors $\rho : \mathcal{D} \rightarrow \mathcal{E}$ of symmetric monoidal categories over $\mathcal{C}$.

The existence of colimits arises from the following assertion:

3.3.2 Proposition (*cf.* [54, Proposition 2.3.5]). *Let I be a small category. Assume that colimits of I-diagram exists in $\mathcal{E}$, as well as reflexive coequalizers. Use the notation $\text{colim}_i^{\mathcal{E}} X_i$ to refer to the colimit formed in the category $\mathcal{E}$, for any I-diagram of objects X_i.*

Let $i \mapsto A_i$ be an I-diagram of P-algebras in $\mathcal{E}$. The colimit of $i \mapsto A_i$ in the category of P-algebras in $\mathcal{E}$ is realized by a reflexive coequalizer of the form:

$$P(\text{colim}_i^{\mathcal{E}} P(A_i)) \overset{s_0}{\underset{d_1}{\overset{d_0}{\rightrightarrows}}} P(\text{colim}_i^{\mathcal{E}} A_i) \longrightarrow \text{colim}_i A_i \ ,$$

where d_0, d_1 are morphisms of free P-algebras in $\mathcal{E}$ and the reflection s_0 is a morphism of $\mathcal{E}$. ☐

This construction occurs in [12, Lemma 1.1.10] in the particular case of coproducts. We refer to the proof of proposition 2.3.5 in [54] for the general case. We recall simply the definition of the morphisms d_0, d_1, s_0 for the sake of completeness. The canonical morphisms $\alpha_i : A_i \to \operatorname{colim}_i^{\mathcal{E}} A_i$ induce

$$\mathrm{P}(\operatorname{colim}_i^{\mathcal{E}} \mathrm{P}(A_i)) \xrightarrow{\mathrm{P}(\alpha_*)} \mathrm{P}(\mathrm{P}(\operatorname{colim}_i^{\mathcal{E}} A_i))$$

The morphism d_0 is defined by the composite of $\mathrm{P}(\alpha_*)$ with the morphism

$$\mathrm{P}(\mathrm{P}(\operatorname{colim}_i^{\mathcal{E}} A_i)) = \mathrm{S}(\mathrm{P} \circ \mathrm{P}, \operatorname{colim}_i^{\mathcal{E}} A_i) \xrightarrow{\mathrm{S}(\mu, \operatorname{colim}_i^{\mathcal{E}} A_i)} \mathrm{S}(\mathrm{P}, \operatorname{colim}_i^{\mathcal{E}} A_i)$$

induced by the operad composition product $\mu : \mathrm{P} \circ \mathrm{P} \to \mathrm{P}$. The P-actions $\lambda_i : \mathrm{P}(A_i) \to A_i$ induce

$$\mathrm{P}(\operatorname{colim}_i^{\mathcal{E}} \mathrm{P}(A_i)) \xrightarrow{\mathrm{P}(\lambda_*)} \mathrm{P}(\operatorname{colim}_i^{\mathcal{E}} A_i)$$

and this morphism defines d_1. The operad unit $\eta : I \to \mathrm{P}$ induces

$$\mathrm{P}(\operatorname{colim}_i^{\mathcal{E}} A_i) \xrightarrow{\mathrm{P}(\eta(A_*))} \mathrm{P}(\operatorname{colim}_i^{\mathcal{E}} \mathrm{P}(A_i))$$

and this morphism defines s_0.

By universality of this construction, we obtain:

3.3.3 Proposition (compare with [54, Corollary 2.4.5]). *Let P be an operad in C. Let $\rho : \mathcal{D} \to \mathcal{E}$ be a functor of symmetric monoidal categories over C. If $\rho : \mathcal{D} \to \mathcal{E}$ preserves colimits, then so does the induced functor $\rho : {}_{\mathrm{P}}\mathcal{D} \to {}_{\mathrm{P}}\mathcal{E}$ on categories of P-algebras.*

Proof. Since ρ preserves colimits by assumption and free P-algebras by observation 3.2.14, we obtain that the functor ρ preserves the coequalizers of the form of proposition 3.3.2, the coequalizers that realize colimits in categories of P-algebras. The conclusion follows. $\qquad\square$

In the case of the functor $\mathrm{S} : \mathcal{M} \to \mathcal{F}(\mathcal{E}, \mathcal{E})$, we obtain as a corollary:

3.3.4 Proposition. *Let P be an operad in C. Let $i \mapsto N_i$ be a diagram of P-algebras in $\mathcal{M}$. We have a natural isomorphism*

$$\mathrm{S}(\operatorname*{colim}_i N_i, X) = \operatorname*{colim}_i \mathrm{S}(N_i, X),$$

for all $X \in \mathcal{E}$, where on the right-hand side we consider the colimit of the P-algebras $\mathrm{S}(N_i, X)$, $i \in I$, associated to $X \in \mathcal{E}$ by the functors $\mathrm{S}(N_i) : \mathcal{E} \to {}_{\mathrm{P}}\mathcal{E}$. $\qquad\square$

This assertion is also a corollary of the adjunction relation of §2.3.3.

3.3.5 Extension and Restriction of Structures for Algebras over Operads.

Recall that a morphism of operads $\phi : \mathsf{P} \to \mathsf{Q}$ yields adjoint functors of *extension and restriction of structures*

$$\phi_! : {}_\mathsf{P}\mathcal{E} \rightleftarrows {}_\mathsf{Q}\mathcal{E} : \phi^*.$$

This classical assertion for algebras in the base category $\mathcal{C}$ can be generalized to algebras in a symmetric monoidal category over $\mathcal{C}$. For the sake of completeness we recall the construction of these functors.

The operad P operates on any Q-algebra B through the morphism $\phi : \mathsf{P} \to \mathsf{Q}$ and this operation defines the P-algebra $\phi^* B$ associated to $B \in {}_\mathsf{Q}\mathcal{E}$ by restriction of structures. The map $B \mapsto \phi^* B$ defines the restriction functor $\phi^* : {}_\mathsf{Q}\mathcal{E} \to {}_\mathsf{P}\mathcal{E}$.

In the converse direction, the Q-algebra $\phi_! A$ associated to a P-algebra $A \in {}_\mathsf{P}\mathcal{E}$ is defined by a reflexive coequalizer of the form:

$$S(\mathsf{Q} \circ \mathsf{P}, A) \overset{s_0}{\underset{\substack{d_0 \\ d_1}}{\rightrightarrows}} S(\mathsf{Q}, A) \longrightarrow \phi_! A \, .$$

The composite

$$S(\mathsf{Q} \circ \mathsf{P}, A) \xrightarrow{S(\mathsf{Q} \circ \phi, A)} S(\mathsf{Q} \circ \mathsf{Q}, A) \xrightarrow{S(\mu, A)} S(\mathsf{Q}, A),$$

where $\mu : \mathsf{Q} \circ \mathsf{Q} \to \mathsf{Q}$ is the composition product of Q, defines $d_0 : S(\mathsf{Q} \circ \mathsf{P}, A) \to S(\mathsf{Q}, A)$. The morphism

$$S(\mathsf{Q} \circ \mathsf{P}, A) = S(\mathsf{Q}, S(\mathsf{P}, A)) \xrightarrow{S(\mathsf{Q}, \lambda)} S(\mathsf{Q}, A)$$

induced by the P-action on A defines $d_1 : S(\mathsf{Q} \circ \mathsf{P}, A) \to S(\mathsf{Q}, A)$. The morphism

$$S(\mathsf{Q}, A) = S(\mathsf{Q} \circ I, A) \xrightarrow{S(\mathsf{Q} \circ \eta, A)} S(\mathsf{Q} \circ \mathsf{P}, A)$$

induced by the operad unit of P gives the reflection $s_0 : S(\mathsf{Q}, A) \to S(\mathsf{Q} \circ \mathsf{P}, A)$. Observe that d_0, d_1 define morphisms of free Q-algebras $d_0, d_1 : \mathsf{Q}(S(\mathsf{P}, A)) \rightrightarrows \mathsf{Q}(A)$. Since the forgetful functor $U : {}_\mathsf{Q}\mathcal{E} \to \mathcal{E}$ creates the coequalizers which are reflexive in $\mathcal{E}$, we obtain that the coequalizer of d_0, d_1 in $\mathcal{E}$ is equipped with a natural Q-algebra structure and represents the coequalizer of d_0, d_1 in the category of Q-algebras. The map $A \mapsto \phi_! A$ defines the extension functor $\phi_! : {}_\mathsf{P}\mathcal{E} \to {}_\mathsf{Q}\mathcal{E}$.

One checks readily that the extension functor $\phi_! : {}_\mathsf{P}\mathcal{E} \to {}_\mathsf{Q}\mathcal{E}$, defined by our coequalizer construction, is left adjoint to the restriction functor $\phi^* : {}_\mathsf{Q}\mathcal{E} \to {}_\mathsf{P}\mathcal{E}$ as required. The adjunction unit $\eta A : A \to \phi^* \phi_! A$ is the morphism

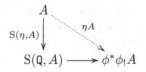

determined by the operad unit $\eta : I \to Q$. The adjunction augmentation $\epsilon B : \phi_! \phi^* B \to B$ is the morphism

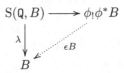

induced by the Q-action on B.

In §9.3, we observe that these functors are instances of functors associated to right modules over operads.

For the unit morphism $\eta : I \to P$, we have $_I\mathcal{E} = \mathcal{E}$, the restriction functor $A \mapsto \eta^* A$ is identified with the forgetful functor $U : {}_P\mathcal{E} \to \mathcal{E}$ and the extension functor $M \mapsto \eta_! M$ represents the free object functor $P(-) : \mathcal{E} \to {}_P\mathcal{E}$. Thus in this case we obtain the adjunction between the forgetful and free object functors.

3.3.6 Extension and Restriction of Left Modules over Operads. In the case $\mathcal{E} = \mathcal{M}$ and $_P\mathcal{E} = {}_P\mathcal{M}$, the category of left P-modules, we obtain that the extension of structures of a left P-module $N \in {}_P\mathcal{E}$ is defined by a reflexive coequalizer of the form:

$$Q \circ P \circ N \underset{d_1}{\overset{d_0}{\underset{\displaystyle\rightrightarrows}{}}} Q \circ N \longrightarrow \phi_! N \ .$$

In §9.3, we identify this coequalizer construction with a relative composition product $Q \circ_P N$, where the operad P acts on Q on the right through the morphism $\phi : P \to Q$ and the operad composition product $\mu : Q \circ Q \to Q$.

3.3.7 Extension and Restriction of Functors. In the case of functor categories $\mathcal{F} = \mathcal{F}(\mathcal{A}, \mathcal{E})$ and $_P\mathcal{F} = \mathcal{F}(\mathcal{A}, {}_P\mathcal{E})$, the extension and restriction functors

$$\phi_! : \mathcal{F}(\mathcal{A}, {}_P\mathcal{E}) \rightleftarrows \mathcal{F}(\mathcal{A}, {}_Q\mathcal{E}) : \phi^*$$

defined by the construction of §3.3.5 are given by the composition of functors with the extension and restriction functors on target categories.

This assertion is straightforward for the restriction functor ϕ^*. To check the assertion for the extension functor $\phi_!$ use simply that colimits are created pointwise in functor categories or check directly that the functor $F \mapsto \phi_! \circ F$ is left adjoint to $F \mapsto \phi^* \circ F$.

The next assertion is an immediate consequence of the universality of the construction of §3.3.5:

3.3.8 Proposition. *Let $\phi : \mathsf{P} \to \mathsf{Q}$ be an operad morphism. Let $\rho : \mathcal{D} \to \mathcal{E}$ be a functor of symmetric monoidal categories over $\mathcal{C}$. If ρ preserves colimits, then the diagram of functors*

$$
\begin{array}{ccc}
{}_{\mathsf{P}}\mathcal{D} & \xrightarrow{\ \rho\ } & {}_{\mathsf{P}}\mathcal{E} \\
\phi^* \Big\uparrow \Big\downarrow \phi_! & & \phi^* \Big\uparrow \Big\downarrow \phi_! \\
{}_{\mathsf{Q}}\mathcal{D} & \xrightarrow{\ \rho\ } & {}_{\mathsf{Q}}\mathcal{E}
\end{array}
,
$$

where we consider the functor on algebras induced by ρ, commutes. ☐

3.4 Endomorphism Operads in Enriched Symmetric Monoidal Categories

In §1.1.12, we observe that any symmetric monoidal category $\mathcal{E}$ over $\mathcal{C}$ is equipped with an external hom-bifunctor $Hom_{\mathcal{E}} : \mathcal{E}^{op} \times \mathcal{E} \to \mathcal{C}$ such that

$$ \mathrm{Mor}_{\mathcal{E}}(C \otimes X, Y) = \mathrm{Mor}_{\mathcal{C}}(C, Hom_{\mathcal{E}}(X, Y)), $$

for all $C \in \mathcal{C}$, $X, Y \in \mathcal{E}$. In this section, we use the existence of these hom-objects to prove the existence of a universal operad acting on X, the *endomorphism operad* of X.

This assertion is standard in the classical setting of operads in a closed symmetric monoidal categories. We extend simply the usual construction to our relative context of a symmetric monoidal category $\mathcal{E}$ enriched over the base $\mathcal{C}$.

3.4.1 The General Construction in Enriched Symmetric Monoidal Categories. The endomorphism operad in $\mathcal{C}$ of an object $X \in \mathcal{E}$ is formed by the hom-objects

$$ End_X(r) = Hom_{\mathcal{E}}(X^{\otimes r}, X). $$

As usual, the symmetric group operates on $End_X(r)$ by permutations of tensors on the source. The operad composition products $\circ_i : End_X(r) \otimes End_X(s) \to End_X(r + s - 1)$ are deduced from the enriched monoidal structure.

The canonical morphisms

$$ \epsilon : Hom_{\mathcal{E}}(X^{\otimes r}, X) \otimes X^{\otimes r} \to X $$

give an action of the endomorphism operad End_X on X so that X forms an algebra over End_X in $\mathcal{E}$.

The next assertion is a formal generalization of the classical universal definition of internal endomorphism operads:

3.4.2 Proposition. *The endomorphism operad End_X is the universal operad in $\mathcal{C}$ acting on X: any action on X of an operad P in $\mathcal{C}$ is the restriction, by a uniquely determined operad morphism $\nabla : \mathsf{P} \to End_X$, of the action of the endomorphism operad End_X.* □

Accordingly:

3.4.3 Proposition. *We have a one-to-one correspondence between operad morphisms $\nabla : \mathsf{P} \to End_X$ and P-algebra structures on X.* □

3.4.4 Endomorphism Operads of Functors. We apply the construction of §3.4.1 to a category of functors $\mathcal{F} = \mathcal{F}(\mathcal{A}, \mathcal{E})$. We have then:

$$End_F(r) = \int_{X \in \mathcal{A}} Hom_{\mathcal{C}}(F(X)^{\otimes r}, F(X)),$$

where we consider the end of the bifunctor $(X, Y) \mapsto Hom_{\mathcal{C}}(F(X)^{\otimes r}, F(Y))$.

Recall that a functor $F : \mathcal{A} \to \mathcal{E}$ forms a P-algebra in $\mathcal{F}$ if and only if the objects $F(X)$, $X \in \mathcal{A}$, are equipped with a P-algebra structure, functorially in X, so that the map $X \mapsto F(X)$ defines a functor from $\mathcal{A}$ to the category of P-algebras. Thus propositions 3.4.2-3.4.3 imply:

3.4.5 Proposition. *We have a one-to-one correspondence between operad morphisms $\nabla : \mathsf{P} \to End_F$ and functorial P-actions on the objects $F(X)$, $X \in \mathcal{A}$, such that $X \mapsto F(X)$ defines a functor from $\mathcal{A}$ to the category of P-algebras.* □

3.4.6 Remark. The functorial action of End_F on the objects $F(X)$, for $X \in \mathcal{A}$, is represented by operad morphisms

$$\epsilon_X : End_F \to End_{F(X)},$$

where $End_{F(X)}$ denotes the endomorphism operad of the object $F(X) \in \mathcal{E}$. This collection of operad morphisms is identified with the universal collections of morphisms associated to the end

$$End_F(r) = \int_{X \in \mathcal{A}} Hom_{\mathcal{C}}(F(X)^{\otimes r}, F(X)).$$

In the remainder of this section, we examine the functoriality of endomorphism operads with respect to morphisms of symmetric monoidal categories. In the general case of a functor of symmetric monoidal categories $\rho : \mathcal{D} \to \mathcal{E}$, we have:

3.4.7 Proposition. *Let $\rho : \mathcal{D} \to \mathcal{E}$ be a functor of symmetric monoidal categories over $\mathcal{C}$. For every object $X \in \mathcal{D}$, we have an operad morphism*

$$\rho_* : End_X \to End_{\rho(X)}$$

determined by ρ.

Proof. In proposition 1.1.15, we observe that the map $f \mapsto \rho(f)$, defined for morphisms in $\mathcal{D}$, extends to a morphism

$$Hom_{\mathcal{D}}(X, Y) \xrightarrow{\rho} Hom_{\mathcal{E}}(\rho(X), \rho(Y)).$$

In the case of an endomorphism operad, we obtain a morphism

$$Hom_{\mathcal{D}}(X^{\otimes r}, X) \xrightarrow{\rho} Hom_{\mathcal{E}}(\rho(X^{\otimes r}), \rho(X)) \xrightarrow{\simeq} Hom_{\mathcal{E}}(\rho(X)^{\otimes r}, \rho(X)),$$

for any object $X \in \mathcal{D}$. One checks readily that this morphism defines an operad morphism $\rho_* : End_X \to End_{\rho X}$. □

In light of propositions 3.4.2-3.4.3, the existence of an operad morphism

$$\rho_* : End_X \to End_{\rho(X)}$$

is equivalent to the construction of §3.2.14, a P-algebra structure on X, where P is an operad in $\mathcal{C}$, gives rise to a P-algebra structure on ρX, for all $X \in \mathcal{E}$.

In the particular case of the functor S : $\mathcal{M} \to \mathcal{F}$, proposition 3.4.7 returns:

3.4.8 Proposition. *Consider the functor S : $\mathcal{M} \to \mathcal{F}(\mathcal{E}, \mathcal{E})$, where $\mathcal{E}$ is any symmetric monoidal category over $\mathcal{C}$. For any $M \in \mathcal{M}$, we have an operad morphism*

$$End_M \xrightarrow{\Theta} End_{S(M)},$$

natural in $\mathcal{E}$. □

By proposition 2.3.10, the morphism S : $Hom_{\mathcal{M}}(M, N) \to Hom_{\mathcal{F}}(S(M), S(N))$ is mono for every $M, N \in \mathcal{M}$ if the category $\mathcal{E}$ is equipped with a faithful functor $\eta : \mathcal{C} \to \mathcal{E}$. Hence, the operad morphism Θ is mono as long as $\mathcal{E}$ is equipped with a faithful functor $\eta : \mathcal{C} \to \mathcal{E}$.

In the case where the underlying categories $\mathcal{E}, \mathcal{C}$ are both the category of modules over a ring $\Bbbk$, we obtain further:

3.4.9 Proposition. *Assume $\mathcal{E} = \mathcal{C} = \Bbbk \, Mod$, the category of modules over a ring $\Bbbk$.*

The operad morphism

$$End_M \xrightarrow{\Theta} End_{S(M)}$$

is an isomorphism if M is a projective Σ_*-module or if the ground ring is an infinite field.

Proof. This proposition is an immediate corollary of proposition 2.3.7 and proposition 2.3.12.   □

In observation 3.2.15, we prove that the structure of a P-algebra on a Σ_*-object M gives rise to a P-algebra structure at the functor level. The existence of an operad isomorphism

$$End_M \xrightarrow[\simeq]{\Theta} End_{S(M)}$$

implies further:

3.4.10 Proposition. *Assume $\mathcal{E} = \mathcal{C} = \Bbbk \operatorname{Mod}$, the category of modules over a ring $\Bbbk$.*

If M is a projective Σ_*-module or if the ground ring is an infinite field, then all functorial P-algebra structures on the objects $S(M, A)$, $A \in \mathcal{E}$, are uniquely determined by a P-algebra structure on the Σ_*-module M.   □

3.4.11 Functoriality in the Case of Functors. We examine applications of proposition 3.4.7 in the context of functor categories. Consider the functor $\alpha^* : \mathcal{F}(\mathcal{B}, \mathcal{E}) \to \mathcal{F}(\mathcal{A}, \mathcal{E})$ induced by a functor $\alpha : \mathcal{A} \to \mathcal{B}$ on source categories.

Recall that $G \mapsto \alpha^*(G)$ forms a functor of symmetric monoidal categories over $\mathcal{C}$. Accordingly, for any $G : \mathcal{B} \to \mathcal{E}$, we have a morphism of endomorphism operads

$$\alpha^* : End_G \to End_{\alpha^* G}$$

induced by $\alpha^* : \mathcal{F}(\mathcal{B}, \mathcal{E}) \to \mathcal{F}(\mathcal{A}, \mathcal{E})$. This morphism can immediately be identified with the natural morphisms

$$\int_{Y \in \mathcal{B}} Hom_{\mathcal{C}}(G(Y)^{\otimes r}, G(Y)) \to \int_{X \in \mathcal{A}} Hom_{\mathcal{C}}(G(\alpha(X))^{\otimes r}, G(\alpha(X)))$$

defined by the restriction of ends.

Chapter 4
Miscellaneous Structures
Associated to Algebras over Operads

Introduction

In this chapter, we recall the definition of miscellaneous structures associated to algebras over operads: *enveloping operads*, which model comma categories of algebras over operads; *enveloping algebras*, which are associative algebras formed from the structure of enveloping operads; *representations*, which are nothing but modules over enveloping algebras; and modules of *Kähler differentials*, which occur in the definition of the homology of algebras over operads. In §10, we address enveloping objects and modules of Kähler differentials as examples of functors associated to right modules over operads. For the moment, we only give the definition of these structures – enveloping operads in §4.1, representations in §4.2, enveloping algebras in §4.3, Kähler differentials in §4.4 – and we study applications of these operadic constructions for the usual commutative, associative and Lie operads.

To simplify, we address our examples within the category of $\Bbbk$-modules, for which we can use the point-set formalism (see §0.4). In the sequel, we use the principle of generalized point-tensors (see §0.5) to apply our results to commutative (respectively, associative, Lie) algebras in categories of dg-modules, in categories of Σ_*-objects, and in categories of right modules over operads (defined next).

4.1 Enveloping Operads

Let P be any operad. The *enveloping operad* of a P-algebra A is introduced in [17] to represent the category of P-algebras under A. The article [13] gives another definition according to which the enveloping operad $U_P(A)$ is the universal operad under P such that $U_P(A)(0) = A$.

B. Fresse, *Modules over Operads and Functors*, Lecture Notes in Mathematics 1967, 77
DOI: 10.1007/978-3-540-89056-0_4, © Springer-Verlag Berlin Heidelberg 2009

The purpose of this section is to review these definitions in the context where the P-algebra A belongs to a symmetric monoidal category $\mathcal{E}$ over the base category $\mathcal{C}$. Though we still assume that the operad P belongs to the base category $\mathcal{C}$, we have to use the category of operads in $\mathcal{E}$ because the enveloping operad of a P-algebra in $\mathcal{E}$ belongs to that category.

4.1.1 The Enveloping Operad. To begin with, we recall the adjoint definition of the enveloping operad.

Recall that we use the notation $\mathcal{O}_{\mathcal{E}}$ to refer to any category of operads in a symmetric monoidal category $\mathcal{E}$ other than the base category $\mathcal{C}$. Form the comma category $P / \mathcal{O}_{\mathcal{E}}$ of operad morphisms $\phi : P \to S$, where we use the functor of symmetric monoidal categories $\eta : \mathcal{C} \to \mathcal{E}$ associated to $\mathcal{E}$ to map P to an operad in $\mathcal{E}$.

The enveloping operad of a P-algebra A is the operad $U_P(A) \in P / \mathcal{O}_{\mathcal{E}}$ defined by the adjunction relation

$$\mathrm{Mor}_{P / \mathcal{O}_{\mathcal{E}}}(U_P(A), S) = \mathrm{Mor}_{P\mathcal{E}}(A, S(0)).$$

To justify this definition, recall that $B = S(0)$ defines the initial object in the category of S-algebras. If S is an operad under P, then $B = S(0)$ forms a P-algebra by restriction of structures.

The functor $S \mapsto S(0)$ preserves limits because limits of operads and algebras over operads are created in the underlying category. The existence of the enveloping operad follows immediately from this observation.

The main purpose of this section is to give a more effective construction of the enveloping operad. Before, we check simply that the enveloping operad defined by the adjunction relation of this paragraph fulfils the objective of [17]:

4.1.2 Proposition (see [13, 17]). *Let A any P-algebra. The comma category of algebras under A, denoted by $A/_P\mathcal{E}$, is isomorphic to the category of $U_P(A)$-algebras. The forgetful functor $U : A/_P\mathcal{E} \to {}_P\mathcal{E}$ is represented by the restriction functor $\eta^* : {}_{U_P(A)}\mathcal{E} \to {}_P\mathcal{E}$, where $\eta : P \to U_P(A)$ is the underlying morphism of the enveloping operad, viewed as an object of the category of operads under P.*

Proof. The $U_P(A)$-algebras form a P-algebra by restriction of structures through $\eta : P \to U_P(A)$. For the endomorphism operad of a P-algebra B, the adjunction relation reads

$$\mathrm{Mor}_{P / \mathcal{O}_{\mathcal{E}}}(U_P(A), End_B) = \mathrm{Mor}_{P\mathcal{E}}(A, B)$$

since we have $End_B(0) = B$. The proposition follows readily from the bijection given by this adjunction relation. □

4.1.3 Toward Explicit Constructions. In the point-set context, the enveloping operad $U_P(A)$ can be defined informally as the object spanned by formal elements

$$u(x_1, \ldots, x_m) = p(x_1, \ldots, x_m, a_1, \ldots, a_r),$$

where $x_1, \ldots, x_m$ are variables, $p \in P(m+r)$, and $a_1, \ldots, a_r \in A$, together with relations of the form

$$p(x_1, \ldots, x_m, a_1, \ldots, a_{e-1}, q(a_e, \ldots, a_{e+s-1}), a_{e+s}, \ldots, a_{r+s-1})$$
$$\equiv p \circ_{m+e} q(x_1, \ldots, x_m, a_1, \ldots, a_{e-1}, a_e, \ldots, a_{e+s-1}, a_{e+s}, \ldots, a_{r+s-1}).$$

The goal of the next paragraphs is to give an effective construction of the enveloping operad in the context of symmetric monoidal categories. In this setting, the pointwise relations are replaced by identities of morphisms and we use colimits to perform the quotient process. To make the construction more conceptual, we replace the relations of this paragraph by equivalent relations which involve all composition products of the operad at once. Roughly, we put relations together to perform the quotient process by a single reflexive coequalizer.

Technically, we define the enveloping operad on the subcategory of free P-algebras first and we use that any P-algebra A is a coequalizer of morphisms of free P-algebras to extend the construction to all P-algebras.

4.1.4 Shifted Σ_*-Objects.

The enveloping operad of a free P-algebra $P(X)$ is yielded by a partial evaluation of the functor $S(P) : \mathcal{E} \to \mathcal{E}$ associated to the operad P. To define this partial evaluation, we use a shift of entries in the underlying Σ_*-object of P.

For any Σ_*-object M, we form a collection of shifted objects $M[m]$, $m \in \mathbb{N}$, such that $M[m](n) = M(m+n)$, for $n \in \mathbb{N}$. The action of Σ_n on $\{m+1, \ldots, m+n\} \subset \{1, \ldots, m+n\}$ determines a morphism $j : \Sigma_n \to \Sigma_{m+n}$ and an action of Σ_n on $M(m+n)$ which provides the collection $M[m](n)$, $n \in \mathbb{N}$, with the structure of a Σ_*-object.

The action of Σ_m on $\{1, \ldots, m\} \subset \{1, \ldots, m+n\}$ determines a morphism $i : \Sigma_m \to \Sigma_{m+n}$ and an action of Σ_m on $M[m](n) = M(m+n)$. This Σ_m-action commutes clearly with the internal Σ_n-action on $M[m](n)$, for all $n \in \mathbb{N}$, so that the collection $M[m]$, $m \in \mathbb{N}$, forms a Σ_*-object in the category of Σ_*-objects.

For any $X \in \mathcal{E}$, the object $S(M[m], X)$ comes equipped with a natural Σ_m-action induced by the action of Σ_m on the Σ_*-object $M[m]$. Consequently, the collection $S[M, X](m) = S(M[m], X)$, $m \in \mathbb{N}$, defines a Σ_*-object in $\mathcal{E}$. For short, we also use the notation $M[X] = S[M, X]$ for the Σ_*-object associated to X by this partial evaluation process.

By construction, we have the relation $S[M, X](0) = S(M, X)$ so that $S(M, X)$ represents the constant part of $S[M, X]$.

4.1.5 Shifted Operads.

For an operad P, the collection $P[\cdot] = \{P[m]\}_{m \in \mathbb{N}}$ forms naturally an operad in the category of Σ_*-objects and the collection $P[X] = \{S(P[m], X)\}_{m \in \mathbb{N}}$, associated to any object $X \in \mathcal{E}$, comes equipped with an induced operad structure. The easiest is to define the composition morphisms of these operads in May's form.

For the shifted collection $P[m]$, $m \in \mathbb{N}$, the composition morphisms

$$\mu : P[r] \otimes P[m_1] \otimes \cdots \otimes P[m_r] \to P[m_1 + \cdots + m_r]$$

are yielded by composites

$$
\begin{aligned}
P(r+s) &\otimes P(m_1 + n_1) \otimes \cdots \otimes P(m_r + n_r) \\
&\simeq P(r+s) \otimes P(m_1 + n_1) \otimes \cdots \otimes P(m_r + n_r) \otimes 1 \otimes \cdots \otimes 1 \\
&\to P(r+s) \otimes P(m_1 + n_1) \otimes \cdots \otimes P(m_r + n_r) \otimes P(1) \otimes \cdots \otimes P(1) \\
&\to P(m_1 + n_1 + \cdots + m_r + n_r + s) \\
&\simeq P(m_1 + \cdots + m_r + s + n_1 + \cdots + n_r),
\end{aligned}
$$

with operad units $\eta: 1 \to P(1)$ at positions $\{r+1, \ldots, r+s\}$ of the composition, and where the latter isomorphism involves the action of a block permutation $w(m_1, \ldots, m_r, n_1, \ldots, n_r, s)$ equivalent to the natural bijection

$$\{1, \ldots, m_1\} \amalg \{1, \ldots, n_1\} \amalg \cdots \amalg \{1, \ldots, m_r\} \amalg \{1, \ldots, n_r\} \amalg \{1, \ldots, s\}$$

$$\downarrow \simeq$$

$$\{1, \ldots, m_1\} \amalg \cdots \amalg \{1, \ldots, m_r\} \amalg \{1, \ldots, s\} \amalg \{1, \ldots, n_1\} \amalg \ldots \amalg \{1, \ldots, n_r\}.$$

Recall that

$$
\begin{aligned}
(P[r] &\otimes P[m_1] \otimes \cdots \otimes P[m_r])(n) \\
&= \bigoplus_{s, n_*} \Sigma_n \otimes_{\Sigma_s \times \Sigma_{n_*}} \{P[r](s) \otimes P[m_1](n_1) \otimes \cdots \otimes P[m_r](n_r)\} \\
&= \bigoplus_{s, n_*} \Sigma_n \otimes_{\Sigma_s \times \Sigma_{n_*}} \{P(r+s) \otimes P(m_1 + n_1) \otimes \cdots \otimes P(m_r + n_r)\},
\end{aligned}
$$

where $\Sigma_s \times \Sigma_{n_*} = \Sigma_s \times \Sigma_{n_1} \times \cdots \times \Sigma_{n_r}$. Our composition morphisms

$$
P(r+s) \otimes P(m_1 + n_1) \otimes \cdots \otimes P(m_r + n_r)
$$
$$
\xrightarrow{\mu} P(m_1 + \cdots + m_r + s + n_1 + \cdots + n_r),
$$

commute with $\Sigma_s \times \Sigma_{n_1} \times \cdots \times \Sigma_{n_r}$-actions and determine morphisms of Σ_*-objects

$$\mu : P[r] \otimes P[m_1] \otimes \cdots \otimes P[m_r] \to P[m_1 + \cdots + m_r]$$

as required. One checks readily that these composition morphisms satisfy May's axioms and hence provide the collection $P[m]$, $m \in \mathbb{N}$, with the structure of an operad in Σ_*-objects.

Since we observe that $S : \mathcal{M} \to \mathcal{E}$ defines a functor of symmetric monoidal categories, we obtain that any collection $P[X](m) = S(P[m], X)$, $m \in \mathbb{N}$, associated to an object $X \in \mathcal{E}$, comes equipped with composition morphisms

$$S(P[r], X) \otimes S(P[m_1], X) \otimes \cdots \otimes S(P[m_r], X)$$
$$\simeq S(P[r] \otimes P[m_1] \otimes \cdots \otimes P[m_r], X) \to S(P[m_1 + \cdots + m_r], X),$$

induced by the composition morphisms of the shifted operad $P[\cdot]$, and hence inherits an operad structure.

Note that $P[X]$ forms an operad in $\mathcal{E}$.

For our needs, we note further:

4.1.6 Observation. *The identity* $P(n) = P[n](0)$, $n \in \mathbb{N}$, *gives a morphism of operads in* Σ_*-*objects*

$$\eta : P(\cdot) \to P[\cdot],$$

where we identify the components of the operad P *with constant* Σ_*-*objects.*

For an object $X \in \mathcal{E}$, *the identity of* $P(n) = P[n](0)$ *with the constant term of* $S(P[n], X)$ *gives a morphism of operads in* $\mathcal{E}$

$$\eta : P \to P[X].$$

Hence $P[X]$ *forms an operad under* P.

4.1.7 Observation. *We have* $P[X](0) = P(X)$, *the free* P-*algebra generated by* X.

Suppose $\phi : R \to S$ is any morphism of operads under P. Then $\phi : R(0) \to S(0)$ defines a morphism of P-algebras. Hence we have a natural map

$$\text{Mor}_{P / \mathcal{O}_{\mathcal{E}}}(R, S) \to \text{Mor}_{P\mathcal{E}}(R(0), S(0)).$$

For the operad $R = P[X]$, we obtain further:

4.1.8 Lemma. *The natural map*

$$\text{Mor}_{P / \mathcal{O}_{\mathcal{E}}}(P[X], S) \to \text{Mor}_{P\mathcal{E}}(P(X), S(0))$$

is an iso and $P[X]$ *satisfies the adjunction relation of the enveloping operad.*

Proof. To check this lemma, we use the operad morphism $\eta : P \to P[X]$ defined in observation 4.1.6 and the natural morphism $\eta[X] : X \to P[X](0)$ defined by the composite

$$X = S(I, X) \xrightarrow{S(\eta, X)} S(P, X) = P[X](0),$$

where $\eta : I \to P$ refers to the unit of the operad P.

Since $P(X)$ is a free P-algebras, we have an isomorphism

$$\text{Mor}_{P\mathcal{E}}(P(X), S(0)) \simeq \text{Mor}_{\mathcal{E}}(X, S(0))$$

and we can use the natural map

$$\rho : \mathrm{Mor}_{\mathrm{P}/\mathcal{O}_{\mathcal{E}}}(\mathrm{P}[X], \mathrm{S}) \to \mathrm{Mor}_{\mathcal{E}}(X, \mathrm{S}(0))$$

which associates to any $\phi : \mathrm{P}[X] \to \mathrm{S}$ the composite

$$X \xrightarrow{\eta[X]} \mathrm{P}[X](0) \xrightarrow{\phi} \mathrm{S}(0).$$

Observe that the canonical morphism

$$\mathrm{P}(m+n) \otimes X^{\otimes n} \to (\mathrm{P}(m+n) \otimes X^{\otimes n})_{\Sigma_n} \hookrightarrow \mathrm{S}(\mathrm{P}[m], X) = \mathrm{P}[X](m)$$

is identified with the composite of the operadic composition product

$$\mathrm{P}[X](m+n) \otimes \mathrm{P}[X](1)^{\otimes m} \otimes \mathrm{P}[X](0)^{\otimes n} \to \mathrm{P}[X](m)$$

with the morphism

$$\mathrm{P}(m+n) \otimes 1^{\otimes m} \otimes X^{\otimes n} \to \mathrm{P}[X](m+n) \otimes \mathrm{P}[X](1)^{\otimes m} \otimes \mathrm{P}[X](0)^{\otimes n}$$

induced by the operad morphism $\eta : \mathrm{P} \to \mathrm{P}[X]$, the operad unit $\eta : 1 \to \mathrm{P}[X](1)$ and $\eta[X] : X \to \mathrm{P}[X](0)$. Consequently, any morphism $\phi : \mathrm{P}[X] \to \mathrm{S}$ of operads under P fits a commutative diagram of the form

$$
\begin{array}{ccc}
\mathrm{P}(m+n) \otimes X^{\otimes n} & \longrightarrow & \mathrm{P}[X](m) \\
{\scriptstyle \simeq} \downarrow & & \\
\mathrm{P}(m+n) \otimes 1^{\otimes m} \otimes X^{\otimes n} & & \Big\downarrow \phi \\
{\scriptstyle \eta \otimes \eta^{\otimes m} \otimes f^{\otimes n}} \downarrow & & \\
\mathrm{S}(m+n) \otimes \mathrm{S}(1)^{\otimes m} \otimes \mathrm{S}(0)^{\otimes n} & \xrightarrow{\ \nu\ } & \mathrm{S}(m),
\end{array}
$$

where $f = \rho(\phi)$ and ν refers to the composition product of S. Thus we obtain that $\phi : \mathrm{P}[X] \to \mathrm{S}$ is determined by the associated morphism $f = \rho(\phi)$.

If we are given a morphism $f : X \to \mathrm{S}(0)$, then straightforward verifications show that the morphisms $\phi : \mathrm{P}[X](m) \to \mathrm{S}(m)$ determined by the diagram define a morphism of operads under P.

Hence we conclude that the map $\phi \mapsto \rho(\phi)$ is one-to-one. $\qquad \square$

In the case $\mathrm{S} = \mathrm{P}[Y]$, we obtain that any morphism of P-algebras $f : \mathrm{P}(X) \to \mathrm{P}(Y)$ determines a morphism $\phi_f : \mathrm{P}[X] \to \mathrm{P}[Y]$ in the category of operads under P. The definition of the correspondence

$$\mathrm{Mor}_{\mathrm{P}/\mathcal{O}_{\mathcal{E}}}(\mathrm{R}, \mathrm{S}) \to \mathrm{Mor}_{\mathrm{P}\mathcal{E}}(\mathrm{R}(0), \mathrm{S}(0))$$

implies that the map $f \mapsto \phi_f$ preserves composites. Hence the map $X \mapsto P[X]$ determines a functor on the full subcategory of $_P\mathcal{E}$ formed by free P-algebras. In addition, we obtain:

4.1.9 Lemma. *The operad morphism* $\phi_f : P[X] \to P[Y]$ *associated to a morphism of free P-algebras* $f : P(X) \to P(Y)$ *fits a commutative diagram*

$$
\begin{array}{ccc}
\mathrm{Mor}_{P/\mathcal{O}_\mathcal{E}}(P[X], S) & \xrightarrow{\simeq} & \mathrm{Mor}_{P\mathcal{E}}(P(X), S(0)) \\
{\scriptstyle \phi_f{}^*} \downarrow & & \downarrow {\scriptstyle f^*} \\
\mathrm{Mor}_{P/\mathcal{O}_\mathcal{E}}(P[Y], S) & \xrightarrow[\simeq]{} & \mathrm{Mor}_{P\mathcal{E}}(P(Y), S(0)). \quad \square
\end{array}
$$

To extend the functor $U_P(P(X)) = P[X]$ to the whole category of P-algebras, we use the following assertion:

4.1.10 Fact. *Any P-algebra A has a natural presentation by a reflexive coequalizer*

$$
P(S(P, A)) \underset{d_1}{\overset{d_0}{\rightrightarrows}} P(A) \xrightarrow{\epsilon} A,
$$

where $d_0, d_1 : P(S(P, A)) \to P(A)$ *and* $s_0 : P(A) \to P(S(P, A))$ *are morphisms of free P-algebras.*

This statement is proved in [44, Chapter 6] in the context of algebras over a monad. For the sake of completeness, we recall simply the definition of the morphisms in this reflexive coequalizer, and we refer to *loc. cit.* for the proof that A is the coequalizer of d_0, d_1.

To define d_0, we use the morphism

$$
S(P, S(P, A)) = S(P \circ P, A) \xrightarrow{S(\mu, A)} S(P, A),
$$

induced by the composition product of the operad $\mu : P \circ P \to P$, which forms obviously a morphism of free P-algebras. To define d_1, we use the morphism of free P-algebras

$$
P(S(P, A)) \xrightarrow{P(\lambda)} P(P, A)
$$

induced by the evaluation morphism of the P-algebra $\lambda : S(P, A) \to A$. To define s_0, we use the morphism of free P-algebras

$$
P(A) = P(S(I, A)) \xrightarrow{P(S(\eta, A))} P(S(P, A))
$$

induced by the operad unit $\eta : I \to P$. To define ϵ, we use simply the evaluation morphism of A:

$$
P(A) = S(P, A) \xrightarrow{\lambda} A.
$$

4.1.11 Proposition. *The enveloping operad of a P-algebra A is constructed by a reflexive coequalizer of the form:*

$$P[S(P, A)] \underset{\phi_{d_1}}{\overset{\phi_{d_0}}{\underset{\longrightarrow}{\rightrightarrows}}} P[A] \dashrightarrow U_P(A),$$

where d_0, d_1, s_0 refer to the morphisms of free P-algebras of fact 4.1.10.

Proof. By lemma 4.1.8, we have an isomorphism

$$\ker\Big\{ \mathrm{Mor}_{P/\mathcal{O}_{\mathcal{E}}}(P[A], S) \underset{\phi_{d_1}^*}{\overset{\phi_{d_0}^*}{\rightrightarrows}} \mathrm{Mor}_{P/\mathcal{O}_{\mathcal{E}}}(P[S(P, A)], S) \Big\}$$

$$\simeq \ker\Big\{ \mathrm{Mor}_{P\mathcal{E}}(P(A), S(0)) \underset{d_1^*}{\overset{d_0^*}{\rightrightarrows}} \mathrm{Mor}_{P\mathcal{E}}(P(S(P, A)), S(0)) \Big\},$$

from which we deduce the relation:

$$\mathrm{Mor}_{P/\mathcal{O}_{\mathcal{E}}}\big(\mathrm{coker}\{ P[S(P, A)] \underset{\phi_{d_1}}{\overset{\phi_{d_0}}{\rightrightarrows}} P[A] \}, S\big)$$

$$\simeq \mathrm{Mor}_{P\mathcal{E}}\big(\mathrm{coker}\{ P(S(P, A)) \underset{d_1}{\overset{d_0}{\rightrightarrows}} P(A) \}, S(0)\big) \simeq \mathrm{Mor}_{P\mathcal{E}}(A, S(0)).$$

Hence we obtain that

$$U_P(A) = \mathrm{coker}\{ P[S(P, A)] \underset{\phi_{d_1}}{\overset{\phi_{d_0}}{\rightrightarrows}} P[A] \}$$

satisfies the adjunction relation of an enveloping operad. $\square$

4.2 Representations of Algebras over Operads

In this section we define the *category of representations associated to an algebra over an operad.* In the classical examples of algebras over the commutative, associative, and Lie operads, we obtain respectively: the category of left modules over an algebra, the category of bimodules, and the usual category of representations of a Lie algebra.

To handle definitions in the setting of symmetric monoidal categories over a base, we borrow a conceptual definition of [20]. In the next section, we prove that representations are left modules over enveloping algebras, associative algebras formed from the structure of enveloping operads.

4.2.1 Functors on Pairs. For the moment, we assume A is any object in a symmetric monoidal category $\mathcal{E}$ over the base category $\mathcal{C}$. For any $E \in \mathcal{E}$, we set

$$(A; E)^{\otimes n} = \bigoplus_{e=1}^{n} A \otimes \cdots \otimes \underset{e}{E} \otimes \cdots \otimes A,$$

where the sum ranges over tensor products $A \otimes \cdots \otimes E \otimes \cdots \otimes A$ with $n - 1$ copies of $A \in \mathcal{E}$ and one copy of $E \in \mathcal{E}$.

To define the notion of a representation, we use functors $S(M, A; -) : \mathcal{E} \to \mathcal{E}$, associated to Σ_*-objects $M \in \mathcal{M}$, defined by a formula of the form:

$$S(M, A; E) = \bigoplus_{n=0}^{\infty} \left(M(n) \otimes (A; E)^{\otimes n} \right)_{\Sigma_n}.$$

For the unit Σ_*-object I, we have an obvious isomorphism

$$S(I, A; E) = 1 \otimes E \simeq E$$

since $I(1) = 1$ and $I(n) = 0$ for $n \neq 1$. For Σ_*-objects $M, N \in \mathcal{M}$, we have a natural isomorphism

$$S(M, S(N, A); S(N, A; E)) \simeq S(M \circ N, A; E)$$

whose definition extends the usual composition isomorphisms of functors associated to Σ_*-objects. These isomorphisms satisfy coherence identities like the usual composition isomorphisms of functors associated to Σ_*-objects.

4.2.2 Representations of Algebras over Operads. By definition, a *representation of an algebra* A over an operad P is an object $E \in \mathcal{E}$ equipped with a morphism $\mu : S(P, A; E) \to E$ so that the following diagrams commute

where $\lambda : S(P, A) \to A$ refers to the evaluation morphism of the P-algebra A. The morphism $\mu : S(P, A; E) \to E$ determines an *action* of the P-algebra A on E.

We use the notation $\mathcal{R}_P(A)$ to refer to the category of representations of a P-algebra A, where a morphism of representations $f : E \to F$ consists obviously of a morphism in $\mathcal{E}$ which commutes with actions on representations.

By definition of the functor $S(P, A; E)$, the structure of a representation of A is equivalent to a collection of morphisms

$$\mu : P(n) \otimes (A \otimes \cdots \otimes E \otimes \cdots \otimes A) \to E,$$

similar to the evaluation morphisms of a P-algebra, but where a module E occurs once in the tensor product of the left-hand side. The unit and associativity relations of the action $\mu : S(P, A; E) \to E$, defined in terms of commutative diagrams, can be written explicitly in terms of these generalized evaluation morphisms.

In the point-set context, the image of a tensor $p \otimes (a_1 \otimes \cdots \otimes x \otimes \cdots \otimes a_n) \in P(n) \otimes (A \otimes \cdots \otimes E \otimes \cdots \otimes A)$ under the evaluation morphism of a representation is denoted by $p(a_1, \ldots, x, \ldots, a_n) \in E$. The unit relation of the action is equivalent to the identity $1(x) = x$, for the unit operation $1 \in P(1)$, and any element $x \in E$. In terms of partial composites, the associativity relation reads:

$$p(a_1, \ldots, a_{e-1}, q(a_e, \ldots, a_{e+n-1}), a_{e+n}, \ldots, x, \ldots, a_{m+n-1})$$
$$= p \circ_e q(a_1, \ldots, a_{e-1}, a_e, \ldots, a_{e+n-1}, a_{e+n}, \ldots, x, \ldots, a_{m+n-1}),$$
$$p(a_1, \ldots, a_{f-1}, q(a_f, \ldots, x, \ldots, a_{f+n-1}), a_{f+n}, \ldots, a_{m+n-1})$$
$$= p \circ_f q(a_1, \ldots, a_{f-1}, a_f, \ldots, x, \ldots, a_{f+n-1}, a_{f+n}, \ldots, a_{m+n-1}),$$

for $p \in P(m)$, $q \in P(n)$, and $a_1 \otimes \cdots \otimes x \otimes \cdots \otimes a_{m+n-1} \in A \otimes \cdots \otimes E \otimes \cdots \otimes A$.

4.2.3 Representations of Algebras over Operads Defined by Generators and Relations.
If $P = F(M)$ is a free operad, then a representation of a P-algebra A is equivalent to an object E equipped with a collection of morphisms

$$\mu : M(n) \otimes (A; E)^{\otimes n} \to E, \ n \in \mathbb{N},$$

with no relation required outside Σ_n-equivariance relations. Roughly, the associativity of actions implies (as in the case of algebras) that the action of formal composites $(\cdots ((\xi_1 \circ_{e_2} \xi_2) \circ_{e_3} \cdots) \circ_{e_r} \xi_r$, which spans the free operad $F(M)$, is determined by the action of the generating operations $\xi_i \in M(n_i)$.

If P is an operad in $\Bbbk$-modules defined by generators and relations $P = F(M)/(R)$, then the structure of a representation on an object E is determined by a collection of morphisms

$$\mu : M(n) \otimes (A; E)^{\otimes n} \to E, \ n \in \mathbb{N},$$

such that we have the relations

$$\rho(a_1, \ldots, x, \ldots, a_n) = 0, \ \forall a_1 \otimes \cdots \otimes x \otimes \cdots \otimes a_n \in A \otimes \cdots \otimes E \otimes \cdots \otimes A,$$

for every generating relation $\rho \in R$.

4.2.4 Examples: The Case of Classical Operads in $\Bbbk$-Modules. In the usual examples of operads in $\Bbbk$-modules, $\mathsf{P} = \mathsf{C}, \mathsf{L}, \mathsf{A}$, the equivalence of §4.2.3 returns the following assertions:

(a) Let A be any commutative algebra in $\Bbbk$-modules. The structure of a representation of A is determined by morphisms $\mu : E \otimes A \to E$ and $\mu : A \otimes E \to E$ so that we have the symmetry relation $\mu(a, x) = \mu(x, a)$, for all $a \in A$, $x \in E$, and the associativity relations

$$\mu(\mu(a, b), x) = \mu(a, \mu(b, x)), \quad \mu(\mu(a, x), b) = \mu(a, \mu(x, b)),$$
$$\mu(\mu(x, a), b) = \mu(x, a \cdot b),$$

for all $a, b \in A$, $x \in E$, where we also use the notation $\mu(a, b) = a \cdot b$ to denote the product of A. As a consequence, the category of representations $\mathcal{R}_\mathsf{C}(A)$, where A is any commutative algebra, is isomorphic to the usual category of left A-modules.

(b) Let A be any associative algebra in $\Bbbk$-modules. The structure of a representation of A is determined by morphisms $\mu : E \otimes A \to E$ and $\mu : A \otimes E \to E$ so that we have the associativity relations

$$\mu(\mu(a, b), x) = \mu(a, \mu(b, x)), \quad \mu(\mu(a, x), b) = \mu(a, \mu(x, b)),$$
$$\mu(\mu(x, a), b) = \mu(x, \mu(a, b)),$$

for all $a, b \in A$, $x \in E$, where we also use the notation $\mu(a, b) = a \cdot b$ to denote the product of A. Hence the category of representations $\mathcal{R}_\mathsf{A}(A)$, where A is any associative algebra, is isomorphic to the usual category of A-bimodules.

(c) Let G be any Lie algebra in $\Bbbk$-modules. The structure of a representation of G is determined by morphisms $\gamma : E \otimes \mathsf{G} \to E$ and $\gamma : \mathsf{G} \otimes E \to E$ so that we have the antisymmetry relation $\gamma(g, x) = \gamma(x, g)$, for all $g \in \mathsf{G}$, $x \in E$, and the Jacobi relations

$$\gamma(\gamma(g, h), x) + \gamma(\gamma(h, x), g) + \gamma(\gamma(x, g), h) = 0$$

for all $g, h \in \mathsf{G}$, $x \in E$, where we also use the notation $\gamma(g, h) = [g, h]$ to denote the Lie bracket of G. Hence the category of representations $\mathcal{R}_\mathsf{L}(\mathsf{G})$, where G is any Lie algebra, is isomorphic to the usual category of representations of the Lie algebra G.

4.3 Enveloping Algebras

The *enveloping algebra* of a P-algebra A is the term $U_\mathsf{P}(A)(1)$ of the enveloping operad $U_\mathsf{P}(A)$. The first purpose of this section is to prove that the category of representations of A is isomorphic to the category of left $U_\mathsf{P}(A)(1)$-modules. Then we determine the enveloping algebra of algebras over the usual commutative, associative, and Lie operads. In all cases, we retrieve usual constructions of classical algebra.

4.3.1 Enveloping Algebras. By definition, the enveloping algebra of a P-algebra A is the term $U_P(A)(1)$ of the enveloping operad $U_P(A)$. The composition product

$$\mu : U_P(A)(1) \otimes U_P(A)(1) \to U_P(A)(1)$$

defines a unital associative product on $U_P(A)(1)$ so that $U_P(A)(1)$ forms a unitary associative algebra in the category $\mathcal{E}$. Since we use no more enveloping operads in this section, we can drop the reference to the arity in the notation of enveloping algebras and, for simplicity, we set $U_P(A) = U_P(A)(1)$.

In the point-set context, the enveloping algebra $U_P(A)$ can be defined informally as the object spanned by formal elements

$$u = p(x, a_1, \ldots, a_n),$$

where x is a variable and $p \in P(1 + n)$, $a_1, \ldots, a_n \in A$, divided out by the relation of §4.1.3:

$$p(x, a_1, \ldots, a_{e-1}, q(a_e, \ldots, a_{e+n-1}), a_{e+s}, \ldots, a_{m+n-1})$$
$$\equiv p \circ_{1+e} q(x, a_1, \ldots, a_{e-1}, a_e, \ldots, a_{e+n-1}, a_{e+n}, \ldots, a_{m+n-1}).$$

The product of $U_P(A)$ is defined by the formula:

$$p(x, a_1, \ldots, a_m) \cdot q(x, b_1, \ldots, b_n) = p \circ_1 q(x, b_1, \ldots, b_n, a_1, \ldots, a_m).$$

The enveloping algebra of an algebra over an operad is defined in these terms in [18] in the context of $\Bbbk$-modules.

The first objective of this section is to prove:

4.3.2 Proposition. *Let A be a P-algebra in any symmetric monoidal category over $\mathcal{C}$. The category of representations $\mathcal{R}_P(A)$ is isomorphic to the category of left $U_P(A)$-modules, where $U_P(A) = U_P(A)(1)$ refers to the enveloping algebra of A.* $\square$

In the point-set context, the proposition is an immediate consequence of this explicit construction of the enveloping algebra and of the explicit definition of the structure of representations in terms of operations $p \otimes (a_1 \otimes \cdots \otimes x \otimes \cdots \otimes a_n) \mapsto p(a_1, \ldots, x, \ldots, a_n)$. Note that any operation of the form $p(a_1, \ldots, a_{i-1}, x, a_i, \ldots, a_n)$ is, by equivariance, equivalent to an operation of the form $wp(x, a_1, \ldots, a_{i-1}, a_i, \ldots, a_n)$, where x is moved to the first position of the tensor product.

The proposition holds in any symmetric monoidal category. To check the generalized statement, we use the explicit construction of the enveloping operad, which generalizes the point-set construction of §4.1.3.

Proof. The construction of proposition 4.1.11 implies that the enveloping algebra of A is defined by a reflexive coequalizer of the form:

$$P[S(P, A)](1) \underset{\phi_{d_1}}{\overset{\phi_{d_0}}{\rightrightarrows}} P[A](1) \dashrightarrow U_P(A)(1),$$

with ϕ_{s_0} above.

For any object E, we have an obvious isomorphism $S(P, A; E) \simeq P[A](1) \otimes E$ and we obtain

$$S(P, S(P, A); S(P, A; E)) \simeq P[S(P, A)](1) \otimes P[A](1) \otimes E.$$

Thus the evaluation morphism of a representation $\rho : S(P, A; E) \to E$ is equivalent to a morphism $\rho : P[A](1) \otimes E \to E$. One checks further, by a straightforward inspection, that $\rho : S(P, A; E) \to E$ satisfies the unit and associativity relation of representations if and only if the equivalent morphism $\rho : P[A](1) \otimes E \to E$ equalizes the morphisms $d_0, d_1 : P[S(P, A)](1) \to P[A](1)$ and induces a unitary and associative action of $U_P(A)(1)$ on E. The conclusion follows. $\square$

As a corollary, we obtain:

4.3.3 Proposition. *The forgetful functor $U : \mathcal{R}_P(A) \to \mathcal{E}$ has a left adjoint $F : \mathcal{E} \to \mathcal{R}_P(A)$ which associates to any object $X \in \mathcal{E}$ the free left $U_P(A)$-module $F(X) = U_P(A) \otimes X$.* $\square$

In other words, the object $F(X) = U_P(A) \otimes X$ represents the free object generated by X in the category of representations $\mathcal{R}_P(A)$.

In the remainder of the section, we determine the operadic enveloping algebras of commutative, associative and Lie algebras. To simplify, we assume $\mathcal{E} = \mathcal{C} = \Bbbk \operatorname{Mod}$ and we use the pointwise representation of tensors in proofs. As explained in the introduction of this chapter, we can use the principle of generalized point-tensors (see §0.5) to extend our results to commutative (respectively, associative, Lie) algebras in categories of dg-modules, in categories of Σ_*-objects, and in categories of right modules over operads.

4.3.4 Proposition. *Let $P = F(M)/(R)$ be an operad in $\Bbbk$-modules defined by generators and relations. The enveloping algebra of any P-algebra A is generated by formal elements $\xi(x_1, a_1, \dots, a_n)$, where $\xi \in M(1 + n)$ ranges over generating relations of P, together with the relations $w\rho(x_1, a_1, \dots, a_n) \equiv 0$, where $\rho \in R(n)$ and $w \in \Sigma_{n+1}$.* $\square$

The next propositions are easy consequences of this statement.

4.3.5 Proposition. *For a commutative algebra without unit A, the operadic enveloping algebra $U_C(A)$ is isomorphic to A_+, the unitary algebra such that $A_+ = 1 \oplus A$.* $\square$

The object $A_+ = 1 \oplus A$ represents the image of A under the extension functor $(-)_+ : {}_C\mathcal{E} \to {}_{C_+}\mathcal{E}$ from commutative algebras without unit to commutative algebras with unit. The product of A_+ extends the product of A on $A \otimes A \subset A_+ \otimes A_+$ and is given by the canonical isomorphisms $1 \otimes 1 \simeq 1$ and $A \otimes 1 \simeq A \simeq 1 \otimes A$ on the other components $1 \otimes 1, A \otimes 1, 1 \otimes A \subset A_+ \otimes A_+$. This construction can also be applied to associative algebras. The result gives an associative algebra with unit A_+ naturally associated to any $A \in {}_A\mathcal{E}$ so that A_+ represents the image of A under the extension functor $(-)_+ : {}_A\mathcal{E} \to {}_{A_+}\mathcal{E}$.

4.3.6 Proposition. *For an associative algebra without unit A, the operadic enveloping algebra $U_A(A)$ is isomorphic to the classical enveloping algebra of A, defined as the tensor product $A_+ \otimes A_+^{op}$, where A_+ is obtained by the addition of a unit to A, as in proposition §4.2.4, and A_+^{op} refers to the opposite algebra of A_+.* □

Recall that the Lie operad L forms a suboperad of the associative operad A: the operad embedding $\iota : L \to A_+$ maps the generating operation of the Lie operad $\gamma = \gamma(x_1, x_2)$ to the commutator $\iota(\gamma) = \mu - \tau\mu = \mu(x_1, x_2) - \mu(x_2, x_1)$. The classical enveloping algebra of a Lie algebra G, usually defined as the algebra generated by elements X_g, $g \in G$, together with the relations $X_g \cdot X_h - X_h \cdot X_g \equiv X_{[g,h]}$, represents the image of G under the extension functor $\iota_! : {}_L\mathcal{E} \to {}_{A_+}\mathcal{E}$.

4.3.7 Proposition. *For a Lie algebra G, the operadic enveloping algebra $U_L(G)$ is isomorphic to the classical enveloping algebra of G, defined as the image of G under the extension functor $\iota_! : {}_L\mathcal{E} \to {}_{A_+}\mathcal{E}$.* □

4.4 Derivations and Kähler Differentials

The *modules of Kähler differentials* $\Omega^1_P(A)$ are representations of algebras over P which appear naturally in the deformation theory of P-algebras and in the definition of a generalized Quillen homology for P-algebras.

To simplify, we take $\mathcal{E} = \mathcal{C} = \Bbbk\,\mathrm{Mod}$ as underlying symmetric monoidal categories and we use the pointwise representation of tensors in $\Bbbk$-modules. Again, we can use the principle of generalized point-tensors to extend the results of this section to the category of dg-modules, to the category of Σ_*-objects, and to categories of right modules over operads.

4.4.1 Derivations and Kähler Differentials. A map $\theta : A \to E$, where A is a P-algebra and E a representation of A, is a derivation if it satisfies the relation

$$\theta(p(a_1, \ldots, a_n)) = \sum_{i=1}^{n} p(a_1, \ldots, \theta(a_i), \ldots, a_n),$$

for all $p \in P(n)$, $a_1, \ldots, a_n \in A$. The module of *Kähler differentials* $\Omega^1_P(A)$ is a representation of A such that

$$\mathrm{Mor}_{\mathcal{R}_P(A)}(\Omega^1_P(A), E) = \mathrm{Der}_P(A, E),$$

for all E, where $\mathrm{Der}_P(A, E)$ denotes the $\Bbbk$-module of derivations $\theta : A \to E$.

Note simply that the functor $\mathrm{Der}_P(A, -) : E \mapsto \mathrm{Der}_P(A, E)$ preserves limits to justify the existence of $\Omega^1_P(A)$.

In the case of a free P-algebra, we obtain:

4.4.2 Observation. *If $A = P(X)$ is a free P-algebra, then any derivation $\theta : P(X) \to E$ is uniquely determined by its restriction to generators. Accordingly, we have an isomorphism*

$$\mathrm{Der}_P(P(X), E) \simeq \mathrm{Mor}_{\mathcal{E}}(X, E),$$

for any representation $E \in \mathcal{R}_P(P(X))$, and the module of Kähler differentials $\Omega^1_P(P(X))$ is isomorphic to the free representation $U_P(P(X)) \otimes X$, where $U_P(P(X)) = U_P(P(X))(1)$ refers to the enveloping algebra of $P(X)$ (recall that representations of A are equivalent to left $U_P(P(X))$-modules).

4.4.3 Explicit Constructions of Kähler Differentials. An explicit construction of the module of Kähler differentials $\Omega^1_P(A)$ can be deduced from observation 4.4.2: as in §4.1, we observe that $U_P(P(X)) \otimes X$ forms a functor on the full subcategory of free P-algebras in $_P\mathcal{E}$ and we use that any P-algebra A has a presentation by a reflexive coequalizer naturally associated to A.

On the other hand, in the context of $\Bbbk$-modules, the module of Kähler differentials $\Omega^1_P(A)$ can be defined easily as the $\Bbbk$-module spanned by formal expressions $p(a_1, \ldots, da_i, \ldots, a_m)$, where $p \in P(m)$, $a_1, \ldots, a_m \in A$, together with relations of the form:

$$p(a_1, \ldots, q(a_i, \ldots, a_{i+n-1}), \ldots, da_j, \ldots, a_{m+n-1})$$
$$\equiv p \circ_i q(a_1, \ldots, a_i, \ldots, a_{i+n-1}, \ldots, da_j, \ldots, a_{m+n-1}), \quad \text{for } i \neq j,$$
$$p(a_1, \ldots, dq(a_i, \ldots, a_{i+n-1}), \ldots, a_{m+n-1})$$
$$\equiv \sum_{j=i}^{i+n-1} p \circ_i q(a_1, \ldots, a_i, \ldots, da_j, \ldots, a_{i+n-1}, \ldots, a_{m+n-1}).$$

The representation structure is given by the formula

$$p(a_1, \ldots, a_{i-1}, q(a_i, \ldots, da_j, \ldots, a_{i+n-1}), a_{i+n}, \ldots, a_{m+n-1})$$
$$= p \circ_i q(a_1, \ldots, a_{i-1}, a_i, \ldots, da_j, \ldots, a_{i+n-1}, a_{i+n}, \ldots, a_{m+n-1}),$$

for every formal element $q(a_i, \ldots, da_j, \ldots, a_{i+n-1}) \in \Omega^1_P(A)$, and where $p \in P(m)$, $q \in P(n)$, and $a_1 \ldots, a_{m+n-1} \in A$. The adjunction relation is immediate from this construction.

In the case of a free P-algebra $A = P(X)$, the natural isomorphism $\theta : U_P(P(X)) \otimes X \xrightarrow{\simeq} \Omega^1_P(P(X))$ identifies a generating element $x \in X$ to a formal differential $dx = 1(dx) \in \Omega^1_P(P(X))$.

4.4.4 Classical Examples. If the operad P is defined by generators and relations $P = F(M)/(R)$, then a map $\theta : A \to E$ forms a derivation if and only if it satisfies the relation

$$\theta(\xi(a_1,\ldots,a_n)) = \sum_{i=1}^{n} \xi(a_1,\ldots,\theta(a_i),\ldots,a_n)$$

with respect to generating operations $\xi \in M(n)$. To prove this assertion, use again that P is spanned by formal composites of generating operations $\xi \in M(n)$ and check the coherence of the derivation relation with respect to operadic composites.

In the context of associative and commutative algebras, we obtain that a map $\theta : A \to E$ forms a derivation if and only if it satisfies the usual identity

$$\theta(a \cdot b) = \theta(a) \cdot b + a \cdot \theta(b)$$

with respect to the product. In the context of Lie algebras, we obtain that a map $\theta : A \to E$ forms a derivation if and only if it satisfies the identity

$$\theta([a,b]) = [\theta(a),b] + [a,\theta(b)]$$

with respect to the Lie bracket. Thus, in all cases, we obtain the standard notions of derivations.

Since derivations of commutative algebras are derivations in the usual sense, we obtain:

4.4.5 Proposition. *For a commutative algebra A, the module of Kähler differentials $\Omega_C^1(A)$ which arises from the theory of operads is isomorphic to the usual module of Kähler differentials of commutative algebra.* ☐

In the case of associative algebras, we obtain:

4.4.6 Proposition. *For an associative algebra (without unit) A, the module of Kähler differentials $\Omega_A^1(A)$ is isomorphic to the tensor product $A_+ \otimes A$ together with the bimodule structure such that*

$$a \cdot (\alpha \otimes \beta) = a\alpha \otimes \beta,$$
$$\text{and} \quad (\alpha \otimes \beta) \cdot b = \alpha \otimes \beta b + \alpha\beta \otimes b,$$

for $a, b \in A$, $\alpha \otimes \beta \in A_+ \otimes A$.

The definition of the module $A_+ \otimes A$ as a non-commutative analogue of Kähler differential appears in non-commutative geometry (see [8, 34, 40]). The proposition asserts that the definition of the theory of operads agrees with the definition of non-commutative geometry. Since the relationship between $A_+ \otimes A$ and derivations does not seem to occur in the literature, we give a proof of the proposition.

Proof. We make explicit inverse isomorphisms

$$\Theta : \mathrm{Mor}_{\mathcal{R}_\mathtt{A}(A)}(A_+ \otimes A, E) \to \mathrm{Der}(A, E)$$
$$\text{and} \quad \Phi : \mathrm{Der}(A, E) \to \mathrm{Mor}_{\mathcal{R}_\mathtt{A}(A)}(A_+ \otimes A, E).$$

In one direction, for a map $\phi : A_+ \otimes A \to E$, we set $\Theta(\phi)(a) = \phi(1 \otimes a)$, for $a \in A$. It is straightforward to check that $\Theta(\phi)$ forms a derivation if ϕ is a morphism of A-bimodules. In the converse direction, for a map $\theta : A \to E$, we set $\Phi(\theta)(a \otimes b) = a \cdot \theta(b)$, for $a \otimes b \in A_+ \otimes A$. It is straightforward to check that $\Phi(\theta)$ forms a morphism of A-bimodules if θ is a derivation. In addition we have clearly $\Theta(\Phi(\theta)) = \theta$, for all maps $\theta : A \to E$. Conversely, for any morphism of A-bimodules $\phi : A_+ \otimes A \to E$, we have

$$\Phi(\Theta(\phi))(a \otimes b) = a \cdot \phi(1 \otimes b) = \phi(a \cdot (1 \otimes b)) = \phi(a \otimes b).$$

Hence we obtain $\Theta\Phi = \mathrm{Id}$ and $\Phi\Theta = \mathrm{Id}$. □

Recall that the operadic enveloping algebra $U_\mathrm{L}(\mathsf{G})$ is identified with the classical enveloping algebra of G, the associative algebra generated by elements X_g, where $g \in \mathsf{G}$, together with the relations $X_g X_h - X_h X_g = X_{[g,h]}$, for $g, h \in \mathsf{G}$. Let $\widetilde{U}_\mathrm{L}(\mathsf{G})$ be the augmentation ideal of $U_\mathrm{L}(\mathsf{G})$, the submodule of $U_\mathrm{L}(\mathsf{G})$ spanned by products $X_{g_1} \cdots X_{g_n}$ of length $n > 0$.

4.4.7 Proposition. *For a Lie algebra G, the module of Kähler differentials $\Omega^1_\mathrm{L}(\mathsf{G})$ is isomorphic to the augmentation ideal $\widetilde{U}_\mathrm{L}(\mathsf{G})$ together with its natural left $U_\mathrm{L}(\mathsf{G})$-module structure.*

Proof. We make explicit inverse isomorphisms

$$\Theta : \mathrm{Mor}_{R_\mathrm{L}(\mathsf{G})}(\widetilde{U}_\mathrm{L}(\mathsf{G}), E) \to \mathrm{Der}(\mathsf{G}, E)$$
$$\text{and} \quad \Phi : \mathrm{Der}(\mathsf{G}, E) \to \mathrm{Mor}_{R_\mathrm{L}(\mathsf{G})}(\widetilde{U}_\mathrm{L}(\mathsf{G}), E).$$

In one direction, for a map $\phi : \widetilde{U}_\mathrm{L}(\mathsf{G}) \to E$, we set $\Theta(\phi)(g) = \phi(X_g)$, for $g \in \mathsf{G}$. The derivation relation $\Theta(\phi)([g, h]) = [\Theta(\phi)(g), h] + [g, \Theta(\phi)(h)]$ follows from the relation $X_{[g,h]} = X_g \cdot X_h - X_h \cdot X_g$ in $U_\mathrm{L}(\mathsf{G})$ and the assumption that ϕ is a morphism of left $U_\mathrm{L}(\mathsf{G})$-module. In the converse direction, for a map $\theta : \mathsf{G} \to E$, we set $\Phi(\theta)(X_{g_1} \cdots X_{g_n}) = X_{g_1} \cdots X_{g_{n-1}} \cdot \theta(g_n)$, for $X_{g_1} \cdots X_{g_n} \in \widetilde{U}_\mathrm{L}(\mathsf{G})$. One checks readily that $\Phi(\theta)$ cancels the ideal generated by the relations $X_g X_h - X_h X_g - X_{[g,h]}$ if θ forms a derivation. Hence the map $\Phi(\theta) : \widetilde{U}_\mathrm{L}(\mathsf{G}) \to E$ is well defined. It is also immediate from the definition that $\Phi(\theta)$ is a morphism of left $U_\mathrm{L}(\mathsf{G})$-modules. We have clearly $\Theta(\Phi(\theta)) = \theta$, for all maps $\theta : \mathsf{G} \to E$. Conversely, for any morphism of left $U_\mathrm{L}(\mathsf{G})$-modules $\phi : \widetilde{U}_\mathrm{L}(\mathsf{G}) \to E$, we have

$$\Phi(\Theta(\phi))(X_{g_1} \cdots X_{g_n}) = X_{g_1} \cdots X_{g_{n-1}} \cdot \phi(X_{g_n}) = \phi(X_{g_1} \cdots X_{g_{n-1}} \cdot X_{g_n}).$$

Hence we obtain $\Theta\Phi = \mathrm{Id}$ and $\Phi\Theta = \mathrm{Id}$. □

Bibliographical Comments on Part I

We refer to [46] for a comprehensive overview of the history of operads. We only give brief indications on matters related to our approach on algebras over operads.

In most applications, authors deal with algebras and operads within a fixed usual category, like the category of topological spaces, the category of dg-modules, the category of simplicial sets, or the category of simplicial modules. Nevertheless, homology theories, and other usual functors which change the underlying category, are often used in the study of operads.

In the original reference [47], the structure of an operad is defined within the category of topological spaces as a device to model the structure of iterated loop spaces. The definition of operads in the framework of symmetric monoidal categories occurs explicitly in [17]. But, as far as we know, the thesis [54] is the first reference which formalizes the naturality of operad structures with respect to functors of symmetric monoidal categories.

In almost all references, the algebras over an operad are defined within the same category as the operad. But, in problems of stable homotopy, authors consider naturally spectra acted on by operads in simplicial sets (see for instance [21]). However, the general axiomatic background of symmetric monoidal categories over a base, which axiomatizes this relationship, has not been formalized in the operadic literature yet.

The composition product ∘ is defined in [32] for Σ_*-objects in sets. The introduction of this operation in the operad literature goes back to [56]. Left modules over operads appear explicitly in [57] in a realization of the cotriple construction of [3, 47] at the operad level. The notion of a left module over operads is studied more thoroughly in [54] and in [14]. The point of view of [54] is to identify algebras over operads with constant left modules. This approach is the usual one of the literature.

The converse point of view adopted in this book, according to which left modules are algebras over operads in a category over the base category, has not been fully exploited yet. However, the equivalence between left modules over operads and algebras in Σ_*-objects is made explicit in [38]. Note further that the generalization of Morita's theorem to modules over operads in [33] uses such an equivalence implicitly.

In the literature, the first examples of left module structures have appeared as generalized algebras: Lie algebras in Σ_*-objects are introduced in [2] (under the name twisted Lie algebras) to study total Hopf invariants. The study of these generalized Lie algebras structures is carried on in [19, 61], and in [11, 12, 13] and [39] in the language of operads.

Representations and enveloping algebras of algebras over an operads are defined in [18]. The idea of the enveloping operad goes back to [17]. The definition adopted in §4.1 is borrowed from [13]. The notion of a Kähler differential for algebras over an operad is introduced in [26] in a definition of the cohomology of algebras over operads.

Part II
The Category of Right Modules over Operads and Functors

Chapter 5
Definitions and Basic Constructions

Introduction

In this chapter, we define the functor $S_R(M) : {}_R\mathcal{E} \to \mathcal{E}$ associated to a right R-module M. Besides, we study the commutation of the functor $S_R(M) : {}_R\mathcal{E} \to \mathcal{E}$ with colimits.

For our purpose, we assume that right modules, like Σ_*-objects and operads, belong to the fixed base category $\mathcal{C}$. But, as usual, we consider algebras in symmetric monoidal categories $\mathcal{E}$ over $\mathcal{C}$.

The construction $S_R(M) : A \mapsto S_R(M, A)$ appears explicitly in [54, §2.3.10] as a particular case of a relative composition product $M \circ_R N$, an operation between right and left modules over an operad R. In §5.1.5, we observe conversely that the relative composition product $M \circ_R N$ is given by a construction of the form $S_R(M, N)$ when N is identified with an R-algebra in Σ_*-objects.

The category of left R-modules is denoted by ${}_R\mathcal{M}$. We adopt the symmetrical notation $\mathcal{M}_R$ for the category of right R-modules.

In §3.2.15, we introduce the notation $\mathcal{F} = \mathcal{F}(\mathcal{E}, \mathcal{E})$ for the category of functors $F : \mathcal{E} \to \mathcal{E}$ and the short notation ${}_R\mathcal{F} = \mathcal{F}(\mathcal{E}, {}_R\mathcal{E})$ for the category of functors $F : \mathcal{E} \to {}_R\mathcal{E}$. If the category $\mathcal{E}$ is clear from the context, then we may use the symmetric short notation $\mathcal{F}_R = \mathcal{F}({}_R\mathcal{E}, \mathcal{E})$ for the category of functors $F : {}_R\mathcal{E} \to \mathcal{E}$. The map $S_R : M \mapsto S_R(M)$ defines a functor $S_R : \mathcal{M}_R \to \mathcal{F}_R$.

Because of our conventions on categories of functors, we should check that $S_R(M) : {}_R\mathcal{E} \to \mathcal{E}$ preserves filtered colimits. The (straightforward) verification of this assertion is addressed in section 5.2 devoted to colimits.

B. Fresse, *Modules over Operads and Functors*, Lecture Notes in Mathematics 1967, 99
DOI: 10.1007/978-3-540-89056-0_5, © Springer-Verlag Berlin Heidelberg 2009

5.1 The Functor Associated to a Right Module over an Operad

We recall the definition of a right module over an operad and we define the functor $S_R(M) : A \mapsto S_R(M, A)$ associated to a right module next.

5.1.1 Right Modules over Operads. In §3.2.9, we recall that a left R-module consists of a Σ_*-object $N \in \mathcal{M}$ equipped with a morphism $\lambda : R \circ N \to N$ that provides N with a left action of the monoid object in $(\mathcal{M}, \circ, I)$ defined by the operad R (see [14, §2.1.5]). Symmetrically, a right R-module consists of a Σ_*-object $M \in \mathcal{M}$ equipped with a morphism $\rho : M \circ R \to M$ that fits commutative diagrams

$$
\begin{array}{ccc}
M \circ R & \xleftarrow{\ M \circ \eta\ } & M \circ I \\
\downarrow{\scriptstyle \rho} & \swarrow{\scriptstyle =} & \\
M & &
\end{array}
\qquad
\begin{array}{ccc}
M \circ R \circ R & \xrightarrow{\ \rho \circ R\ } & M \circ R \\
\downarrow{\scriptstyle M \circ \mu} & & \downarrow{\scriptstyle \rho} \\
M \circ R & \xrightarrow{\ \rho\ } & M,
\end{array}
$$

where as usual $\eta : I \to R$ (respectively, $\mu : R \circ R \to R$) refers to the unit (respectively, product) of the operad R. In short, the morphism $\rho : M \circ R \to M$ provides M with a right action of the monoid object in $(\mathcal{M}, \circ, I)$ defined by the operad R (see [14, §2.1.5]).

Note that the composition product of Σ_*-objects is not symmetric. Therefore, in general, we can not recover properties of left R-modules by symmetry with right R-modules.

The category of right R-modules is denoted by $\mathcal{M}_R$, a morphism of right R-modules consists obviously of a morphism of Σ_*-objects $f : M \to N$ that preserves operad actions.

Like an operad structure, the action of an operad R on a right R-module M is equivalent to composition products

$$M(r) \otimes R(n_1) \otimes \cdots \otimes R(n_r) \to M(n_1 + \cdots + n_r)$$

that satisfy natural equivariance properties. The unit and associativity relations of operad actions can be expressed in terms of these composition products by an obvious generalization of May's axioms. The partial compositions product of an operad (see §3.1) have an obvious generalization in the context of right-modules so that a right R-module structure is fully determined by collections of morphisms

$$\circ_i : M(m) \otimes R(n) \to M(m + n - 1), \ i = 1, \ldots, m,$$

that satisfy natural equivariance properties, as well as unit and associativity relations (see [46]).

In the point-set context, the composite of an element $\xi \in M(r)$ with operations $q_1 \in R(n_1), \ldots, q_r \in R(n_r)$ is denoted by $\xi(q_1, \ldots, q_r) \in M(n_1 +$

$\cdots + n_r$), like a composite of operad operations. Similarly, we use the notation $\xi \circ_e q$ for the partial composite of $\xi \in M(m)$ with $q \in R(n)$. The definition of partial composites reads explicitly $\xi \circ_e q = \xi(1, \ldots, q, \ldots, 1)$, where operad units occur at positions $i \neq e$ in the composite of the right-hand side.

5.1.2 Colimits and Limits. Since left R-modules form a category of R-algebras in a symmetric monoidal category over $\mathcal{C}$, proposition 3.3.1 implies that the forgetful functor $U : {}_R\mathcal{M} \to \mathcal{M}$ creates filtered colimits and reflexive coequalizers in ${}_R\mathcal{M}$.

In the case of right R-modules, we obtain that the forgetful functor $U : \mathcal{M}_R \to \mathcal{M}$ creates all colimits in $\mathcal{M}_R$, because the composition product of Σ_*-objects preserves colimits on the left: since we have an isomorphism $(\mathrm{colim}_{i \in I} M_i) \circ R \xleftarrow{\simeq} \mathrm{colim}_{i \in I}(M_i \circ R)$, we can form a composite

$$(\operatorname*{colim}_{i \in I} M_i) \circ R \xleftarrow{\simeq} \operatorname*{colim}_{i \in I}(M_i \circ R) \to \operatorname*{colim}_{i \in I} M_i$$

to provide every colimit of right R-modules M_i with a natural right R-module structure and the conclusion follows.

The forgetful functor $U : \mathcal{M}_R \to \mathcal{M}$ creates all limits, as in the case of left R-modules, because we have a natural morphism $(\lim_{i \in I} M_i) \circ R \to \lim_{i \in I}(M_i \circ R)$ in the good direction to provide $\lim_{i \in I} M_i$ with a right R-module structure.

5.1.3 The Functor Associated to a Right Module over an Operad. The purpose of this paragraph is to define the functor $S_R(M) : {}_R\mathcal{E} \to \mathcal{E}$ associated to a right R-module M, where $\mathcal{E}$ is any symmetric monoidal category over $\mathcal{C}$. The image of an R-algebra $A \in {}_R\mathcal{E}$ under this functor $S_R(M) : A \mapsto S_R(M, A)$ is defined by a reflexive coequalizer of the form

$$S(M \circ R, A) \underset{d_1}{\overset{d_0}{\underset{\longrightarrow}{\rightrightarrows}}} \;\;\;\; {}^{s_0} \!\! \text{ } S(M, A) \longrightarrow S_R(M, A) \, .$$

On one hand, the functor $S(M) : \mathcal{E} \to \mathcal{E}$ associated to a right R-module M is equipped with a natural transformation

$$S(M \circ R, A) \xrightarrow{S(\rho, A)} S(M, A)$$

induced by the morphism $\rho : M \circ R \to M$ that defines the right R-action on M. On the other hand, the structure of an R-algebra is determined by a morphism $\lambda : S(R, A) \to A$ in $\mathcal{E}$. Therefore, for an R-algebra A, we have another natural morphism

$$S(M \circ R, A) = S(M, S(R, A)) \xrightarrow{S(M, \lambda)} S(M, A)$$

induced by the morphism $\lambda : S(R, A) \to A$. In the converse direction, we have a natural morphism

$$S(M, A) = S(M \circ I, A) \xrightarrow{S(M \circ \eta, A)} S(M \circ R, A)$$

induced by the operad unit $\eta : I \to R$. The object $S_R(M, A) \in \mathcal{E}$ associated to $A \in {}_R\mathcal{E}$ is defined by the coequalizer of $d_0 = S(\rho, A)$ and $d_1 = S(M, \lambda)$ in $\mathcal{E}$ together with the reflection $s_0 = S(M \circ \eta, A)$.

In the point-set context, the object $S_R(M, A)$ can be defined intuitively as the object spanned by tensors $x(a_1, \ldots, a_r)$, where $x \in M(r)$, $a_1, \ldots, a_r \in A$, divided out by the coinvariant relations of $S(M, A)$

$$\sigma x(a_1, \ldots, a_r) \equiv x(a_{\sigma(1)}, \ldots, a_{\sigma(r)})$$

and relations

$$x \circ_i p(a_1, \ldots, a_{r+s-1}) \equiv x(a_1, \ldots, p(a_i, \ldots, a_{i+s-1}), \ldots, a_{r+s-1})$$

yielded by the coequalizer construction.

Clearly, the map $S_R(M) : A \mapsto S_R(M, A)$ defines a functor $S_R(M) : {}_R\mathcal{E} \to \mathcal{E}$, from the category of R-algebras in $\mathcal{E}$ to the category $\mathcal{E}$, and the map $S_R : M \mapsto S_R(M)$ defines a functor $S_R : \mathcal{M}_R \to \mathcal{F}({}_R\mathcal{E}, \mathcal{E})$, from the category of right R-modules to the category of functors $F : {}_R\mathcal{E} \to \mathcal{E}$. For short, we can adopt the notation $\mathcal{F}_R$ for this functor category $\mathcal{F}_R = \mathcal{F}({}_R\mathcal{E}, \mathcal{E})$.

The construction $S_R : M \mapsto S_R(M)$ is also functorial in $\mathcal{E}$. Explicitly, for any functor $\rho : \mathcal{D} \to \mathcal{E}$ of symmetric monoidal categories over $\mathcal{C}$, the diagram of functors

$$
\begin{array}{ccc}
{}_R\mathcal{D} & \xrightarrow{\rho} & {}_R\mathcal{E} \\
\scriptstyle S_R(M) \downarrow & & \downarrow \scriptstyle S_R(M) \\
\mathcal{D} & \xrightarrow{\rho} & \mathcal{E}
\end{array}
$$

commutes up to natural isomorphisms.

5.1.4 Constant Modules and Constant Functors. Recall that the base category $\mathcal{C}$ is isomorphic to the full subcategory of $\mathcal{M}$ formed by constant Σ_*-objects. By proposition 2.1.6, we also have a splitting $\mathcal{M} = \mathcal{C} \times \mathcal{M}^0$, where $\mathcal{M}^0$ is the category of connected Σ_*-objects in $\mathcal{C}$.

Any constant Σ_*-object has an obvious right R-module structure: all partial composites $\circ_i : M(r) \otimes R(s) \to M(r + s - 1)$ are necessarily trivial if we assume $M(r) = 0$ for $r > 0$. Hence the base category $\mathcal{C}$ is also isomorphic to the full subcategory of $\mathcal{M}_R$ formed by constant right R-modules. Observe that the map $S_R : M \mapsto S_R(M)$ associates constant functors $F(A) \equiv C$ to constant right R-modules $C \in \mathcal{C}$.

In the case of right R-modules, we have a splitting

$$\mathcal{M}_R = \mathcal{C} \times \mathcal{M}_R^0,$$

where $\mathcal{M}_R^0$ denotes the full subcategory of $\mathcal{M}_R$ formed by connected right R-modules, only if R is a non-unitary operad.

5.1.5 Relative Composition Products. For a right R-module M and a left R-module N, we have a *relative composition product* $M \circ_R N$ defined by a reflexive coequalizer

$$M \circ R \circ N \underset{d_1}{\overset{d_0}{\rightrightarrows}} \overset{s_0}{\overbrace{}} M \circ N \longrightarrow M \circ_R N \,,$$

where the morphism $d_0 = \rho \circ N$ is induced by the right R-action on M, the morphism $d_1 = M \circ \lambda$ is induced by the left R-action on N, and $s_0 = M \circ \eta \circ N$ is induced by the operad unit. This general construction makes sense in any monoidal category in which reflexive coequalizers exist.

Recall that $M \circ N$ is identified with the image of $N \in \mathcal{M}$ under the functor $S(M) : \mathcal{E} \to \mathcal{E}$ defined by $M \in \mathcal{M}$ for the category $\mathcal{E} = \mathcal{M}$. As a byproduct, in the case $\mathcal{E} = \mathcal{M}$, we have a formal identification of the construction of §5.1 with the relative composition product:

$$S_R(M, N) = M \circ_R N.$$

In §9.1 we observe that the operad R forms a left module over itself. In this case, we obtain $S_R(M, R) = M \circ_R R \simeq M$. In §9.2.6, we note further that $S_R(M, R) = M \circ_R R$ forms naturally a right R-module so that M is isomorphic to $S_R(M, R)$ as a right R-module. In the sequel, we use this observation to recover the right R-module M from the associated collection of functors $S_R(M) : {}_R\mathcal{E} \to \mathcal{E}$, where $\mathcal{E}$ ranges over symmetric monoidal categories over $\mathcal{C}$.

5.1.6 The Example of Right Modules over the Commutative Operad. In this paragraph, we examine the structure of right modules over the operad C_+ associated to the category of unitary commutative algebras. Recall that this operad is given by the trivial representations $C_+(n) = *$, for every $n \in \mathbb{N}$ (see §3.1.10).

We check that a right C_+-module is equivalent to a contravariant functor $F : Fin^{op} \to \mathcal{C}$, where Fin refers to the category formed by finite sets and all maps between them (this observation comes from [33]).

For a Σ_*-object M and a finite set with n elements I, we set

$$M(I) = Bij(\{1, \ldots, n\}, I) \otimes_{\Sigma_n} M(n),$$

where $Bij(\{1, \ldots, n\}, I)$ refers to the set of bijections $i_* : \{1, \ldots, n\} \to I$. The coinvariants make the internal Σ_n-action on $M(n)$ agree with the action of Σ_n by right translations on $Bij(\{1, \ldots, n\}, I)$. In this manner we obtain a functor $I \mapsto M(I)$ on the category of finite sets and bijections between them.

For a composite $M \circ C_+$, we have an identity of the form:

$$M \circ C_+(I) = \bigoplus_{I_1 \amalg \cdots \amalg I_r = I} M(r) \otimes C_+(I_1) \otimes \cdots \otimes C_+(I_r)/ \equiv,$$

where the sum ranges over all partitions $I_1 \amalg \cdots \amalg I_r = I$ and is divided out by a natural action of the symmetric groups Σ_r. Since $C_+(I_k) = \Bbbk$, we also have

$$M \circ C_+(I) = \bigoplus_{I_1 \amalg \cdots \amalg I_r = I} M(r)/\equiv,$$

and we can use the natural correspondence between partitions $I_1 \amalg \cdots \amalg I_r = I$ and functions $f : I \to \{1, \ldots, r\}$, to obtain an expansion of the form:

$$M \circ C_+(I) = \bigoplus_{f : I \to J} M(J)/\equiv .$$

According to this expansion, we obtain readily that a right C_+-action $\rho : M \circ C_+ \to M$ determines morphisms $f^* : M(J) \to M(I)$ for every $f \in Fin$, so that a right C_+-module determines a contravariant functor on the category Fin. For details, we refer to [33].

On the other hand, one checks that the map $n \mapsto A^{\otimes n}$, where A is any commutative algebra in a symmetric monoidal category $\mathcal{E}$ extends to a covariant functor $A^{\otimes -} : Fin \to \mathcal{E}$. In the point-set context, the morphism $f_* : A^{\otimes m} \to A^{\otimes n}$ associated to a map $f : \{1, \ldots, m\} \to \{1, \ldots, n\}$ is defined by the formula:

$$f_*\left(\bigotimes_{i=1}^{m} a_i\right) = \bigotimes_{j=1}^{n}\left\{ \prod_{i \in f^{-1}(j)} a_i\right\}.$$

The functor $S_{C_+}(M) : {}_{C_+}\mathcal{E} \to \mathcal{E}$ associated to a right C_+-module M is also determined by the coend

$$S_{C_+}(M, A) = \int^{Fin} M(I) \otimes A^{\otimes I}.$$

This observation follows from a straightforward inspection of definitions.

5.1.7 The Example of Right Modules over Initial Unitary Operads.

Recall that the initial unitary operad $*_C$ associated to an object $C \in \mathcal{C}$ is the operad such that $*_C(0) = C$, $*_C(1) = 1$, and $*_C(n) = 0$ for $n > 0$.

In §5.1.1, we mention that the structure of a right module over an operad R is fully determined by partial composites

$$\circ_i : M(m) \otimes R(n) \to M(m + n - 1), \ i = 1, \ldots, m,$$

together with natural relations, for which we refer to [46]. In the case of an initial unitary operad $R = *_C$, we obtain that a right $*_C$-module structure is determined by morphisms

$$\partial_i : M(m) \otimes C \to M(m - 1), \ i = 1, \ldots, m,$$

which represent the partial composites $\circ_i : M(m) \otimes *_C(0) \to M(m - 1)$, because the term $*_C(n)$ of the operad is reduced to the unit for $n = 1$ and vanishes for $n > 1$. Finally, the relations of partial composites imply that he structure of a right $*_C$-module is equivalent to a collection of morphisms $\partial_i : M(m) \otimes C \to M(m-1)$, $i = 1, \ldots, m$, that satisfy the equivariance relation

$$\partial_i \cdot w = w \circ_i * \cdot \partial_{w^{-1}(i)},$$

for every $w \in \Sigma_m$, where $w \circ_i * \in \Sigma_{m-1}$ refers to a partial composite with a unitary operation $*$ in the permutation operad (see §3.1.9), and so that we have commutative diagrams

$$
\begin{array}{ccc}
M(m) \otimes C \otimes C & \xrightarrow{M(r) \otimes \tau} M(m) \otimes C \otimes C \xrightarrow{\partial_i} & M(m-1) \otimes C \\
\downarrow{\scriptstyle \partial_j} & & \downarrow{\scriptstyle \partial_{j-1}} \\
M(m-1) \otimes C & \xrightarrow[\partial_i]{\hspace{4cm}} & M(m-2),
\end{array}
$$

for $i < j$, where $\tau : C \otimes C \to C \otimes C$ refers to the commutation of tensors.

5.1.8 The Example of Λ_*-Objects and Generalized James's Construction.

In the case of the operad $* = *_1$ associated to the unit object $C = 1$, a right $*$-module is equipped with morphisms $\partial_i : M(n) \to M(n-1)$, $i = 1, \ldots, n$, since $M(n) \otimes 1 \simeq M(n)$. Thus we obtain that a right $*$-module is equivalent to a Σ_*-object M equipped with operations $\partial_i : M(n) \to M(n-1)$, $i = 1, \ldots, n$, so that $\partial_i \cdot w = w \circ_i * \cdot \partial_{w(i)}$, for every $w \in \Sigma_n$, and $\partial_i \partial_j = \partial_{j-1} \partial_i$, for $i < j$. According to [7], this structure is equivalent to a covariant functor $M : \Lambda_*^{op} \to \mathcal{C}$, where Λ_* refers to the category formed by sets $\{1, \ldots, n\}$, $n \in \mathbb{N}$, and injective maps between them. The operation $\partial_i : M(n) \to M(n-1)$ is associated to the injection $d^i : \{1, \ldots, n-1\} \to \{1, \ldots, n\}$ that avoids $i \in \{1, \ldots, n\}$. The action of a permutation $w : M(n) \to M(n)$ is associated to the bijection $w^{-1} : \{1, \ldots, n\} \to \{1, \ldots, n\}$ defined by the inverse of this permutation.

In the topological context, an algebra over the initial unitary operad $*$ is equivalent to a pointed space $X \in \mathrm{Top}_*$ since a morphism $* : \mathrm{pt} \to X$ defines a base point $* \in X$. By definition, the functor $\mathrm{S}_*(M)$ associated to a Λ_*-space M is given by the quotient

$$\mathrm{S}_*(M, X) = \coprod_{n=0}^{\infty} (M(n) \times X^{\times n})_{\Sigma_n} / \equiv$$

of the coinvariant modules $(M(n) \times X^{\times n})_{\Sigma_n}$ under the relation

$$\xi(x_1, \ldots, *, \ldots, x_n) = \xi \circ_i *(x_1, \ldots, \widehat{*}, \ldots, x_n).$$

Equivalently, the map $n \mapsto X^n$ determines a covariant functor on the category Λ_* and $\mathrm{S}_*(M, X)$ is given by the coend

$$\mathrm{S}_*(M, X) = \int^{\Lambda_*} M(n) \times X^{\times n}.$$

Examples of this construction are the classical James's model of $\Omega\Sigma X$ (see [31]) and its generalizations to iterated loop spaces $\Omega^n\Sigma^n X$ (see [7, 47]). Actually, in the original definition of [47, Construction 2.4], motivated by recognition problems for iterated loop spaces, the monad associated to a topological operad is defined by a functor of the form $S_*(P)$.

5.2 Colimits

In this section we examine the commutation of our construction with colimits. On one hand, we obtain readily:

5.2.1 Proposition. *The functor* $S_R : \mathcal{M}_R \to \mathcal{F}_R$ *preserves all colimits.*

Proof. The proposition follows from an immediate interchange of colimits.
□

Recall that the forgetful functor $U : \mathcal{M}_R \to \mathcal{M}$ creates all colimits. Recall also that colimits in functor categories are obtained pointwise. Thus proposition 5.2.1 asserts that, for any R-algebra A, we have the identity:

$$S_R(\operatorname*{colim}_i M_i, A) = \operatorname*{colim}_i S_R(M_i, A).$$

In contrast, one can observe that the functor $S_R(M) : A \mapsto S_R(M, A)$ does not preserve colimits in general just because: the functor $S(M) : X \mapsto S(M, X)$ does not preserve all colimits, the forgetful functor $U : {}_R\mathcal{E} \to \mathcal{E}$ does not preserve all colimits. On the other hand: in §2.4 we observe that the functor $S(M) : X \mapsto S(M, X)$ preserves filtered colimits and reflexive coequalizers; in §3.3 we observe that the forgetful functor $U : {}_R\mathcal{E} \to \mathcal{E}$ creates filtered colimits and the coequalizers which are reflexive in the underlying category $\mathcal{E}$. Therefore, we obtain:

5.2.2 Proposition. *The functor* $S_R(M) : {}_R\mathcal{E} \to \mathcal{E}$ *associated to a right R-module* $M \in \mathcal{M}_R$ *preserves filtered colimits and the coequalizers of R-algebras which are reflexive in the underlying category* $\mathcal{E}$.

Proof. Immediate consequence of the results recalled above the proposition.
□

Chapter 6
Tensor Products

Introduction

The goal of this chapter is to generalize the assertions of proposition 2.1.5 to right modules over an operad:

Theorem 6.A. *Let* R *be an operad in* $\mathcal{C}$. *The category of right* R-*modules* $\mathcal{M}_R$ *is equipped with the structure of a symmetric monoidal category over* $\mathcal{C}$ *so that the map* $S_R : M \mapsto S_R(M)$ *defines a functor of symmetric monoidal categories over* $\mathcal{C}$

$$S_R : (\mathcal{M}_R, \otimes, 1) \to (\mathcal{F}(_R\mathcal{E}, \mathcal{E}), \otimes, 1),$$

functorially in $\mathcal{E}$, *for every symmetric monoidal category* $\mathcal{E}$ *over* $\mathcal{C}$.

We have further:

Theorem 6.B. *Let* R *be a non-unitary operad in* $\mathcal{C}$.
  The category $\mathcal{M}_R^0$ *of connected right* R-*modules forms a reduced symmetric monoidal category over* $\mathcal{C}$. *The category splitting* $\mathcal{M}_R = \mathcal{C} \times \mathcal{M}_R^0$ *is compatible with symmetric monoidal structures and identifies* $\mathcal{M}_R$ *with the symmetric monoidal category over* $\mathcal{C}$ *associated to the reduced category* $\mathcal{M}_R^0$.
  Moreover, the functor $S_R : M \mapsto S_R(M)$ *fits a diagram of symmetric monoidal categories over* $\mathcal{C}$:

$$
\begin{array}{ccc}
\mathcal{M}_R & \xrightarrow{\;\;S_R\;\;} & \mathcal{F}(_R\mathcal{E}, \mathcal{E}) \\
{\scriptstyle \simeq}\big\uparrow & & \big\uparrow{\scriptstyle \simeq} \\
\mathcal{C} \times \mathcal{M}_R^0 & \xrightarrow[\mathrm{Id} \times S_R^0]{} & \mathcal{C} \times \mathcal{F}(_R\mathcal{E}, \mathcal{E})^0.
\end{array}
$$

We prove in [13, §3] that the category of right R-modules inherits a symmetric monoidal structure from Σ_*-objects. We survey this construction

B. Fresse, *Modules over Operads and Functors*, Lecture Notes in Mathematics 1967, 107
DOI: 10.1007/978-3-540-89056-0_6, © Springer-Verlag Berlin Heidelberg 2009

in §6.1. We prove that this symmetric monoidal structure fits the claims of theorems 6.A-6.B in §6.2. We record the definition of internal hom-objects for right modules over an operad in §6.3.

6.1 The Symmetric Monoidal Category of Right Modules over an Operad

We survey the definition of the symmetric monoidal structure of [13, §3] for right modules over operad. We use observations of §2.2 to make this construction more conceptual.

6.1.1 The Tensor Product of Right Modules over an Operad.
To define the tensor product of right R-modules M, N, we observe that the tensor product of M, N in the category of Σ_*-objects is equipped with a natural R-module structure. The right R-action on $M \otimes N$ is given explicitly by the composite of the distribution isomorphism of §2.2.3

$$(M \otimes N) \circ R \simeq (M \circ R) \otimes (N \circ R)$$

with the morphism

$$(M \circ R) \otimes (N \circ R) \xrightarrow{\rho \otimes \rho} M \otimes N$$

induced by right R-actions on M and N. The external tensor product of a right R-module N with an object $C \in C$ in the category of Σ_*-objects comes equipped with a similar natural R-module structure.

Thus we have an internal tensor product $\otimes : \mathcal{M}_R \times \mathcal{M}_R \to \mathcal{M}_R$ and an external tensor product $\otimes : C \times \mathcal{M}_R \to \mathcal{M}_R$ on the category of right R-modules. The external tensor product can also be identified with the tensor product of a right R-module with a constant right R-module.

The associativity and symmetry isomorphisms of the tensor product of Σ_*-modules define morphisms of right R-modules. This assertion follows from a straightforward inspection. The unit object 1 is equipped with a trivial right R-module structure, like any constant Σ_*-module, and defines a unit with respect to the tensor product of right R-modules as well. Hence we conclude that the category of right R-modules inherits a symmetric monoidal structure, as asserted in theorem 6.A.

In [13] we prove that the category of right R-modules comes also equipped with an internal hom $\mathrm{HOM}_R(M, N)$ such that

$$\mathrm{Mor}_{\mathcal{M}_R}(L \otimes M, N) = \mathrm{Mor}_{\mathcal{M}_R}(L, \mathrm{HOM}_R(M, N)),$$

but we do not use this structure in this book.

6.1.2 The Tensor Product of Connected Right Modules over an Operad. The tensor product of connected right R-modules forms obviously a connected right R-module, and similarly as regards the tensor product of a connected right R-module with a constant right R-module. Thus the category of connected right R-modules $\mathcal{M}_R^0$ is preserved by the symmetric monoidal structure of $\mathcal{M}$ and forms a reduced symmetric monoidal category over $\mathcal{C}$.

Suppose R is a connected operad. The isomorphism $\sigma : \mathcal{C} \times \mathcal{M}_R^0 \xrightarrow{\simeq} \mathcal{M}_R$ maps a pair $(C, M) \in \mathcal{C} \times \mathcal{M}_R^0$ to the direct sum $C \oplus M \in \mathcal{M}_R$. The canonical functor $\eta : \mathcal{C} \to \mathcal{M}_R$ defines tautologically a functor of symmetric monoidal categories

$$\eta : (\mathcal{C}, \otimes, 1) \to (\mathcal{M}_R, \otimes, 1)$$

since we provide $\mathcal{M}_R$ with the structure of a symmetric monoidal category over $\mathcal{C}$. As the tensor product is assumed to preserve colimits, we conclude that σ defines an equivalence of symmetric monoidal categories over $\mathcal{C}$ so that $\mathcal{M}_R$ is equivalent to the canonical symmetric monoidal category over $\mathcal{C}$ associated to the reduced category $\mathcal{M}_R^0$. Thus we obtain the first assertion of theorem 6.B.

6.1.3 The Pointwise Representation of Tensors in Right R-Modules. The tensor product of R-modules inherits an obvious pointwise representation from Σ_*-objects. By construction, the underlying Σ_*-object of $M \otimes N$ is given by the tensor product of M, N in the category of Σ_*-objects. Naturally, we take the tensors of §2.1.9

$$x \otimes y \in M(p) \otimes N(q)$$

to span the tensor product of M, N in right R-modules. The pointwise expression of the right R-action on a generating tensor $x \otimes y \in M(p) \otimes N(q)$ is given by the formula:

$$(x \otimes y)(\rho_1, \ldots, \rho_r) = x(\rho_1, \ldots, \rho_p) \otimes y(\rho_{p+1}, \ldots, \rho_r),$$

for every $\rho_1 \in R(n_1), \ldots, \rho_r \in R(n_r)$, where $r = p + q$.

Recall that $(M \otimes N)(r)$ is spanned as a $\mathcal{C}$-object by tensors $w \cdot x \otimes y$, where $w \in \Sigma_r$. By equivariance, the action of R on $w \cdot x \otimes y$ reads:

$$(w \cdot x \otimes y)(\rho_1, \ldots, \rho_r)$$
$$= w(n_1, \ldots, n_r) \cdot x(\rho_{w(1)}, \ldots, \rho_{w(p)}) \otimes y(\rho_{w(p+1)}, \ldots, \rho_{w(r)}).$$

In [13], the right R-action on $M \otimes N$ is defined by this pointwise formula.

By fact §2.1.10, a morphism of Σ_*-objects $f : M \otimes N \to T$ is equivalent to a collection of $\Sigma_p \times \Sigma_q$-equivariant morphisms $f : M(p) \otimes N(q) \to T(r)$. Clearly, a morphism of right R-modules $f : M \otimes N \to T$ is equivalent to a collection of $\Sigma_p \times \Sigma_q$-equivariant morphisms $f : M(p) \otimes N(q) \to T(r)$ such that:

$$f(x \otimes y)(\rho_1, \ldots, \rho_r) = f(x(\rho_1, \ldots, \rho_p) \otimes y(\rho_{p+1}, \ldots, \rho_r))$$

for every generating tensors $x \otimes y \in M(p) \otimes N(q)$, and every $\rho_1 \in R(n_1), \ldots, \rho_r \in R(n_r)$.

6.2 Tensor Products of Modules and Functors

We achieve the proof of theorems 6.A-6.B in this chapter. We have essentially to check:

6.2.1 Lemma. *For right R-modules M, N, we have a natural functor isomorphism $S_R(M \otimes N) \simeq S_R(M) \otimes S_R(N)$ which commutes with the unit, associativity and symmetry isomorphisms of tensor products.*

Proof. In the construction of §5.1, we observe that

$$S(M \circ R, A) \underset{d_1}{\overset{d_0}{\rightrightarrows}} S(M, A) \longrightarrow S_R(M, A)$$

forms a reflexive coequalizer. Accordingly, by proposition 1.2.1, the tensor product $S_R(M, A) \otimes S_R(N, A)$ can be identified with the coequalizer of the tensor product

$$S(M \circ R, A) \otimes S(N \circ R, A) \underset{d_1 \otimes d_1}{\overset{d_0 \otimes d_0}{\rightrightarrows}} S(M, A) \otimes S(N, A) .$$

The statements of §§2.2.4-2.2.5 imply the existence of a natural isomorphism of coequalizer diagrams

$$
\begin{array}{ccccc}
S((M \otimes N) \circ R, A) & \rightrightarrows & S(M \otimes N, A) & \longrightarrow & S_R(M \otimes N, A) \\
{\scriptstyle \simeq} \downarrow & & {\scriptstyle \simeq} \downarrow & & \vdots \downarrow \\
S(M \circ R, A) \otimes S(N \circ R, A) & \rightrightarrows & S(M, A) \otimes S(N, A) & \longrightarrow & S_R(M, A) \otimes S_R(N, A)
\end{array}
$$

from which we deduce the required isomorphism $S_R(M \otimes N, A) \simeq S_R(M, A) \otimes S_R(N, A)$.

This isomorphism inherits the commutation with the unit, associativity and symmetry isomorphism of tensor products from the isomorphism $S(M \otimes N, X) \simeq S(M, X) \otimes S(N, X)$ of the symmetric monoidal functor $S : M \mapsto S(M)$.

Note that the isomorphism $S_R(M \otimes N) \simeq S_R(M) \otimes S_R(N)$ is natural with respect to the underlying category $\mathcal{E}$. $\square$

Then we obtain:

6.2.2 Proposition. *The map* $S_R : M \mapsto S_R(M)$ *defines a symmetric monoidal functor*

$$S_R : (\mathcal{M}_R, \otimes, 1) \to (\mathcal{F}_R, \otimes, 1),$$

functorially in $\mathcal{E}$*, where* $\mathcal{F}_R = \mathcal{F}(_R\mathcal{E}, \mathcal{E})$*, and we have a commutative triangle of symmetric monoidal categories*

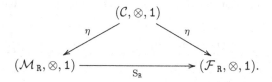

Proof. Lemma 6.2.1 gives the first assertion of the proposition. The commutativity of the triangle is checked by an immediate inspection.   □

The claims of theorems 6.A-theorem 6.B are immediate consequences of this proposition.   □

6.3 On Enriched Category Structures

If $\mathcal{C}$ is a closed symmetric monoidal category, then we obtain further that the category of right modules over an operad R inherits a hom-bifunctor

$$Hom_{\mathcal{M}_R}(-, -) : (\mathcal{M}_R)^{op} \times (\mathcal{M}_R) \to \mathcal{C}$$

from the category of Σ_*-objects.
formally, the hom-object $Hom_{\mathcal{M}_R}(M, N)$, for $M, N \in \mathcal{M}_R$, is defined by a reflexive equalizer of the form

$$Hom_{\mathcal{M}_R}(M, N) \longrightarrow Hom_{\mathcal{M}}(M, N) \underset{d^1}{\overset{d^0}{\rightrightarrows}} Hom_{\mathcal{M}}(M \circ R, N) .$$

The morphism

$$Hom_{\mathcal{M}}(M, N) \xrightarrow{Hom_{\mathcal{M}}(\rho, N)} Hom_{\mathcal{M}}(M \circ R, N)$$

induced by the R-action on M defines d^0. Since we observe in §2.2.3 that the map $M \mapsto M \circ R$ defines a functor of symmetric monoidal categories over $\mathcal{C}$

$$- \circ R : \mathcal{M} \to \mathcal{M},$$

we have a natural morphism on hom-objects

$$Hom_{\mathcal{M}}(M, N) \xrightarrow{-\circ R} Hom_{\mathcal{M}}(M \circ R, N \circ R).$$

The composite of this morphism with the morphism

$$Hom_{\mathcal{M}}(M \circ R, N \circ R) \xrightarrow{Hom_{\mathcal{M}}(M \circ R, \rho)} Hom_{\mathcal{M}}(M \circ R, N)$$

induced by the R-action on N defines d^1. The reflection s^0 is defined by the morphism

$$Hom_{\mathcal{M}}(M \circ R, N) \xrightarrow{Hom_{\mathcal{M}}(M \circ \eta, N)} Hom_{\mathcal{M}}(M \circ I, N) = Hom_{\mathcal{M}}(M, N)$$

induced by the operad unit $\eta : I \to R$.

The morphism set $Mor_{\mathcal{M}_R}(M, N)$ can be defined by the same equalizer

$$Mor_{\mathcal{M}_R}(M, N) \longrightarrow Mor_{\mathcal{M}}(M, N) \underset{d^1}{\overset{d^0}{\rightrightarrows}} Mor_{\mathcal{M}}(M \circ R, N),$$

with the reflection s^0.

because for a morphism $\phi : M \to N$ the commutation with operad actions is formally equivalent to the equation $d^0(\phi) = d^1(\phi)$. The adjunction relation

$$Mor_{\mathcal{M}_R}(C \otimes M, N) = Mor_C(C, Hom_{\mathcal{M}_R}(M, N)),$$

is an immediate consequence of this observation and of the adjunction relation at the Σ_*-object level.

Chapter 7
Universal Constructions
on Right Modules over Operads

Introduction

The usual constructions of module categories (namely free objects, extension and restriction of structures) make sense in the context of right modules over operads. In this chapter we check that these constructions correspond to natural operations on functors.

In §7.1, we determine the functor associated to free objects in right modules over operads. Besides, we observe that the category of right modules over an operad R is equipped with small projective generators defined by the free objects associated to the generating Σ_*-objects $F_r = I^{\otimes r}$, $r \in \mathbb{N}$, and we determine the associated functors on R-algebras.

In §7.2, we define the extension and restriction functors for right modules over operads and we make explicit the corresponding operations on functors.

7.1 Free Objects and Generators

In §3.2.13, we recall that the free left R-module generated by $L \in \mathcal{M}$ is represented by the composite Σ_*-object $N = R \circ L$ equipped with the left R-action defined by the morphism $\mu \circ L : R \circ R \circ L \to R \circ L$, where $\mu : R \circ R \to R$ denotes the composition product of R. Symmetrically, the free right R-module generated by $L \in \mathcal{M}$ is defined by the composite Σ_*-object $M = L \circ R$ equipped with the right R-action defined by the morphism $L \circ \mu : L \circ R \circ R \to L \circ R$.

At the functor level, the free object functor $- \circ R : \mathcal{M} \to \mathcal{M}_R$ and the forgetful functor $U : \mathcal{M}_R \to \mathcal{M}$ have the following interpretation:

7.1.1 Theorem.

(a) For a free right R-module $M = L \circ R$, the diagram

B. Fresse, *Modules over Operads and Functors*, Lecture Notes in Mathematics 1967, 113
DOI: 10.1007/978-3-540-89056-0_7, © Springer-Verlag Berlin Heidelberg 2009

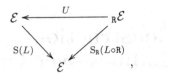

in which $U : {}_R\mathcal{E} \to \mathcal{E}$ denotes the forgetful functor, commutes up to a natural isomorphism of functors.

(b) For any right R-module $M \in \mathcal{M}_R$, the diagram

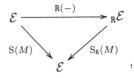

in which $R(-) : \mathcal{E} \to {}_R\mathcal{E}$ denotes the free R-algebra functor, commutes up to a natural isomorphism of functors.

In other words, we have functor isomorphisms $S_R(L \circ R) \simeq S(L) \circ U$, for all free right R-modules $M = L \circ R$, and $S(M) \simeq S_R(M) \circ R(-)$, for all right R-modules $M \in \mathcal{M}_R$. The isomorphisms that give the functor relations of theorem 7.1.1 are also natural with respect to the symmetric monoidal category $\mathcal{E}$ on which functors are defined.

This theorem occurs as the particular case of an operad unit $\eta : I \to R$ in theorem 7.2.2 proved next. Note that we prove directly the general case of theorem 7.2.2 and our arguments do not rely on theorem 7.1.1. Therefore we can refer to theorem 7.2.2 for the proof of this proposition.

7.1.2 Generating Objects in Right Modules over Operads. Consider the free right R-modules $F_r \circ R$, $r \in \mathbb{N}$, associated to the generators F_r, $r \in \mathbb{N}$ of the category of Σ_*-objects.

As $F_r = I^{\otimes r}$, we have $F_1 \circ R = I \circ R \simeq R$, the operad R considered as a right module over itself, and, for all $r \in \mathbb{N}$, we obtain $F_r \circ R = (I^{\otimes r}) \circ R \simeq R^{\otimes r}$, the rth tensor power of R in the category of right R-modules.

Since $S(F_r) \simeq \text{Id}^{\otimes r}$, the rth tensor power of the identity functor $\text{Id} : \mathcal{E} \to \mathcal{E}$, theorem 7.1.1 returns:

7.1.3 Proposition. *We have* $S_R(F_1 \circ R) = S_R(R) = U$, *the forgetful functor* $U : {}_R\mathcal{E} \to \mathcal{E}$, *and* $S_R(F_r \circ R) = S_R(R)^{\otimes r} = U^{\otimes r}$, *the* rth *tensor power of the forgetful functor* $U : {}_R\mathcal{E} \to \mathcal{E}$. $\qquad\square$

The next proposition is an immediate consequence of proposition 2.1.13 and of the adjunction relation $Hom_{\mathcal{M}_R}(L \circ R, M) \simeq Hom_{\mathcal{M}}(L, M)$.

7.1.4 Proposition. *We have a natural isomorphism*

$$\omega_r(M) : M(r) \xrightarrow{\simeq} Hom_{\mathcal{M}_R}(F_r \circ R, M)$$

for every $M \in \mathcal{M}_R$. $\qquad\square$

This proposition and the discussion of §7.1.2 imply further:

7.1.5 Proposition. *The objects $F_r \circ R$, $r \in \mathbb{N}$ define small projective generators of $\mathcal{M}_R$ in the sense of enriched categories. Explicitly, the functors*

$$M \mapsto Hom_{\mathcal{M}_R}(F_r \circ R, M)$$

preserve filtered colimits and coequalizers and the canonical morphism

$$\bigoplus_{r=0}^{\infty} Hom_{\mathcal{M}_R}(F_r \circ R, M) \otimes F_r \circ R \to M$$

is a regular epi, for every $M \in \mathcal{M}_R$. □

This proposition is stated for the sake of completeness. Therefore we do not give more details on these assertions.

7.2 Extension and Restriction of Structures

Let $\psi : R \to S$ be an operad morphism. In §3.3.5, we recall the definition of extension and restriction functors $\psi_! : {}_R\mathcal{E} \rightleftarrows {}_S\mathcal{E} : \psi^*$. In the case $\mathcal{E} = \mathcal{M}$, we obtain extension and restriction functors on left modules over operads $\psi_! : {}_R\mathcal{M} \rightleftarrows {}_S\mathcal{M} : \psi^*$.

Symmetrically, we have extension and restriction functors on right module categories

$$\psi_! : \mathcal{M}_R \rightleftarrows \mathcal{M}_S : \psi^*.$$

In this section, we recall the definition of these operations and we prove that they reflect natural operations on functors.

7.2.1 Extension and Restriction of Right-Modules over Operads.
The operad R operates on any right S-module N through the morphism $\psi : R \to S$. This R-action defines the right R-module $\psi^* N$ associated to N by restriction of structures. In the converse direction, the right S-module $\psi_! M$ obtained by extension of structures from a right R-module M is defined by the relative composition product $\psi_! M = M \circ_R S$.

To justify the definition of $\psi_! M$, recall that the relative composition product $\psi_! M = M \circ_R S$ is defined by a reflexive coequalizer of the form:

$$M \circ R \circ S \underset{d_1}{\overset{d_0}{\rightrightarrows}} M \circ S \longrightarrow \psi_! M \ .$$

with s_0 labeling the reflexive arrow.

(This reflexive coequalizer is symmetric to the reflexive coequalizer of §3.3.6.) One observes that d_0, d_1 are morphisms of free right S-modules, so that

$\psi_! M$ inherits a natural S-module structure, and an immediate inspection of definitions proves that $M \mapsto \psi_! M$ is left adjoint to the restriction functor $N \mapsto \psi^* N$. The adjunction unit $\eta M : M \to \psi^* \psi_! M$ is identified with the morphism $\eta M : M \to M \circ_R S$ induced by

$$M = M \circ I \xrightarrow{M \circ \eta} M \circ S$$

where $\eta : I \to S$ denotes the unit of the operad S. The adjunction augmentation $\epsilon N : \psi_! \psi^* N \to N$ is identified with the morphism $\epsilon N : N \circ_R S \to N$ induced by the right S-action on N:

$$N \circ S \xrightarrow{\rho} N.$$

For the unit morphism $\eta : I \to R$ of an operad R, we have $\mathcal{M}_I = \mathcal{M}$, the restriction functor $M \mapsto \eta^* M$ represents the forgetful functor $U : \mathcal{M}_R \to \mathcal{M}$ and the extension functor $L \mapsto \eta_! L$ is isomorphic to the free object functor $L \mapsto L \circ R$. Thus in this case we obtain the adjunction between the forgetful and free object functors.

7.2.2 Theorem. *Let $\psi : R \to S$ be any operad morphism.*

(a) For any right R-module M, the diagram of functors

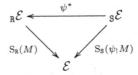

commutes up to a natural functor isomorphism.

(b) For any right S-module N, the diagram of functors

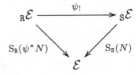

commutes up to a natural functor isomorphism.

The isomorphisms that give these functor relations are also natural with respect to the symmetric monoidal category $\mathcal{E}$ on which functors are defined.

Proof. In the case of a right R-module M and an S-algebra B, we obtain that the morphism

$$S(M, B) = S(M \circ I, A) \xrightarrow{S(M \circ \eta, A)} S(M \circ S, A)$$

induced by the operad unit $\eta : I \to S$ and the morphism

$$S(M \circ S, B) = S(M, S(S, B)) \xrightarrow{S(M, \lambda)} S(M, B)$$

induced by the operad action on B induce inverse natural isomorphisms between $S_R(M, \psi^* B)$ and $S_S(\psi_! M, B)$.

In the case of a right S-module N and an R-algebra A, we obtain that the morphism

$$S(N, A) \xrightarrow{S(N \circ \eta, A)} S(N \circ S, A)$$

induced by the operad unit $\eta : I \to S$ and the morphism

$$S(N, S(A)) = S(N \circ S, A) \xrightarrow{S(\rho, A)} S(N, A)$$

induced by the right operad action on N induce inverse natural isomorphisms between $S_S(N, \psi_! A)$ and $S_R(\psi^* N, A)$. $\qquad\square$

7.2.3 Extension and Restriction of Functors on the Right. The composition on the right with a functor $\alpha : \mathcal{A} \to \mathcal{B}$ induces a functor $\alpha^* : \mathcal{F}(\mathcal{B}, \mathcal{X}) \to \mathcal{F}(\mathcal{A}, \mathcal{X})$. In this section, we study compositions with adjoint functors

$$\alpha_! : \mathcal{A} \rightleftarrows \mathcal{B} : \alpha^*,$$

for instance the extension and restriction functors $\psi_! : {}_R\mathcal{E} \rightleftarrows {}_S\mathcal{E} : \psi^*$ associated to an operad morphism $\psi : R \to S$. In this situation, we define extension and restriction functors (on the right) on functor categories

$$\alpha_! : \mathcal{F}(\mathcal{A}, \mathcal{X}) \rightleftarrows \mathcal{F}(\mathcal{B}, \mathcal{X}) : \alpha^*$$

by $\alpha_! F(B) = F(\alpha^* B)$, for $B \in \mathcal{B}$, respectively $\alpha^* G(A) = G(\alpha_! A)$, for $A \in \mathcal{A}$.

7.2.4 Proposition. *The extension and restriction functors*

$$\alpha_! : \mathcal{F}(\mathcal{A}, \mathcal{X}) \rightleftarrows \mathcal{F}(\mathcal{B}, \mathcal{X}) : \alpha^*$$

define adjoint functors, for any target category $\mathcal{X}$.

Proof. Observe simply that the natural transformations $F(\eta A) : F(A) \to F(\alpha^* \alpha_! A)$ and $G(\epsilon B) : G(\alpha_! \alpha^* B) \to G(B)$ induced by the adjunction unit and the adjunction augmentation of $\alpha_! : \mathcal{A} \rightleftarrows \mathcal{B} : \alpha^*$ define an adjunction unit and an adjunction augmentation at the functor level. $\qquad\square$

Theorem 7.2.2 asserts that we have natural functor isomorphisms $S_S(\psi_! M) \simeq \psi_! S_R(M)$, for every $M \in \mathcal{M}_R$, and $S_R(\psi^* N) \simeq \psi^* S_S(N)$, for every $N \in \mathcal{M}_S$. By a direct and straightforward inspection, we obtain further:

7.2.5 Observation. *The isomorphisms* $S_S(\psi_! M) \simeq \psi_! S_R(M)$ *and* $S_R(\psi^* N) \simeq \psi^* S_S(N)$ *commute with adjunction units and adjunction augmentations.* $\qquad\square$

And, as a byproduct, we obtain:

7.2.6 Proposition. *The diagram*

$$
\begin{array}{ccc}
Hom_{\mathcal{M}_S}(\psi_! M, N) & \xrightarrow{\;\;\simeq\;\;} & Hom_{\mathcal{M}_R}(M, \psi^* N) \\
{\scriptstyle S_S}\downarrow & & \downarrow{\scriptstyle S_R} \\
Hom_{\mathcal{F}_S}(S_S(\psi_! M), S_S(N)) & & Hom_{\mathcal{F}_R}(S_R(M), \psi^* S_S(N)) \\
\end{array}
$$

$$
Hom_{\mathcal{F}_S}(\psi_! S_R(M), S_S(N)) \xrightarrow[\simeq]{} Hom_{\mathcal{F}_R}(S_R(M), S_R(\psi^* N))
$$

commutes, for every $M \in \mathcal{M}_R$ and $N \in \mathcal{M}_S$. □

This proposition asserts roughly that the functor $S_R : M \mapsto S_R(M)$ preserves the adjunction relation of the extension and restriction of structures up to functor isomorphisms.

To complete the results of this section, we inspect the commutation of extension and restriction functors with tensor structures. At the module level, we obtain:

7.2.7 Proposition. *Let $\psi : R \to S$ be an operad morphism. The extension functor $\psi_! : \mathcal{M}_R \to \mathcal{M}_S$ and the restriction functor $\psi^* : \mathcal{M}_S \to \mathcal{M}_R$ are functors of symmetric monoidal categories over $\mathcal{C}$ and define an adjunction relation in the 2-category of symmetric monoidal categories over $\mathcal{C}$.*

Proof. For the restriction functor, we have obviously an identity $\psi^*(M \otimes N) = \psi^*(M) \otimes \psi^*(N)$ and $\psi^* : \mathcal{M}_S \to \mathcal{M}_R$ forms clearly a functor of symmetric monoidal categories.

Concerning the extension functor, since we have identifications $\psi_!(M) = M \circ_R S = S_R(M, S)$, lemma 6.2.1 gives a natural isomorphism of Σ_*-objects $\psi_!(M \otimes N) \simeq \psi_!(M) \otimes \psi_!(N)$. One checks easily that this isomorphism defines a morphism in the category of right R-modules. Hence we conclude that $\psi_! : \mathcal{M}_R \to \mathcal{M}_S$ forms a functor of symmetric monoidal categories as well.

One checks further that the adjunction unit $\eta M : M \to \psi^* \psi_! M$ and the adjunction augmentation $\epsilon N : \psi_! \psi^* N \to N$, made explicit in §7.2.1, are natural transformations of symmetric monoidal categories. Hence we obtain finally that $\psi_! : \mathcal{M}_R \to \mathcal{M}_S$ and $\psi^* : \mathcal{M}_R \rightleftarrows \mathcal{M}_S$ define an adjunction of symmetric monoidal categories over $\mathcal{C}$. □

At the functor level, we obtain:

7.2.8 Proposition. *Let $\alpha_! : \mathcal{A} \to \mathcal{B} : \alpha^*$ be adjoint functors. The associated extension and restriction functors*

$$
\alpha_! : \mathcal{F}(\mathcal{A}, \mathcal{E}) \rightleftarrows \mathcal{F}(\mathcal{B}, \mathcal{E}) : \alpha^*,
$$

where $\mathcal{E}$ is a symmetric monoidal category over C, are functors of symmetric monoidal categories over C and define an adjunction relation in the 2-category of symmetric monoidal categories over C.

Proof. In §3.2.16, we already observe that functors of the form $F \mapsto \alpha_! F$ and $G \mapsto \alpha^* G$ are functors of symmetric monoidal categories over C. One checks by an immediate inspection that the adjunction unit $\eta F : F \to \alpha^* \alpha_! F$ and the adjunction augmentation $\epsilon G : \alpha_! \alpha^* G \to G$ are natural transformations of symmetric monoidal categories over C.   □

By a direct and straightforward inspection, we obtain:

7.2.9 Observation. *The isomorphisms $S_S(\psi_! M) \simeq \psi_! S_R(M)$ and $S_R(\psi^* N) \simeq \psi^* S_S(N)$ are natural equivalences of symmetric monoidal categories over C.*
                                                                                     □

To summarize:

7.2.10 Proposition. *Let $\psi : R \to S$ be an operad morphism. The diagram*

$$
\begin{array}{ccc}
\mathcal{M}_R & \underset{\psi^*}{\overset{\psi_!}{\rightleftarrows}} & \mathcal{M}_S \\
{\scriptstyle S_R} \downarrow & & \downarrow {\scriptstyle S_S} \\
\mathcal{F}_R & \underset{\psi^*}{\overset{\psi_!}{\rightleftarrows}} & \mathcal{F}_S
\end{array}
$$

commutes up to natural equivalences of symmetric monoidal categories over C.
                                                                                     □

Chapter 8
Adjunction and Embedding Properties

Introduction

Let R be an operad in $\mathcal{C}$. In this chapter, we generalize results of §2.3 to the functor $S_R : \mathcal{M}_R \to \mathcal{F}_R$ from the category of right R-modules to the category of functors $F : {}_R\mathcal{E} \to \mathcal{E}$, where $\mathcal{E}$ is a fixed symmetric monoidal category over $\mathcal{C}$.

Again, since we observe in §5.2 that the functor $S_R : \mathcal{M}_R \to \mathcal{F}_R$ preserves colimits, we obtain that this functor has a right adjoint $\Gamma_R : \mathcal{F}_R \to \mathcal{M}_R$. In §8.1, we generalize the construction of §2.3.4 to give an explicit definition of this functor $\Gamma_R : G \mapsto \Gamma_R(G)$.

In §8.2, we prove that the functor $S_R : \mathcal{M}_R \mapsto \mathcal{F}_R$ is faithful in an enriched sense (like $S : \mathcal{M} \mapsto \mathcal{F}$) if the category $\mathcal{E}$ is equipped with a faithful functor $\eta : \mathcal{C} \to \mathcal{E}$. Equivalently, we obtain that the adjunction unit $\eta(M) : M \to \Gamma_R(S_R(M))$ forms a monomorphism.

In the case $\mathcal{E} = \mathcal{C} = \Bbbk\,\mathrm{Mod}$, we observe that the unit $\eta(M) : M \to \Gamma(S(M))$ of the adjunction $S : \mathcal{M} \rightleftarrows \mathcal{F} : \Gamma$ forms an isomorphism if M is a projective Σ_*-module or if the ground ring is an infinite field. In §8.3, we extend these results to the context of right R-modules:

Theorem 8.A. *In the case $\mathcal{E} = \mathcal{C} = \Bbbk\,\mathrm{Mod}$, the adjunction unit $\eta_R(M) : M \to \Gamma_R(S_R(M))$ defines an isomorphism if M is a projective Σ_*-module or if the ground ring is an infinite field.*

As a corollary, we obtain that the functor $S_R : \mathcal{M}_R \to \mathcal{F}_R$ is full and faithful in the case $\mathcal{E} = \mathcal{C} = \Bbbk\,\mathrm{Mod}$, where $\Bbbk$ is an infinite field.

To prove this theorem, we observe that the underlying Σ_*-object of $\Gamma_R(G)$, for a functor $G : {}_R\mathcal{E} \to \mathcal{E}$, is identified with the Σ_*-object $\Gamma(G \circ R(-))$ associated to the composite of $G : {}_R\mathcal{E} \to \mathcal{E}$ with the free R-algebra functor $R(-) : \mathcal{E} \to {}_R\mathcal{E}$. Then we use the relation $S_R(M) \circ R(-) \simeq S(M)$, for a functor of the form $G = S_R(M)$, to deduce theorem 8.A from the corresponding assertions about the unit of the adjunction $S : \mathcal{M} \rightleftarrows \mathcal{F} : \Gamma$.

B. Fresse, *Modules over Operads and Functors*, Lecture Notes in Mathematics 1967, 121
DOI: 10.1007/978-3-540-89056-0_8, © Springer-Verlag Berlin Heidelberg 2009

The results of this chapter are not used elsewhere in this book and we give an account of these results essentially for the sake of completeness.

8.1 The Explicit Construction of the Adjoint Functor $\Gamma_R : \mathcal{F}_R \to \mathcal{M}_R$

The explicit construction of the adjoint functor $\Gamma_R : \mathcal{F}_R \to \mathcal{M}_R$ is parallel to the construction of §2.3 for the functor $\Gamma : \mathcal{F} \to \mathcal{M}$ to the category of Σ_*-objects $\mathcal{M}$. First, we prove the existence of a pointwise adjunction relation

$$\mathrm{Mor}_\mathcal{E}(S_R(M, A), X) \simeq \mathrm{Mor}_{\mathcal{M}_R}(M, End_{A,X}),$$

for fixed objects $A \in {}_R\mathcal{E}$, $X \in \mathcal{E}$, where $End_{A,X}$ is the endomorphism module defined in §2.3.1. For this aim, we observe that the endomorphism module $End_{A,X}$ is equipped with a right R-module structure if A is an R-algebra and we have essentially to check that the adjunction relation of §2.3.2

$$\mathrm{Mor}_\mathcal{E}(S(M, A), X) \simeq \mathrm{Mor}_\mathcal{M}(M, End_{A,X}),$$

makes correspond morphisms of right R-modules $g : M \to End_{A,X}$ to morphisms $f : S_R(M, A) \to X$.

Then, we apply this pointwise adjunction relation to a category of functors $\mathcal{E} = \mathcal{F}_R$ to define the right adjoint $\Gamma_R : \mathcal{F}_R \to \mathcal{M}_R$ of $S_R : \mathcal{M}_R \to \mathcal{F}_R$.

8.1.1 On Endomorphism Modules and Endomorphism Operads.
Recall that the endomorphism module of a pair objects $X, Y \in \mathcal{E}$ is defined by the collection

$$End_{X,Y}(r) = Hom_\mathcal{E}(X^{\otimes r}, Y)$$

on which symmetric groups act by tensor permutations on the source. In the case $X = Y$, we obtain $End_{X,Y} = End_X$, the endomorphism operad of X, defined in §3.4.

We have natural composition products

$$\circ_i : Hom_\mathcal{E}(X^{\otimes r}, Y) \otimes Hom_\mathcal{E}(X^{\otimes s}, X) \to Hom_\mathcal{E}(X^{\otimes r+s-1}, Y)$$

which generalize the composition products of the endomorphism operad End_X and which provide $End_{X,Y}$ with the structure of a right module over this endomorphism operad End_X.

Recall that the structure of an R-algebra A is equivalent to an operad morphism $\nabla : R \to End_A$. Hence, in the case where $X = A$ is an R-algebra, we obtain that $End_{A,Y}$ forms naturally a right R-module by restriction of structures.

Our first goal is to check:

8.1.2 Proposition. *We have a natural adjunction relation*

$$\mathrm{Mor}_{\mathcal{E}}(\mathrm{S}_R(M, A), Y) \simeq \mathrm{Mor}_{\mathcal{M}_R}(M, End_{A,Y})$$

for every $M \in \mathcal{M}_R$, $A \in {}_R\mathcal{E}$ and $Y \in \mathcal{E}$.

Proof. By definition, the object $\mathrm{S}_R(M, A)$ is defined by a reflexive coequalizer

$$\mathrm{S}(M \circ R, A) \underset{d_1}{\overset{d_0}{\rightrightarrows}} \mathrm{S}(M, A) \longrightarrow \mathrm{S}_R(M, A) \ ,$$

where d_0 is induced by the right R-action on M and d_1 is induced by the left R-action on A. As a consequence, we obtain that the morphism set $\mathrm{Mor}_{\mathcal{E}}(\mathrm{S}_R(M, A), Y)$ is determined by an equalizer of the form

$$\mathrm{Mor}_{\mathcal{E}}(\mathrm{S}_R(M, A), Y) \longrightarrow \mathrm{Mor}_{\mathcal{E}}(\mathrm{S}(M, A), Y) \underset{d^1}{\overset{d^0}{\rightrightarrows}} \mathrm{Mor}_{\mathcal{E}}(\mathrm{S}(M \circ R, A), Y) \ .$$

By proposition 2.3.2, we have adjunction relations

$$\mathrm{Mor}_{\mathcal{E}}(\mathrm{S}(M, A), Y) \simeq \mathrm{Mor}_{\mathcal{M}}(M, End_{A,Y})$$
$$\text{and} \quad \mathrm{Mor}_{\mathcal{E}}(\mathrm{S}(M \circ R, A), Y) \simeq \mathrm{Mor}_{\mathcal{M}}(M \circ R, End_{A,Y})$$

that transport this equalizer to an isomorphic equalizer of the form:

$$\ker(d^0, d^1) \dashrightarrow \mathrm{Mor}_{\mathcal{M}}(M, End_{A,Y}) \underset{d^1}{\overset{d^0}{\rightrightarrows}} \mathrm{Mor}_{\mathcal{M}}(M \circ R, End_{A,Y}) \ .$$

By functoriality, we obtain that the map d^0 in this equalizer is induced by the morphism $\rho : M \circ R \to M$ that determines the right R-action on M. Thus we have $d^0(f) = f \cdot \rho$, for all $f : M \to End_{A,Y}$, where ρ denotes the right R-action on M. On the other hand, by going through the construction of proposition 2.3.2, we check readily that $d^1(f) = \rho \cdot f \circ R$, where in this formula ρ denotes the right R-action on $End_{A,Y}$. Hence we have $f \in \ker(d^0, d^1)$ if and only if $f : M \to End_{A,Y}$ forms a morphism of right R-modules.

To conclude, we obtain that the adjunction relation of proposition 2.3.2 restricts to an isomorphism

$$\mathrm{Mor}_{\mathcal{E}}(\mathrm{S}_R(M, A), Y) \simeq \mathrm{Mor}_{\mathcal{M}_R}(M, End_{A,Y})$$

and this achieves the proof of proposition 8.1.2. $\qquad\square$

8.1.3 Remark. In §2.3.3, we mention that the endomorphism module $End_{X,Y}$ forms also a left module over End_Y, the endomorphism operad of Y. Explicitly, we have natural composition products

$$Hom_{\mathcal{E}}(Y^{\otimes r}, Y) \otimes Hom_{\mathcal{E}}(X^{\otimes n_1}, Y) \otimes \cdots \otimes Hom_{\mathcal{E}}(X^{\otimes n_r}, Y)$$
$$\to Hom_{\mathcal{E}}(X^{\otimes n_1 + \cdots + n_r}, Y)$$

which generalize the composition products of the endomorphism operad End_Y and which provide $End_{X,Y}$ with the structure of a left module over this endomorphism operad End_Y.

Obviously, the left End_Y-action commutes with the right End_X-action on $End_{X,Y}$ so that $End_{X,Y}$ forms a End_Y-End_X-bimodule (see §9.1 for this notion). As a corollary, if $X = A$ is an algebra over an operad R and $Y = B$ is an algebra over an operad P, then we obtain that $End_{A,B}$ forms a P-R-bimodule by restriction of structures on the left and on the right (see §9.3).

Next (see §9.2) we observe that the map $S_R(N) : A \mapsto S_R(N, A)$ defines a functor $S_R(N) : {}_R\mathcal{E} \to {}_P\mathcal{E}$ when N is equipped with the structure of a P-R-bimodule. In this context, we have an adjunction relation

$$Mor_{{}_P\mathcal{E}}(S_R(N, A), B) \simeq Mor_{{}_P\mathcal{M}_R}(N, End_{A,B})$$

for all $N \in {}_P\mathcal{M}_R$, $A \in {}_R\mathcal{E}$ and $B \in {}_P\mathcal{E}$, where ${}_P\mathcal{M}_R$ refers to the category of P-R-bimodules.

8.1.4 Definition of the Adjoint Functor $\Gamma_R : \mathcal{F}_R \to \mathcal{M}_R$. To define the functor $\Gamma_R : \mathcal{F}_R \to \mathcal{M}_R$, we apply the pointwise adjunction relation of proposition 8.1.2 to the category of functors $\mathcal{F}_R$.

Observe that the forgetful functor $U : {}_R\mathcal{E} \to \mathcal{E}$ can be equipped with the structure an R-algebra in the category of functors $\mathcal{F}_R = \mathcal{F}({}_R\mathcal{E}, \mathcal{E})$. This assertion is tautological since we observe in §3.2.8 that R-algebras in categories of functors $\mathcal{F}_R = \mathcal{F}(\mathcal{X}, \mathcal{E})$ are equivalent to functors $F : \mathcal{X} \to {}_R\mathcal{E}$. The forgetful functor $U : {}_R\mathcal{E} \to \mathcal{E}$ together with its R-algebra structure corresponds to the identity functor $Id : {}_R\mathcal{E} \to {}_R\mathcal{E}$.

Equivalently, the evaluation products $R(r) \otimes A^{\otimes r} \to A$, where A ranges over the category of R-algebras, are equivalent to functor morphisms $R(r) \otimes U^{\otimes r} \to U$ that provide the forgetful functor $U : {}_R\mathcal{E} \to \mathcal{E}$ with the structure of an R-algebra in $\mathcal{F}_R$.

We have a tautological identity $S_R(M, U(A)) = S_R(M, A)$ from which we deduce the functor relation $S_R(M, U) = S_R(M)$. As a consequence, if we set $\Gamma_R(G) = End_{U,G}$ for $G \in \mathcal{F}_R$, then proposition 8.1.2 returns:

8.1.5 Proposition. *The functor $\Gamma_R : \mathcal{F}_R \to \mathcal{M}_R$ defined by the map $G \mapsto End_{U,G}$ is right adjoint to $S_R : \mathcal{M}_R \to \mathcal{F}_R$.* □

By proposition 1.1.16, we have as well:

8.1.6 Proposition. *The functors $S_R : \mathcal{M}_R \rightleftarrows \mathcal{F}_R : \Gamma_R$ satisfies an enriched adjunction relation*

$$Hom_{\mathcal{M}_R}(S_R(M), G) \simeq Hom_{\mathcal{F}_R}(M, \Gamma_R(G)),$$

where we replace morphism sets by hom-objects over $\mathcal{C}$. □

8.2 On Adjunction Units and Enriched Embedding Properties

The unit of the adjunction $S_R : \mathcal{M}_R \to \mathcal{F}_R : \Gamma_R$ is denoted by $\eta_R(N) : N \to \Gamma_R(S_R(N))$.

By proposition 1.1.15, the map $S_R : M \mapsto S_R(M)$ defines a functor $S_R : \mathcal{M}_R \to \mathcal{F}_R$ in the sense of $\mathcal{C}$-enriched categories. In this section, we extend the observations of §§2.3.7-2.3.11 to the morphism on hom-objects

$$S_R : Hom_{\mathcal{M}_R}(M, N) \to Hom_{\mathcal{F}}(S_R(M), S_R(N))$$

induced by the functor $S_R : \mathcal{M}_R \to \mathcal{F}_R$ and to the adjunction unit $\eta_R(N) : N \to \Gamma_R(S_R(N))$.

By proposition 1.1.16, we have:

8.2.1 Observation. *We have a commutative diagram:*

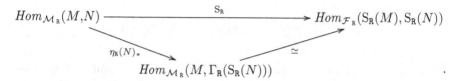

In addition:

8.2.2 Proposition. *The component $\eta_R(N) : N(r) \to Hom_{\mathcal{F}_R}(U^{\otimes r}, S_R(N))$ of the adjunction unit $\eta_R(N) : N \to \Gamma_R(S_R(N))$ coincides with the morphism*

$$N(r) \xrightarrow{\simeq} Hom_{\mathcal{M}_R}(F_r \circ R, N) \xrightarrow{S_R} Hom_{\mathcal{F}_R}(S_R(F_r \circ R), S_R(N))$$
$$\xrightarrow{\simeq} Hom_{\mathcal{F}_R}(U^{\otimes r}, S_R(N))$$

formed by the composite of the isomorphism $\omega_r(N) : N(r) \xrightarrow{\simeq} Hom_{\mathcal{M}_R}(F_r \circ R, N)$ of proposition 7.1.4, the morphism induced by the functor $S_R : \mathcal{M}_R \to \mathcal{F}_R$ on hom-objects, and the isomorphism induced by the relation $S_R(F_r \circ R) \simeq U^{\otimes r}$.

Proof. Similar to proposition 2.3.8.   □

The next assertion is a generalization of proposition 2.3.10:

8.2.3 Proposition. *The functor $S_R : \mathcal{M}_R \to \mathcal{F}(_R\mathcal{E}, \mathcal{E})$ is faithful for every symmetric monoidal category over $\mathcal{C}$ equipped with a faithful functor $\eta : \mathcal{C} \to \mathcal{E}$.*

If $\mathcal{E}$ is also enriched over $\mathcal{C}$, then the functor $S_R : \mathcal{M}_R \to \mathcal{F}(_R\mathcal{E}, \mathcal{E})$ is faithful in an enriched sense. Explicitly, the morphism induced by S_R on hom-objects

$$Hom_{\mathcal{M}_R}(M, N) \xrightarrow{S_R} Hom_{\mathcal{F}}(S_R(M), S_R(N))$$

is mono in $\mathcal{C}$, for every $M, N \in \mathcal{M}_R$.

Proof. To prove this result, we use the diagram of functors

$$
\begin{array}{ccc}
\mathcal{M}_{R} & \xrightarrow{\ S_R\ } & \mathcal{F}(_R\mathcal{E}, \mathcal{E}) \\
{\scriptstyle U}\big\downarrow & & \big\downarrow{\scriptstyle -\circ R(-)} \\
\mathcal{M} & \xrightarrow{\ \ S\ \ } & \mathcal{F}(\mathcal{E}, \mathcal{E})
\end{array}
$$

where $U : \mathcal{M}_R \to \mathcal{M}$ denotes the forgetful functor and $G \mapsto G \circ R(-)$ maps a functor $G : {}_R\mathcal{E} \to \mathcal{E}$ to the composite of G with the free R-algebra functor $R(-) : \mathcal{E} \to {}_R\mathcal{E}$.

In theorem 7.1.1, we observe that this diagram commutes up to a natural isomorphism. Accordingly, for $M, N \in \mathcal{M}_R$, we have a commutative diagram

$$
\begin{array}{ccc}
\mathrm{Mor}_{\mathcal{M}_R}(M, N) & \xrightarrow{\ S_R\ } & \mathrm{Mor}_{\mathcal{F}(_R\mathcal{E}, \mathcal{E})}(S_R(M), S_R(N)) \\
{\scriptstyle U}\big\downarrow & & \big\downarrow \\
\mathrm{Mor}_{\mathcal{M}}(M, N) & \xrightarrow[\ \ S\ \ ]{} & \mathrm{Mor}_{\mathcal{F}(\mathcal{E}, \mathcal{E})}(S(M), S(N)).
\end{array}
$$

By proposition 2.3.10, the functor $S : \mathcal{M} \to \mathcal{F}$ is faithful and induces an injective map on morphism sets. Since the map $U : \mathrm{Mor}_{\mathcal{M}_R}(M, N) \to \mathrm{Mor}_{\mathcal{M}}(M, N)$ is tautologically injective, we can conclude that S induces an injective map on morphism sets as well and hence defines a faithful functor.

These arguments extend immediately to the enriched category setting. $\square$

By propositions 8.2.1-8.2.2, we have equivalently:

8.2.4 Proposition. *The adjunction unit* $\eta_R(N) : N \to \Gamma_R(S_R(N))$ *is mono in* $\mathcal{M}_R$, *for every* $N \in \mathcal{M}_R$. $\square$

8.3 Comparison of Adjunctions and Application

In general, we do not have more than the results of propositions 8.2.3-8.2.4, but, in the introduction of this chapter, we announce that, in the case $\mathcal{E} = \mathcal{C} = \Bbbk\mathrm{Mod}$, the adjunction unit $\eta_R(N) : N \to \Gamma_R(S_R(N))$ forms an isomorphism if N a projective Σ_*-module or if the ground ring $\Bbbk$ is an infinite field. The purpose of this section is to prove this claim.

For this aim, we compare the adjunction units of $S_R : \mathcal{M}_R \rightleftarrows \mathcal{F}_R : \Gamma_R$ and $S : \mathcal{M} \rightleftarrows \mathcal{F} : \Gamma$. Our comparison result holds for every symmetric monoidal category over a base category $\mathcal{C}$ and not necessarily for the category of $\Bbbk$-modules. Observe first:

8.3.1 Lemma. *For any functor* $G : {}_R\mathcal{E} \to \mathcal{E}$, *the underlying* Σ_*-*object of* $\Gamma_R(G)$ *is isomorphic to* $\Gamma(G \circ R(-))$, *the* Σ_*-*object associated to the composite of* $G : {}_R\mathcal{E} \to \mathcal{E}$ *with the free* R-*algebra functor* $R(-) : \mathcal{E} \to {}_R\mathcal{E}$.

Proof. The adjunction relation of proposition 7.2.4 gives isomorphisms

$$Hom_{\mathcal{F}_R}(U^{\otimes r}, G) \simeq Hom_{\mathcal{F}}(\mathrm{Id}^{\otimes r}, G \circ R(-)),$$

for all $r \in \mathbb{N}$. The lemma follows.   □

In the case $G = S_R(M)$, we have an isomorphism $S_R(M) \circ R(-) \simeq S(M)$ by theorem 7.1.1. Hence, in this case, we obtain:

8.3.2 Observation. *For any right* R-*module* M, *we have a natural isomorphism* $\Gamma_R(S_R(M)) \simeq \Gamma(S(M))$ *in* $\mathcal{M}$.

We check further:

8.3.3 Lemma. *The units of the adjunctions*

$$S : \mathcal{M} \rightleftarrows \mathcal{F} : \Gamma \quad and \quad S_R : \mathcal{M}_R \rightleftarrows \mathcal{F}_R : \Gamma_R$$

fit a commutative diagram

$$
\begin{array}{ccc}
 & N & \\
{}^{\eta_R(N)}\swarrow & & \searrow^{\eta(N)} \\
\Gamma_R(S_R(N)) & \xrightarrow{\;\simeq\;} & \Gamma(S(N)),
\end{array}
$$

for every $N \in \mathcal{M}_R$.

Proof. We apply proposition 7.2.6 to the unit morphism $\eta : I \to R$ of the operad R. Recall that ${}_I\mathcal{E} = \mathcal{E}$ and the extension and restriction functors $\eta_! : {}_I\mathcal{E} \rightleftarrows {}_R\mathcal{E} : \eta^*$ are identified with the free object and forgetful functors $R(-) : \mathcal{E} \rightleftarrows {}_R\mathcal{E} : U$. Similarly, we have $\mathcal{M}_I = \mathcal{M}$ and the extension and restriction functors $\eta_! : \mathcal{M}_I \rightleftarrows \mathcal{M}_R : \eta^*$ are identified with the free object and forgetful functors for right R-modules. Accordingly, by proposition 7.2.6, we have a commutative hexagon:

$$
\begin{array}{ccc}
Hom_{\mathcal{M}_R}(L \circ R, N) & \xrightarrow{\;\;\simeq\;\;} & Hom_{\mathcal{M}}(L, N) \\
{}_{S_R}\downarrow & & \downarrow^{S} \\
Hom_{\mathcal{F}_R}(S_R(L \circ R), S_R(N)) & & Hom_{\mathcal{F}}(S(L), S(N)) \\
{}^{\simeq}\searrow & & \swarrow^{\simeq} \\
Hom_{\mathcal{F}_R}(S(L) \circ U, S_R(N)) & \xrightarrow{\;\simeq\;} & Hom_{\mathcal{F}}(S(L), S_R(N) \circ R(-))
\end{array}
\quad,
$$

for every free object $M = L \circ R$ and all right R-modules N.

We apply this result to the generating objects $M = F_r \circ R = R^{\otimes r}$. We form a commutative diagram:

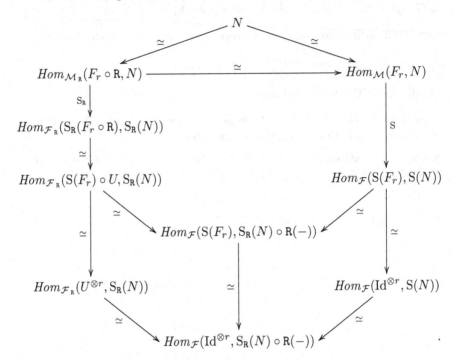

By proposition 8.2.2, the left-hand side composite represents the adjunction unit $\eta_R(N) : N(r) \to \Gamma_R(S_R(N))$. By proposition 2.3.8, the right-hand side composite represents the adjunction unit $\eta(N) : N(r) \to \Gamma(S(N))$. The bottom isomorphisms represent the isomorphisms $\Gamma_R(S_R(N)) \simeq \Gamma(S_R(N) \circ R(-)) \simeq \Gamma(S(N))$ of lemma 8.3.1 and observation 8.3.2. Hence the commutativity of the diagram implies lemma 8.3.3.   □

The conclusion of theorem 8.A is an immediate consequence of this lemma and of proposition 2.3.12.   □

Chapter 9
Algebras in Right Modules over Operads

Introduction

The purpose of this chapter is to survey applications of the general statements of §3 to the category of right modules over an operad R.

In §2, we prove that the functor S : $M \mapsto S(M)$ defines a functor of symmetric monoidal categories over $\mathcal{C}$

$$S : (\mathcal{M}, \otimes, 1) \to (\mathcal{F}, \otimes, 1)$$

and we use this statement in §3.2 in order to model functors $F : \mathcal{E} \to {}_P\mathcal{E}$, where P is an operad in $\mathcal{C}$, by P-algebras in the category of Σ_*-objects. In a similar way, we can use that the functor $S_R : M \mapsto S_R(M)$ defines a functor of symmetric monoidal categories over $\mathcal{C}$

$$S_R : (\mathcal{M}_R, \otimes, 1) \to (\mathcal{F}_R, \otimes, 1),$$

to model functors $F : {}_R\mathcal{E} \to {}_P\mathcal{E}$ by P-algebras in the category of right R-modules.

In this chapter, we also check that the structure of a P-algebra in right R-modules, for P an operad in $\mathcal{C}$, is equivalent to the structure of a P-R-*bimodule*, an object equipped with both a left P-action and a right R-action in the monoidal category $(\mathcal{M}, \circ, I)$, and we give an analogue of our constructions in the bimodule formalism. In the sequel, we aim to extend results of the theory of algebras over operads. For that reason we use the language of algebras in symmetric monoidal categories rather than the bimodule language.

B. Fresse, *Modules over Operads and Functors*, Lecture Notes in Mathematics 1967, 129
DOI: 10.1007/978-3-540-89056-0_9, © Springer-Verlag Berlin Heidelberg 2009

9.1 Algebras in Right Modules over Operads and Bimodules

Formally, a bimodule over operads P and R (in the sense of [14, §2.1] and [54, §2.1.19]) consists of a Σ_*-object N equipped with both a left P-action $\lambda : P \circ N \to N$ and a right R-action $\lambda : N \circ R \to R$ such that the diagram

$$
\begin{array}{ccc}
P \circ N \circ R & \xrightarrow{P \circ \rho} & P \circ N \\
\lambda \circ R \downarrow & & \downarrow \lambda \\
N \circ R & \xrightarrow{\rho} & N
\end{array}
$$

commutes. (This notion of a bimodule is also used in [25] for another purpose, namely to model A_∞-morphisms.) The category of P-R-bimodules is denoted by ${}_P \mathcal{M}_R$.

In §§3.2.9-3.2.10, we observe that the structure of a P-algebra in the category of Σ_*-objects, where P is any operad in $\mathcal{C}$, is equivalent to the structure of a left P-module. In this section, we extend this equivalence to obtain that a P-algebra in right R-modules is equivalent to a P-R-bimodule. Because of this observation, the notation ${}_P \mathcal{M}_R$ for the category of P-R-bimodules is coherent with our conventions for categories of algebras over operads.

In §2.2, we observe that the functor $N \mapsto S(M, N)$, where N ranges over $\mathcal{E} = \mathcal{M}$, is identified with the composition product of Σ_*-objects: $S(M, N) = M \circ N$. In the case $\mathcal{E} = \mathcal{M}_R$, we obtain readily:

9.1.1 Observation. *Let $M \in \mathcal{M}$. The image of a right R-module $N \in \mathcal{M}_R$, under the functor $S(M) : \mathcal{E} \to \mathcal{E}$ defined by a Σ_*-object $M \in \mathcal{M}$ for $\mathcal{E} = \mathcal{M}_R$ is identified with the composite Σ_*-object $S(M, N) = M \circ N$ equipped with the right R-action $M \circ \rho : M \circ N \circ R \to M \circ N$ induced by the right R-action of N.*

In the case $\mathcal{E} = \mathcal{M}$, we observe that the structure of a P-algebra in Σ_*-object is determined by a morphism of Σ_*-objects

$$
S(P, N) = P \circ N \xrightarrow{\lambda} N.
$$

In the case of right R-modules $\mathcal{E} = \mathcal{M}_R$, we assume simply that $\lambda : P \circ N \to N$ defines a morphism in the category of right R-modules. Clearly, this assertion holds if and only if the morphism $\lambda : P \circ N \to N$ defines a left P-action on N that commutes with the right R-action. As a consequence, we obtain:

9.1.2 Proposition. *The structure of a P-algebra in right R-modules is equivalent to the structure of a P-R-bimodule.*

Thus we have a category identity ${}_P(\mathcal{M}_R) = {}_P \mathcal{M}_R$.

Since an operad P forms a bimodule over itself in an obvious way, proposition 9.1.2 returns immediately:

9.1.3 Observation. *Any operad forms naturally an algebra over itself in the category of right modules over itself.*

9.1.4 Free Objects. In the case $\mathcal{E} = \mathcal{M}$, the free P-algebra generated by a Σ_*-object M is identified with the composite $P(M) = S(P, M) = P \circ M$ equipped with the left P-action defined by the morphism

$$P \circ P \circ M \xrightarrow{\mu \circ M} P \circ M$$

induced by the operad composition product $\mu : P \circ P \to P$. In the context of right R-modules $\mathcal{E} = \mathcal{M}_R$, the free P-algebra $P(M) = S(P, M) = P \circ M$ comes also equipped with the right R-action induced by the right R-action on M.

9.2 Algebras in Right Modules over Operads and Functors

In §3.2.15, we use the notation $\mathcal{F}$ for the category of functors $F : \mathcal{E} \to \mathcal{E}$ and the notation $_P\mathcal{F}$ for the category of functors $F : \mathcal{E} \to {_P\mathcal{E}}$, which correspond to left modules over operads. In parallel, we use the notation $\mathcal{F}_R$ for the category of functors $F : {_R\mathcal{E}} \to \mathcal{E}$, which correspond to right modules over operads, and we may use the notation $_P\mathcal{F}_R$ for the category of functors $F : {_R\mathcal{E}} \to {_P\mathcal{E}}$. Again, recall that a P-algebra in the category of functors $F : {_R\mathcal{E}} \to \mathcal{E}$ is equivalent to a functor $F : {_R\mathcal{E}} \to {_P\mathcal{E}}$. Therefore, this notation $_P\mathcal{F}_R$ is coherent with our conventions for categories of algebras over operads.

Since we prove that the functor $S_R : M \mapsto S_R(M)$ defines a functor of symmetric monoidal categories over $\mathcal{C}$

$$S_R : (\mathcal{M}_R, \otimes, 1) \to (\mathcal{F}_R, \otimes, 1),$$

we obtain by constructions of §3.2.14 that the functor $S_R : \mathcal{M}_R \to \mathcal{F}_R$ restricts to a functor

$$S_R : {_P\mathcal{M}_R} \to {_P\mathcal{F}_R}.$$

This construction is functorial with respect to the symmetric monoidal category $\mathcal{E}$. Explicitly, if $\rho : \mathcal{D} \to \mathcal{E}$ is a functor of symmetric monoidal categories over $\mathcal{C}$, then the diagram

$$
\begin{array}{ccc}
_R\mathcal{D} & \xrightarrow{\rho} & _R\mathcal{E} \\
{\scriptstyle S_R(N)}\big\downarrow & & \big\downarrow{\scriptstyle S_R(N)} \\
_P\mathcal{D} & \xrightarrow{\rho} & _P\mathcal{E}
\end{array}
$$

commutes up to natural functor isomorphisms, for every $N \in {_P\mathcal{M}_R}$. Moreover we have:

9.2.1 Proposition.

(a) *For a free* P*-algebra* $N = P(M)$, *where* $M \in \mathcal{M}_R$, *we have*

$$S_R(P(M), A) = P(S_R(M, A)),$$

for all $A \in {}_R\mathcal{E}$, *where on the right-hand side we consider the free* P*-algebra in* $\mathcal{E}$ *generated by the object* $S_R(M, A) \in \mathcal{E}$ *associated to* $A \in {}_R\mathcal{E}$ *by the functor* $S_R(M) : {}_R\mathcal{E} \to \mathcal{E}$.

(b) *For a diagram* $i \mapsto N_i$ *of* P*-algebras in* $\mathcal{M}_R$, *we have*

$$S_R(\operatorname*{colim}_i N_i, A) = \operatorname*{colim}_i S_R(N_i, A),$$

for all $A \in {}_R\mathcal{E}$, *where on the right-hand side we consider the colimit of the diagram of* P*-algebras associated to* $A \in {}_R\mathcal{E}$ *by the functors* $S_R(N_i) : {}_R\mathcal{E} \to {}_P\mathcal{E}$. □

This proposition is also a corollary of the adjunction relation of §8.1.3.

For the P-algebra in right P-modules defined by the operad itself, we obtain:

9.2.2 Proposition. *The functor* $S_P(P) : {}_P\mathcal{E} \to {}_P\mathcal{E}$ *is the identity functor of the category of* P*-algebras.* □

9.2.3 Functors on Algebras in Σ_*-Objects and Right Modules over Operads. Let Q be an operad in $\mathcal{C}$. Since $\mathcal{E} = \mathcal{M}$ and $\mathcal{E} = \mathcal{M}_Q$ form symmetric monoidal categories over $\mathcal{C}$, we can apply the constructions $S_R : \mathcal{M}_R \to \mathcal{F}({}_R\mathcal{E}, \mathcal{E})$ and $S_R : {}_P\mathcal{M}_R \to \mathcal{F}({}_R\mathcal{E}, {}_P\mathcal{E})$ to these categories $\mathcal{E} = \mathcal{M}$ and $\mathcal{E} = \mathcal{M}_Q$:

- In the case $M \in \mathcal{M}_R$ and $N \in {}_R\mathcal{M}_Q$, we obtain an object $S_R(M, N) \in \mathcal{M}_Q$.
- In the case $M \in {}_P\mathcal{M}_R$ and $N \in {}_R\mathcal{M}$, we have $S_R(M, N) \in {}_P\mathcal{M}$.
- In the case $M \in {}_P\mathcal{M}_R$ and $N \in {}_R\mathcal{M}_Q$, we have $S_R(M, N) \in {}_P\mathcal{M}_Q$.

In §5.1.5, we observe that the functor $N \mapsto S_R(M, N)$, for $M \in \mathcal{M}_R$ and $N \in {}_R\mathcal{M}$ is identified with the relative composition product $S_R(M, N) = M \circ_R N$. In the examples $\mathcal{E} = \mathcal{M}$ and $\mathcal{E} = \mathcal{M}_Q$, the structure of $S_R(M, N) = M \circ_R N$ can also be deduced from:

9.2.4 Observation. *Let* $M \in \mathcal{M}_R$ *and* $N \in {}_R\mathcal{M}$.

- *If* $M \in {}_P\mathcal{M}_R$, *then the morphism* $\lambda \circ N : P \circ M \circ N \to M \circ N$ *induces a left* P*-action on* $M \circ_R N$ *so that* $M \circ_R N$ *forms a left* P*-module.*
- *If* $N \in {}_R\mathcal{M}_Q$, *then the morphism* $M \circ \rho : M \circ N \circ Q \to M \circ N$ *induces a right* Q*-action on* $M \circ_R N$ *so that* $M \circ_R N$ *forms a right* Q*-module*
- *If* $M \in {}_P\mathcal{M}_R$ *and* $N \in {}_R\mathcal{M}_Q$, *then we obtain that* $M \circ_R N$ *forms a* P*-*Q*-bimodule.*

These observations are standard for a relative tensor product in a monoidal category whose tensor product preserves reflexive coequalizers. For the sake of completeness, we give an interpretation of the operation $(M, N) \mapsto M \circ_R N$ at the functor level:

9.2.5 Proposition (*cf.* [54, Proposition 2.3.13]). *Let $M \in {}_P\mathcal{M}_R$ and $N \in {}_R\mathcal{M}_Q$. The diagram of functors*

$$
\begin{array}{ccc}
 & \xrightarrow{\ S_Q(M \circ_R N)\ } & \\
{}_Q\mathcal{E} & & {}_P\mathcal{E} \\
 S_Q(N) \searrow & & \nearrow S_R(M) \\
 & {}_R\mathcal{E} &
\end{array}
$$

commutes up to a natural functor isomorphism.

Proof. For an R-Q-bimodule N, the reflexive coequalizer of 5.1.3

$$
S(N \circ Q, A) \underset{d_1}{\overset{d_0}{\rightrightarrows}} S(N, A) \longrightarrow S_Q(N, A) \ .
$$

with the reflection s_0,

defines a coequalizer in the category of R-algebras, reflexive in the underlying category $\mathcal{E}$. By proposition 5.2.2, this coequalizer is preserved by the functor $S_R(M) : {}_R\mathcal{E} \to \mathcal{E}$. As a consequence, we obtain that the image of a Q-algebra A under the composite functor $S_R(M) \circ S_Q(N)$ is given by a reflexive coequalizer of the form

$$
S_R(M, S(N \circ Q, A)) \underset{d_1}{\overset{d_0}{\rightrightarrows}} S_R(M, S(N, A)) \longrightarrow S_R(M, S_Q(N, A)) \ .
$$

On the other hand, the image of A under the functor $S_Q(M \circ_R N)$ associated to the relative composite $M \circ_R N$ is defined by a reflexive coequalizer of the form

$$
S((M \circ_R N) \circ Q, A)) \underset{d_1}{\overset{d_0}{\rightrightarrows}} S(M \circ_R N, A) \longrightarrow S_Q(M \circ_R N, A) \ .
$$

The relation $S_Q(M \circ_R N, A) \simeq S_R(M, S_Q(N, A))$ is deduced from these identifications by an interchange of colimits in a multidiagram of the form

$$
\begin{array}{ccc}
S(M \circ R \circ N \circ Q, A) & \rightrightarrows & S(M \circ R \circ N, A) \ . \\
\Downarrow\Downarrow & & \Downarrow\Downarrow \\
S(M \circ N \circ Q, A) & \rightrightarrows & S(M \circ N, A)
\end{array}
$$

If we perform horizontal colimits first, then we obtain coequalizer (1). If we perform vertical colimits first, then we obtain coequalizer (2).

Use also that the forgetful functor $U : {}_P\mathcal{E} \to \mathcal{E}$ creates reflexive coequalizers to prove that $S_R(M, S_Q(N, A))$ and $S_Q(M \circ_R N, A)$ are naturally isomorphic as P-algebras. $\square$

9.2.6 How to Recover a Right Module over an Operad R from the Associated Functors $S_R(M) : {}_R\mathcal{E} \to \mathcal{E}$.

In the case $N = R$, we have a natural isomorphism $S_R(M, R) = M \circ_R R \simeq M$. To prove this, one can apply theorem 7.1.1 and use that $N = R$ represents the free R-algebra on the unit object I in ${}_R\mathcal{M}$. Then check directly that $S_R(M, R) = M \circ_R R \simeq M$ defines an isomorphism of right R-modules (respectively, of P-R-bimodules if M is so). In the sequel, we use this observation (rather than the results of §8) to recover the object M from the associated collection of functors $S_R(M) : {}_R\mathcal{E} \to \mathcal{E}$ where $\mathcal{E}$ ranges over symmetric monoidal categories over $\mathcal{C}$.

In fact, for any collection of functors $G : {}_R\mathcal{E} \to \mathcal{E}$ we can form a right R-module by applying the functor $G : {}_R\mathcal{E} \to \mathcal{E}$ for $\mathcal{E} = \mathcal{M}_R$ to the R-algebra in $\mathcal{M}_R$ formed by the operad itself.

9.3 Extension and Restriction Functors

In §3.3.6, we recall that an operad morphism $\phi : P \to Q$ yields extension and restriction functors on left module categories

$$\phi_! : {}_P\mathcal{M} \rightleftarrows {}_Q\mathcal{M} : \phi^*$$

and we have an identification $\phi_! N = Q \circ_P N$, for every $N \in {}_P\mathcal{M}$. In §7.2.1, we recall symmetrically that an operad morphism $\psi : R \to S$ yields extension and restriction functors on right module categories

$$\psi_! : \mathcal{M}_R \rightleftarrows \mathcal{M}_S : \psi^*$$

and we have an identification $\psi_! M = M \circ_R S$, for every $M \in \mathcal{M}_R$.

In the case of bimodules over operads, we obtain functors of extension and restriction of structures on the left $\phi_! : {}_P\mathcal{M}_R \rightleftarrows {}_Q\mathcal{M}_R : \phi^*$ and on the right $\psi_! : {}_P\mathcal{M}_R \rightleftarrows {}_P\mathcal{M}_S : \psi^*$. In addition, extensions and restrictions on the left commute with extensions and restrictions on the right up to coherent functor isomorphisms. In the bimodule formalism, the structure of the relative composition product $\phi_! N = Q \circ_P N$, respectively $\psi_! M = M \circ_R S$, can be deduced from observation 9.2.4.

To obtain an interpretation at the functor level of the extension and restriction of structures on the left, we apply proposition 3.3.8. Indeed, in the case of the functor

$$S_R : (\mathcal{M}_R, \otimes, 1) \to (\mathcal{F}_R, \otimes, 1),$$

this proposition returns:

9.3.1 Proposition. *Let* $\phi : P \to Q$ *be an operad morphism. Let* R *be an operad.*

(a) *For any* P*-algebra* A *in right* R*-modules, the diagram of functors*

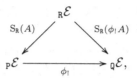

where $\phi_! A$ *is obtained from* A *by extension of structures on the left, commutes up to a natural functor isomorphism.*

(b) *For any* Q*-algebra* B *in right* R*-modules, the diagram of functors*

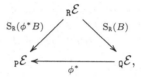

where $\phi^* B$ *is obtained from* B *by extension of structures on the left, commutes up to a natural functor isomorphism.*   □

Symmetrically, for extension and restriction of structures on the right we obtain:

9.3.2 Proposition. *Let* P *be an operad. Let* $\psi : R \to S$ *be an operad morphism.*

(a) *For any* P*-algebra* A *in right* R*-modules, the diagram of functors*

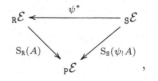

where $\psi_! A$ *is obtained from* A *by extension of structures on the right, commutes up to a natural functor isomorphism.*

(b) *For any* P*-algebra* B *in right* S*-modules, the diagram of functors*

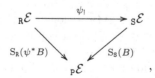

where $\psi^* B$ *is obtained from* B *by restriction of structures on the right, commutes up to a natural functor isomorphism.*

Proof. In theorem 7.2.2 we prove the existence of an isomorphism $\psi_! S_R(A) \simeq S_S(\psi_! A)$, respectively $\psi^* S_S(B) = S_R(\psi^* B)$, in the functor category $\mathcal{F}_S = \mathcal{F}(_S\mathcal{E}, \mathcal{E})$, respectively $\mathcal{F}_R = \mathcal{F}(_R\mathcal{E}, \mathcal{E})$.

Check simply that these functor isomorphisms commute with the left P-action and the proposition follows. This assertion is a formal consequence of the commutation requirement between left and right operad actions. $\square$

Recall that an operad Q forms an algebra over itself in the category of right modules over itself. Propositions 9.1.3, 9.3.1 and 9.3.2 give as a corollary:

9.3.3 Proposition. *Let $\phi : P \to Q$ be an operad morphism.*

(a) The functor of restriction of structures $\phi^* : {}_Q\mathcal{E} \to {}_P\mathcal{E}$ is identified with the functor $S_Q(Q) : {}_Q\mathcal{E} \to {}_P\mathcal{E}$ associated to $A = Q$, the P-algebra in right Q-modules obtained from Q by restriction of structures on the left.

(b) The functor of extension of structures $\phi_! : {}_P\mathcal{E} \to {}_Q\mathcal{E}$ is identified with the functor $S_P(Q) : {}_P\mathcal{E} \to {}_Q\mathcal{E}$ associated to $B = Q$, the Q-algebra in right P-modules obtained from Q by restriction of structures on the right. $\square$

9.4 Endomorphism Operads

In §9.2, we observe that the structure of a P-algebra on a right R-module M gives rise to a natural P-algebra structure on the associated functor $S_R(M)$: $A \mapsto S_R(M, A)$. In this section we use general constructions of §3.4 to review this correspondence of structures in light of endomorphism operads.

In §6, we prove that the map $S_R : M \mapsto S_R(M)$ defines a functor of symmetric monoidal categories $S_R : (\mathcal{M}_R, \otimes, 1) \to (\mathcal{F}_R, \otimes, 1)$. Accordingly, by proposition 3.4.7, we obtain:

9.4.1 Proposition. *For a right R-module $M \in \mathcal{M}_R$ and the associated functor $S_R(M) \in \mathcal{F}(_R\mathcal{E}, \mathcal{E})$, where $\mathcal{E}$ is any symmetric monoidal category over $\mathcal{C}$, we have an operad morphism*

$$End_M \xrightarrow{\Theta_R} End_{S_R(M)},$$

natural in $\mathcal{E}$. $\square$

By proposition 8.2.3, the morphism $S_R : Hom_{\mathcal{M}_R}(M, N) \to Hom_{\mathcal{F}_R}(S_R(M), S_R(N))$ is mono for all $M, N \in \mathcal{M}_R$ as long as the category $\mathcal{E}$ is equipped with a faithful functor $\eta : \mathcal{C} \to \mathcal{E}$. As a corollary, we obtain that Θ_R is a monomorphism as well under this assumption.

In the case where the underlying categories $\mathcal{E}$ and $\mathcal{C}$ are both the category of modules over a ring $\Bbbk$, we obtain further:

9.4.2 Proposition. *Assume $\mathcal{E} = \mathcal{C} = \Bbbk\,\mathrm{Mod}$, the category of modules over a ring $\Bbbk$.*

The operad morphism

$$End_M \xrightarrow{\Theta_R} End_{S_R(M)}$$

is an isomorphism if the right R-module M forms a projective Σ_*-module or if the ground ring is an infinite field.   □

The existence of an operad morphism $\Theta_R : End_M \to End_{S_R(M)}$ is equivalent to the construction of §9.2: an R-algebra structure on M gives rise to a functorial R-algebra structure on the objects $S_R(M, A)$, for $A \in {}_R\mathcal{E}$. The existence of an operad isomorphism $\Theta_R : End_M \xrightarrow{\simeq} End_{S_R(M)}$ implies that this construction defines a one-to-one correspondence. Hence, the assertion of proposition 9.4.2 implies further:

9.4.3 Theorem. *Assume $\mathcal{E} = \mathcal{C} = \Bbbk\,\mathrm{Mod}$, the category of modules over a ring $\Bbbk$.*

If a right R-module M forms a projective Σ_*-module or if the ground ring is an infinite field, then all functorial P-algebra structures on the objects $S_R(M, A)$, $A \in {}_R\mathcal{E}$, are uniquely determined by a P-algebra structure on M.   □

To conclude this chapter, we check that the morphism Θ_R satisfies a functoriality statement with respect to operad changes. To be explicit, let $\psi : R \to S$ be an operad morphism. In §7.2, we observe that the extension and restriction of structures of right modules over operads define functors of symmetric monoidal categories $\psi_! : \mathcal{M}_R \rightleftarrows \mathcal{M}_S : \psi^*$. Accordingly, by proposition 3.4.7, we obtain:

9.4.4 Proposition. *Let $\psi : R \to S$ be an operad morphism.*

(a) *For any right R-module M, we have a natural operad morphism*

$$\psi_! : End_M \to End_{\psi_! M}$$

yielded by the extension functor $\psi_! : \mathcal{M}_R \to \mathcal{M}_S$.

(b) *For any right S-module M, we have a natural operad morphism*

$$\psi^* : End_N \to End_{\psi^* N}$$

yielded by the restriction functor $\psi^* : \mathcal{M}_S \to \mathcal{M}_R$.

At the functor level, we have extension and restriction functors $\psi_! : \mathcal{F}_R \to \mathcal{F}_S : \psi^*$ defined by the composition with extension and restriction functors on the source. By §3.4.11, these functors yield natural operad morphisms

$$\psi_! : End_F \to End_{\psi_! F} \quad \text{and} \quad \psi^* : End_F \to End_{\psi^* F}.$$

In §7.2.10, we record that

$$\begin{array}{ccc}
\mathcal{M}_R & \underset{\psi^*}{\overset{\psi_!}{\rightleftarrows}} & \mathcal{M}_S \\
{\scriptstyle S_R}\downarrow & & \downarrow{\scriptstyle S_S} \\
\mathcal{F}_R & \underset{\psi^*}{\overset{\psi_!}{\rightleftarrows}} & \mathcal{F}_S
\end{array}$$

forms a diagram of functors of symmetric monoidal categories over $\mathcal{C}$ that commutes up to a natural equivalence of symmetric monoidal categories over $\mathcal{C}$. As a consequence, by naturality of our constructions, we obtain:

9.4.5 Proposition. *Let $\psi : R \to S$ be an operad morphism.*

(a) *The diagram*

$$\begin{array}{ccc}
End_M & \overset{\Theta_R}{\longrightarrow} & End_{S_R(M)} \\
{\scriptstyle \psi_!}\downarrow & & \downarrow{\scriptstyle \psi_!} \\
End_{\psi_! M} \overset{\Theta_S}{\longrightarrow} End_{S_S(\psi_! M)} & \overset{\simeq}{\longrightarrow} & End_{\psi_! S_R(M)} \cdot
\end{array}$$

commutes, for every right R-module M.

(b) *The diagram*

$$\begin{array}{ccc}
End_N & \overset{\Theta_S}{\longrightarrow} & End_{S_S(N)} \\
{\scriptstyle \psi^*}\downarrow & & \downarrow{\scriptstyle \psi^*} \\
End_{\psi^* N} \overset{\Theta_R}{\longrightarrow} End_{S_R(\psi^* N)} & \overset{\simeq}{\longrightarrow} & End_{\psi^* S_S(N)} \cdot
\end{array}$$

commutes, for every right S-module N. $\square$

Chapter 10
Miscellaneous Examples

Introduction

Many usual functors are defined by right modules over operads. In this chapter we study the universal constructions of §4: enveloping operads (§10.1), enveloping algebras (§10.2), and Kähler differentials (§10.3).

In §§10.2-10.3, we apply the principle of generalized point-tensors to extend constructions of §4 in order to obtain structure results on the modules which represent the enveloping algebra and Kähler differential functors. The examples of these sections are intended as illustrations of our constructions.

In §17, we study other constructions which occur in the homotopy theory of algebras over operads: cofibrant replacements, simplicial resolutions, and cotangent complexes. New instances of functors associated to right modules over operads can be derived from these examples by using categorical operations of §6, §7 and §9.

10.1 Enveloping Operads and Algebras

Recall that the enveloping operad of an algebra A over an operad R is the object $U_R(A)$ of the category of operads under R defined by the adjunction relation

$$\mathrm{Mor}_{R/\mathcal{O}}(U_R(A), S) = \mathrm{Mor}_{R\mathcal{E}}(A, S(0)).$$

The goal of this section is to prove that the functor $A \mapsto U_R(A)$ is associated to an operad in right R-modules, the shifted operad $R[\,\cdot\,]$ introduced in the construction of §4.1.

In §4.1, we only prove that $R[\,\cdot\,]$ forms an operad in Σ_*-objects. First, we review the definition of the shifted operad $R[\,\cdot\,]$ to check that the components $R[m]$, $m \in \mathbb{N}$, come equipped with the structure of a right R-module so that $R[\,\cdot\,]$ forms an operad in that category. Then we observe that the construction of §4.1 identifies the enveloping operad $U_R(A)$ with the functor $S_R(R[\,\cdot\,], A)$ associated to the shifted operad $R[\,\cdot\,]$.

B. Fresse, *Modules over Operads and Functors*, Lecture Notes in Mathematics 1967, 139
DOI: 10.1007/978-3-540-89056-0_10, © Springer-Verlag Berlin Heidelberg 2009

10.1.1 Shifted Operads. Recall that the shifted operad $R[\,\cdot\,]$ consists of the collection of objects $R[m](n) = R(m+n)$, $m, n \in \mathbb{N}$.

In §4.1.4, we use the action of Σ_n on $\{m+1, \ldots, m+n\} \subset \{1, \ldots, m+n\}$ to define an action of Σ_n on $R(m+n)$, for every $n \in \mathbb{N}$, and to give to the collection $R[m](n)$, $n \in \mathbb{N}$, the structure of a Σ_*-object. In the case of an operad, the composites at last positions

$$R(r+s) \otimes R(n_1) \otimes \cdots \otimes R(n_s) \to R(r+n_1+\cdots+n_s)$$

(insert operad units $\eta : 1 \to R(1)$ at positions $1, \ldots, r$) provide each object $R[r]$, $r \in \mathbb{N}$, with the additional structure of a right R-module.

In §4.1.5, we also observe that the collection $\{R[r]\}_{r \in \mathbb{N}}$ is equipped with the structure of an operad: the action of Σ_m on $\{1, \ldots, m\} \subset \{1, \ldots, m+n\}$ determines an action of Σ_m on $R[m](n) = R(m+n)$, for every $m, n \in \mathbb{N}$, and we use operadic composites at first positions

$$R(r+s) \otimes R(m_1+n_1) \otimes \cdots \otimes R(m_r+n_r) \to R(m_1+n_1+\cdots+m_r+n_r+s)$$

together with the action of appropriate bloc permutations

$$R(m_1+n_1+\cdots+m_r+n_r+s) \xrightarrow{\simeq} R(m_1+\cdots+m_r+n_1+\cdots+n_r+s)$$

to define composition products:

$$R[r] \otimes R[m_1] \otimes \cdots \otimes R[m_r] \to R[m_1+\cdots+m_r].$$

The axioms of operads (in May's form) imply immediately that the action of permutations $w \in \Sigma_m$ preserves the internal right R-module structure of $R[m]$ and similarly regarding the composition products of $R[\,\cdot\,]$. Finally, we obtain that the collection $\{R[r]\}_{r \in \mathbb{N}}$ forms an operad in the symmetric monoidal category of right R-modules.

The morphism of §4.1.6

$$\eta : R \to R[\,\cdot\,],$$

which identifies $R(m)$ with the constant part of $R[m]$, forms a morphism of operads in right R-modules.

In the sequel, we use the notation $\mathcal{O}_R$ to refer to the category of operads in right R-modules, and the notation $R \,/\, \mathcal{O}_R$ to refer to the comma category of objects under $R \in \mathcal{O}$, where we identify the objects $P(m)$ underlying an operad $P \in \mathcal{O}$ with constant right R-modules to form a functor $\mathcal{O} \to \mathcal{O}_R$. Our definition makes the operad $R[\,\cdot\,]$ an object of this category $R/\mathcal{O}_R$.

10.1.2 Functors on Operads. In §4.1.5, we use that $S : M \mapsto S(M)$ defines a functor of symmetric monoidal categories $S : \mathcal{M} \to \mathcal{F}$ to obtain that the collection $\{S(R[r], X)\}_{r \in \mathbb{N}}$, associated to any object $X \in \mathcal{E}$, forms an operad in $\mathcal{E}$. We obtain similarly that the collection $\{S_R(R[r], A)\}_{r \in \mathbb{N}}$, associated to any R-algebra $A \in {}_R\mathcal{E}$, forms an operad in $\mathcal{E}$.

Furthermore, the morphism $\eta : R \to R[\,\cdot\,]$ induces a morphism of operads

$$\eta : R \to S_R(R[\,\cdot\,], A)$$

so that $S_R(R[\,\cdot\,], A)$ forms an operad under R. As a conclusion, we obtain that the map $S_R(R[\,\cdot\,]) : A \mapsto S_R(R[\,\cdot\,], A)$ defines a functor $S_R(R[\,\cdot\,]) : {}_R\mathcal{E} \to R/\mathcal{O}_{\mathcal{E}}$.

Our goal, announced in the introduction, is to prove:

10.1.3 Theorem. *For every R-algebra $A \in {}_R\mathcal{E}$, the enveloping operad $U_R(A)$ is isomorphic to the operad $S_R(R[\,\cdot\,], A) \in R/\mathcal{O}_{\mathcal{E}}$ associated to A by the functor $S_R(R[\,\cdot\,]) : {}_R\mathcal{E} \to R/\mathcal{O}_{\mathcal{E}}$.*

Proof. By lemma 4.1.11, the enveloping operad of an R-algebra A is realized by a reflexive coequalizer of the form:

$$R[S(R, A)] \underset{\phi_{d_1}}{\overset{\phi_{d_0}}{\rightrightarrows}} R[A] \dashrightarrow U_R(A),$$

where we use the notation $R[X] = S(R[\,\cdot\,], X)$. To define the morphisms $\phi_{d_0}, \phi_{d_1}, \phi_{s_0}$ we use that $R[X]$ represents the enveloping operad of the free R-algebra $A = R(X)$ and we apply the adjunction relation of enveloping operads (see §§4.1.8-4.1.11). But a straightforward inspection of constructions shows that these morphisms are identified with the morphisms

$$S(R[\,\cdot\,], S(R, A)) \underset{d_1}{\overset{d_0}{\rightrightarrows}} S(R[\,\cdot\,], A)$$

which occur in the definition of the functor $S(R[\,\cdot\,], A)$ associated to the right R-module $R[\,\cdot\,]$. Therefore we have an identity $U_R(A) = S_R(R[\,\cdot\,], A)$, for every $A \in {}_R\mathcal{E}$. □

10.1.4 Remark. In the point-set context, the relation $U_R(A) = S_R(R[\,\cdot\,], A)$ asserts that $U_R(A)$ is spanned by elements of the form

$$u(x_1, \ldots, x_m) = p(x_1, \ldots, x_m, a_1, \ldots, a_n),$$

where $x_1, \ldots, x_m$ are variables and $a_1, \ldots, a_n \in A$, together with the relations

$$p(x_1, \ldots, x_m, a_1, \ldots, a_{e-1}, q(a_e, \ldots, a_{e+s-1}), a_{e+s}, \ldots, a_{n+s-1})$$
$$\equiv p \circ_{m+e} q(x_1, \ldots, x_m, a_1, \ldots, a_{e-1}, a_e, \ldots, a_{e+s-1}, a_{e+s}, \ldots, a_{n+s-1}).$$

Thus we recover the pointwise construction of §4.1.3.

10.1.5 Functoriality of Enveloping Operads. To complete this section, we study the functoriality of enveloping operads.

Let $\psi : R \to S$ be any operad morphism. For an S-algebra B, we have a natural morphism of operads

$$\psi_\flat : U_R(\psi^* B) \to U_S(B)$$

that corresponds to the morphism of R-algebras $\psi^* B \to U_S(B)(0) = B$ defined by the identity of B. In the converse direction, for an R-algebra A, we have a natural morphism of operads

$$\psi_\sharp : U_R(A) \to U_S(\psi_! A)$$

that corresponds to the morphism of R-algebras $A \to U_S(\psi_! A)(0) = \psi^* \psi_! A$ defined by the unit of the adjunction between extension and restriction functors $\psi_! : {}_R\mathcal{E} \rightleftarrows {}_S\mathcal{E} : \psi^*$.

These natural transformations are realized by morphisms of operads of the form $\psi_\flat : \psi_! R[\,\cdot\,] \to S[\,\cdot\,]$ and $\psi_\sharp : R[\,\cdot\,] \to \psi^* S[\,\cdot\,]$. To explain this assertion, recall that, by proposition 7.2.7, the extension and restriction functors

$$\psi_! : \mathcal{M}_R \rightleftarrows \mathcal{M}_S : \psi^*$$

are functors of symmetric monoidal categories. As a consequence, if P is an operad in right R-modules, then $\psi_! P$ forms an operad in right S-modules, if Q is an operad in right S-modules, then $\psi^* Q$ forms an operad in right R-modules. Furthermore, these extension and restriction functors on operads

$$\psi_! : \mathcal{O}_R \rightleftarrows \mathcal{O}_S : \psi^*$$

are adjoint to each other. The obvious morphism $\psi : R[\,\cdot\,] \to S[\,\cdot\,]$ induced by ψ defines a morphisms

$$\psi_\sharp : R[\,\cdot\,] \to \psi^* S[\,\cdot\,]$$

in $\mathcal{O}_R$. Let

$$\psi_\flat : \psi_! R[\,\cdot\,] \to S[\,\cdot\,]$$

be the associated adjoint morphism. By theorem 7.2.2, we have $S_S(\psi_! R[\,\cdot\,], B) = S_R(R[\,\cdot\,], \psi^* B) = U_R(\psi^* B)$ and $S_R(\psi^* S[\,\cdot\,], A) = S_S(R[\,\cdot\,], \psi_! A) = U_S(\psi_! A)$.

It is straightforward to check that $\psi_\flat : \psi_! R[\,\cdot\,] \to S[\,\cdot\,]$ is the morphism that represents $\psi_\flat : U_R(\psi^* B) \to U_S(B)$ for the algebra in right S-modules B formed by the operad itself $B = S$. As a byproduct, we obtain further that $\psi_\flat : U_R(\psi^* B) \to U_S(B)$ is identified with the morphism of functors

$$S_S(\psi_\flat, B) : S_S(\psi_! R[\,\cdot\,], B) \to S_S(S[\,\cdot\,], B),$$

for all $B \in {}_S\mathcal{E}$. We check similarly that $\psi_\sharp : U_R(A) \to U_S(\psi_! A)$ is identified with the morphism of functors

$$S_R(\psi_\sharp, A) : S_R(R[\,\cdot\,], A) \to S_R(\psi^* S[\,\cdot\,], A),$$

for all $A \in {}_R\mathcal{E}$.

10.2 Enveloping Algebras

Recall that the enveloping algebra of an algebra A over an operad R is the term $U_R(A)(1)$ of its enveloping operad $U_R(A)$. Theorem 10.1.3 implies that $A \mapsto U_R(A)(1)$ is the functor associated to an associative algebra in right R-modules formed by the shifted object $R[1]$.

In this section, we study the right R-module $R[1]$ associated to the classical operads $R = C, A, L$, of commutative algebras, associative algebras, and Lie algebras. Since we use no more enveloping operads in this section, we can drop the reference to the term in the notation of enveloping algebras and, for simplicity, we set $U_R(A) = U_R(A)(1)$.

To determine the structure of the right R-modules $R[1]$, where $R = C, A, L$, we use the identity $R[1] = U_R(R)$, deduced from the principle of §9.2.6, and the results of §4.3 about enveloping algebras over classical operads. The principle of generalized pointwise tensors of §0.5 is applied to extend these results to the category of right R-modules.

Regarding commutative algebras, proposition 4.3.5 gives an isomorphism $U_C(A) \simeq A_+$, where A_+ represents the unitary algebra $A_+ = 1 \oplus A$. The next lemma reflects this relation:

10.2.1 Lemma. *For the commutative operad C, we have an isomorphism of associative algebras in right C-modules $C[1] \simeq C_+$.*

Proof. Apply the isomorphism $U_C(A) \simeq A_+$ to the commutative algebra in right C-modules formed by the commutative operad itself. $\square$

This lemma implies immediately:

10.2.2 Proposition. *In the case of the commutative operad C, the object $C[1]$ forms a free right C-module.*

Proof. The direct sum $C_+ = 1 \oplus C$ forms obviously a free right C-modules since we have the relation $C_+ = 1 \circ C \oplus I \circ C = (1 \oplus I) \circ C$. (Observe that $1 \circ M = 1$, for every Σ_*-object M.) $\square$

Regarding associative algebras, proposition 4.3.6 gives an isomorphism $U_A(A) \simeq A_+ \otimes A_+^{op}$, The next lemma reflects this relation:

10.2.3 Lemma. *For the associative operad A, we have an isomorphism of associative algebras in right A-modules $A[1] \simeq A_+ \otimes A_+^{op}$.*

Proof. Apply the isomorphism $U_A(A) \simeq A_+ \otimes A_+^{op}$ to the associative algebra in right A-modules A formed by the associative operad itself $A = A$. $\square$

As a byproduct, we obtain:

10.2.4 Proposition. *In the case of the associative operad A, the object $A[1]$ forms a free right A-module.*

Proof. The object A_+^{op} has the same underlying right A-module as the operad A_+. Hence we obtain isomorphisms of right A-modules:

$$A_+[1] \simeq A_+ \otimes A_+^{op} \simeq A_+ \otimes A_+ \simeq 1 \oplus A \oplus A \oplus A \otimes A \simeq (1 \oplus I \oplus I \oplus I \otimes I) \circ A$$

from which we conclude that $A[1]$ forms a free object in the category of right A-modules.   □

Recall that the operadic enveloping algebra $U_L(G)$ of a Lie algebra G can be identified with the classical enveloping algebra of G, defined as the image of G under the extension functor $\iota_! : {}_L\mathcal{E} \to {}_{A_+}\mathcal{E}$, from Lie algebras to unitary associative algebras. At the module level, the relation $U_L(G) \simeq \iota_! G$ is reflected by:

10.2.5 Lemma. *For the Lie operad* L, *we have an isomorphism* $L[1] \simeq \iota^* A_+$, *where* $\iota^* A_+$ *is the associative algebra in right L-modules obtained by restriction of structures on the right from the unitary associative operad* A_+.

Proof. Again, apply the isomorphism $U_{L_+}(G) \simeq \iota_! G$ to the Lie algebra in right L-modules G formed by the Lie operad itself $G = L$. In this case, we have a canonical isomorphism $\iota_! L \simeq A_+$. To check this assertion, observe first that the operad morphism $\iota : L \to A_+$ defines a morphism $\iota_\sharp : L \to \iota^* A_+$ in the category of Lie algebras in right L-modules, where $\iota^* A_+$ refers to the two-sided restriction of structures of As_+. By adjunction, we have a morphism $\iota_\flat : \iota_! L \to \iota^* A_+$ in the category of unitary associative algebras in right L-modules, where $\iota_! L$ is the unitary associative algebra obtained by extension of structures on the left from L and $\iota^* A_+$ is the unitary associative algebras in right L-modules obtained by restriction of structures on the right from A_+. If we forget module structures, then we obtain readily that $\iota_\flat : \iota_! L \to \iota^* A_+$ forms an isomorphism, because the Lie operad L defines a free Lie algebra in the category of Σ_*-objects. The conclusion follows.   □

The classical Poincaré-Birkhoff-Witt theorem asserts that $U_L(G)$ is naturally isomorphic to the module of symmetric tensors $S(G)$ as long as the ground ring is a field of characteristic 0. We also have $S(X) = S(C_+, X)$, where C_+ is the Σ_*-object, underlying the operad of unitary commutative algebras, defined by the trivial representations of the symmetric group. Accordingly, by theorem 7.1.1, we have $S(G) = S_L(C_+ \circ L, G)$ for all Lie algebras G. At the module level, the relation $U_L(G) \simeq S(G)$ is reflected by:

10.2.6 Lemma. *In characteristic* 0, *we have an isomorphism of right L-modules* $A_+ \simeq C_+ \circ L$.

Proof. Apply the Poincaré-Birkhoff-Witt theorem to the Lie algebra in right L-modules defined by the operad itself.   □

As a corollary, we obtain:

10.2.7 Proposition. *In the case of the Lie operad* L, *the object* L[1] *forms a free right* L-*module as long as the ground ring is a field of characteristic* 0.   □

In positive characteristic, the Poincaré-Birkhoff-Witt theorem gives only isomorphisms $\mathrm{gr}_n U_\mathsf{L}(\mathsf{G}) \simeq \mathsf{S}_n(\mathsf{G})$, for subquotients of a natural filtration on the enveloping algebra $U_\mathsf{L}(\mathsf{G})$. Thus, in positive characteristic, we have only an isomorphism of the form $\mathrm{gr}\, \mathsf{A}_+ \simeq \mathsf{C}_+ \circ \mathsf{L}$.

10.3 Kähler Differentials

In this section, we prove that the module of Kähler differentials $\Omega^1_\mathsf{R}(A)$, defined in §4.4 for any algebra A over an operad R, is a functor associated to a certain right module Ω^1_R over the operad R.

For simplicity, we take $\mathcal{E} = \mathcal{C} = \Bbbk\,\mathrm{Mod}$ as underlying symmetric monoidal categories. Again, we can apply the principle of generalized pointwise tensors to extend the definition of Ω^1_R in the context of dg-modules and so. The relation $\Omega^1_\mathsf{R}(A) \simeq \mathsf{S}_\mathsf{R}(\Omega^1_\mathsf{R}, A)$ holds for R-algebras in dg-modules, in Σ_*-objects, or in any category of right modules over an operad S.

Recall that the module of Kähler differentials $\Omega^1_\mathsf{R}(A)$ is the $\Bbbk$-module spanned by formal expressions $p(a_1, \ldots, da_i, \ldots, a_m)$, where $p \in \mathsf{R}(m)$, $a_1, \ldots, a_m \in A$, together with relations of the form:

$$p(a_1, \ldots, q(a_i, \ldots, a_{i+n-1}), \ldots, da_j, \ldots, a_{m+n-1})$$
$$\equiv p \circ_i q(a_1, \ldots, a_i, \ldots, a_{i+n-1}, \ldots, da_j, \ldots, a_{m+n-1}), \quad \text{for } i \neq j,$$
$$p(a_1, \ldots, dq(a_i, \ldots, a_{i+n-1}), \ldots, a_{m+n-1})$$
$$\equiv \sum_{j=i}^{i+n-1} p \circ_i q(a_1, \ldots, a_i, \ldots, da_j, \ldots, a_{i+n-1}, \ldots, a_{m+n-1}).$$

From this definition, we deduce immediately:

10.3.1 Proposition. *We have* $\Omega^1_\mathsf{R}(A) = \mathsf{S}_\mathsf{R}(\Omega^1_\mathsf{R}, A)$ *for the right* R-*module* Ω^1_R *spanned by formal expressions* $p(x_1, \ldots, dx_i, \ldots, x_m)$, $p \in \mathsf{R}(m)$, *where* $(x_1, \ldots, x_m)$ *are variables, together with the right* R-*action such that:*

$$p(x_1, \ldots, dx_i, \ldots, x_m) \circ_k q$$
$$= \begin{cases} p\circ_k q(x_1, \ldots, x_k, \ldots, x_{k+n-1}, \ldots, dx_{i+n-1}, \ldots, x_{m+n-1}), & \text{for } k < i, \\ \sum_{j=k}^{k+n-1} p\circ_i q(x_1, \ldots, x_k, \ldots, dx_j, \ldots, x_{k+n-1}, \ldots, x_{m+n-1}), & \text{for } k = i, \\ p\circ_k q(x_1, \ldots, dx_i, \ldots, x_k, \ldots, dx_{k+n-1}, \ldots, x_{m+n-1}), & \text{for } k > i, \end{cases}$$

for all $q \in \mathsf{R}(n)$.   □

According to this proposition, we have $\Omega^1_\mathsf{R}(n) = \Sigma_n \otimes_{\Sigma_{n-1}} \mathsf{R}[1](n-1)$ if we forget the right R-action on Ω^1_R. The object Ω^1_R comes equipped with the structure of a left R[1]-module that reflects the definition of $\Omega^1_\mathsf{R}(A)$ as a left module over the enveloping algebra $U_\mathsf{R}(A) = \mathsf{S}_\mathsf{R}(\mathsf{R}[1], A)$.

In the next statements, we study the structure of the modules of Kähler differentials Ω_R^1 associated to the classical operads $R = A, L$. The case of the commutative operad $R = C$ is more complicated and is addressed only in a remark.

In the associative case, proposition 4.4.6 gives an isomorphism $\Omega_A^1(A) = A_+ \otimes A$. The next lemma reflects this relation:

10.3.2 Lemma. *For the associative operad* A, *we have an isomorphism of right* A-*modules* $\Omega_A^1 \simeq A_+ \otimes A$.

Proof. Apply the isomorphism $\Omega_A^1(A) = A_+ \otimes A$ to the associative algebra in right A-modules A formed by the associative operad itself $A = A$. □

As a byproduct, we obtain:

10.3.3 Proposition. *For* $R = A$, *the operad of associative algebras, the object* Ω_A^1 *forms a free right* A-*module.*

Proof. We have isomorphisms of right A-modules

$$\Omega_A^1 \simeq A_+ \otimes A \simeq 1 \oplus A \oplus A \otimes A \simeq (1 \oplus I \oplus I \otimes I) \circ A$$

from which the conclusion follows.   □

In the next paragraphs, we study the module of Kähler differentials Ω_L^1 associated to the Lie operad L. Recall that the operadic enveloping algebra $U_L(G)$ is identified with the classical enveloping algebra of G, defined by the image of G under the extension functor $\iota_! : {}_L\mathcal{E} \to {}_{A_+}\mathcal{E}$ from Lie algebras to unitary associative algebras. By proposition 4.4.7, we have a natural isomorphism $\widetilde{U}_L(G) \simeq \Omega_L^1(G)$, where $\widetilde{U}_L(G)$ denotes the augmentation ideal of $U_L(G)$. At the module level, we obtain:

10.3.4 Lemma. *For* $R = L$, *the operad of Lie algebras, we have an isomorphism of right* L-*modules* $A \simeq \Omega_L^1$, *where the Lie operad* L *acts on* A *through the morphism* $\iota : L \to A$.

Proof. Apply the isomorphism $\Theta : \widetilde{U}_L(G) \simeq \Omega_L^1(G)$ to the Lie algebra in right L-modules defined by the operad itself. By lemma 10.2.5, we have $U_L(L) \simeq A_+$ and hence $\widetilde{U}_L(L) \simeq A$. The lemma follows.   □

As a byproduct, we obtain:

10.3.5 Proposition. *For* $R = L$, *the operad of Lie algebras, the object* Ω_L^1 *forms a free right* L-*module as long as the ground ring is a field of characteristic* 0.

Proof. By lemma 10.2.6, we have an identity $A_+ = C_+ \circ L$. If we remove units, then we obtain $\Omega_L^1 = A = C \circ L$ and the conclusion follows.   □

In positive characteristic, we only have an isomorphism of the form gr $\Omega_L^1 \simeq C \circ L$.

10.3.6 Remark. The module of Kähler differentials $\Omega^1_{\mathsf{C}}(A)$ associated to the commutative operad C does not form a projective right C-module. This negative result follows from theorem 15.1.A, according to which the functor $S_R(M) : A \mapsto S_R(M, A)$ associated to a projective right R-module M preserves all weak-equivalences of differential graded R-algebras. The functor of commutative Kähler differentials $\Omega^1_{\mathsf{C}}(A)$ does not preserve weak-equivalences between differential graded commutative algebras. For more details, we refer to §17.3 where we address applications of the results of §15 to the homology of algebras over operads.

Bibliographical Comments on Part II

The notion of a right module over an operad is introduced in Smirnov's papers [56, 57] as a left coefficient for the operadic version of the cotriple construction of Beck [3] and May [47]. But, as far as we know, the first thorough studies of categories of modules over operads appear in [11] and [54]. The connection between modules over operads and other classical structures, like Γ-objects, appears in [33].

The objective of the work [11], published in [12, 13], is to prove structure and classification results for formal groups over operads. The symmetric monoidal category of right modules over operads is introduced as a background for the operadic generalization of the Hopf algebra of differential operators of a formal group.

In [54], the operadic cotriple construction occurs as a tool to compute the homotopy of moduli spaces of R-algebra structures, for a given operad R.

The relative composition product $\circ_R$ is also introduced in the papers [56, 57]. The functor $S_R(M)$ associated to a right module over an operad R is defined in [54] as a particular case of the relative composition product $M \circ_R N$. The relative composition product $M \circ_R N$ is used in [33], together with endomorphism operads of right R-modules, in order to define generalized Morita equivalences in the context of operads.

In §5, §6, §7 and §9, we essentially unify results of the literature. The only original idea is to identify the relative composition product $M \circ_R N$ with a particular case of a functor of the form $S_R(M, N)$ rather than the contrary. The statements of §8 and §10 are new.

Part III
Homotopical Background

Chapter 11
Symmetric Monoidal Model Categories for Operads

Introduction

Our next purpose is to study the homotopy of functors associated to modules over operads. To deal with homotopy problems in a general setting, we use the language of *model categories*.

The aim of this part is to recall applications of model categories to operads and algebras over operads. In this chapter, we review the definition of a model category in the context of symmetric monoidal categories over a base and we study the homotopy properties of functors $S(M) : \mathcal{E} \to \mathcal{E}$ associated to Σ_*-objects.

To begin with, in §11.1, we recall briefly basic definitions of model categories (we refer to [27, 28] for a modern and comprehensive account of the theory). For our needs, we also recall the notion of a cofibrantly generated model category, the structure used to define model categories by adjunctions (the model categories defined in this book are obtained as such). In §11.2 we survey basic examples of model categories in dg-modules. In the sequel, we illustrate our results by constructions taken in these model categories.

In §11.3 we review the axioms of monoidal model categories. We use these axioms in the relative context of symmetric monoidal model categories over a base. In §11.4, we prove that the category of Σ_*-objects inherits the structure of a cofibrantly generated monoidal model category from the base category.

In §11.5 we use the model structure of Σ_*-objects to study the homotopy of functors $S(M) : \mathcal{E} \to \mathcal{E}$. Essentially, we observe that the bifunctor $(M, X) \mapsto S(M, X)$ satisfies a pushout-product axiom, like the tensor product of a symmetric monoidal model category. Usual homotopy invariance properties are consequences of this axiom. For instance, we check that the functor $S(M) : \mathcal{E} \to \mathcal{E}$ preserves weak-equivalences between cofibrant objects as long as M is cofibrant in the category of Σ_*-object.

¶ In many usual examples, the functor $S(M) : \mathcal{E} \to \mathcal{E}$ preserves weak-equivalences for a larger class of Σ_*-objects. Therefore we introduce an

B. Fresse, *Modules over Operads and Functors*, Lecture Notes in Mathematics 1967, 153
DOI: 10.1007/978-3-540-89056-0_11, © Springer-Verlag Berlin Heidelberg 2009

axiomatic setting, the notion of a symmetric monoidal category with regular tensor powers, to handle finer homotopy invariance properties. These refinements are addressed in §11.6. Usual model categories of spectra give examples of categories in which the homotopy invariance of functors $S(M) : \mathcal{E} \to \mathcal{E}$ can be refined.

The pushout-product property of the bifunctor $(M, X) \mapsto S(M, X)$ is the base of many subsequent constructions. In the sequel, the possible improvements which hold for categories with regular tensor powers and spectra are deferred to remarks (marked by the symbol ¶).

11.1 Recollections: The Language of Model Categories

In this section, we recall the definition of a model category and the notion of a cofibrantly generated model category with the aim to address homotopy problems in the context of operads.

11.1.1 Basic Notions. Model categories help to handle categories of fractions in which a class of morphisms, usually called *weak-equivalences*, is formally inverted to yield actual isomorphisms. In the context of a model category $\mathcal{A}$, the category of fractions is realized by a *homotopy category* $\mathrm{Ho}\,\mathcal{A}$ and its morphisms are represented by homotopy classes of morphisms of $\mathcal{A}$, a natural notion of homotopy being associated to $\mathcal{A}$. Since we do not use explicitly homotopy categories in this book, we refer to the literature for this motivating application of model categories. For our needs we recall only the axioms of model categories.

The core idea of model categories is to characterize classes of morphisms by lifting properties which occur naturally in homotopy. For the moment, recall simply that a morphism $i : A \to B$ has the *left lifting property* with respect to $p : X \to Y$, and $p : X \to Y$ has the *right lifting property* with respect to $i : A \to B$, if every solid diagram of the form

$$
\begin{array}{ccc}
A & \longrightarrow & X \\
\downarrow{\scriptstyle i} & \nearrow & \downarrow{\scriptstyle p} \\
B & \longrightarrow & Y
\end{array}
$$

can be filled out by a dotted morphism, the *lifting*. Recall that a morphism f is a *retract* of g if we have morphisms

$$
\begin{array}{ccccc}
A & \xrightarrow{\ i\ } & C & \xrightarrow{\ r\ } & A \\
\downarrow{\scriptstyle f} & & \downarrow{\scriptstyle g} & & \downarrow{\scriptstyle f} \\
B & \xrightarrow[\ j\ ]{} & D & \xrightarrow[\ s\ ]{} & B
\end{array}
$$

which make the whole diagram commutes and so that $ri = $ id and $sj = $ id.
The retract f inherits every left lifting property satisfied by g, and similarly
as regards right lifting properties.

11.1.2 The Axioms of Model Categories. The structure of a model category consists of a category $\mathcal{A}$ equipped with three classes of morphisms, called
weak-equivalences (denoted by $\xrightarrow{\sim}$), *cofibrations* (denoted $\rightarrowtail$) and *fibrations*
(denoted $\twoheadrightarrow$), so that the following axioms M1-5 hold:

M1 (*completeness axiom*): Limits and colimits exist in $\mathcal{A}$.

M2 (*two-out-of-three axiom*): Let f and g be composable morphisms. If any
   two among f, g and fg are weak-equivalences, then so is the third.

M3 (*retract axiom*): Suppose f is a retract of g. If g is a weak-equivalence
   (respectively a cofibration, a fibration), then so is f.

M4 (*lifting axioms*):

   i. The cofibrations have the left lifting property with respect to acyclic
      fibrations, where an *acyclic fibration* refers to a morphism which is both
      a weak-equivalence and a fibration.

   ii. The fibrations have the right lifting property with respect to acyclic
      cofibrations, where an *acyclic cofibration* refers to a morphism which is
      both a weak-equivalence and a cofibration.

M5 (*factorization axioms*):

   i. Any morphism has a factorization $f = pi$ such that i is a cofibration
      and p is an acyclic fibration.

   ii. Any morphism has a factorization $f = pj$ such that j is an acyclic
      cofibration and q is a fibration.

By convention, an object X in a model category $\mathcal{A}$ is *cofibrant* if the initial
morphism $0 \to X$ is a cofibration, *fibrant* if the terminal morphism $X \to *$ is
a fibration.

Axiom M5.i implies that, for every object $X \in \mathcal{A}$, we can pick a cofibrant
object $0 \rightarrowtail A$ together with a weak-equivalence $A \xrightarrow{\sim} X$. Any such object is
called a *cofibrant replacement* of X. Axiom M5.ii implies symmetrically that,
for every object $X \in \mathcal{A}$, we can pick a fibrant object $A \twoheadrightarrow *$ together with
a weak-equivalence $X \xrightarrow{\sim} A$. Any such object is called a *fibrant replacement*
of X.

11.1.3 Pushouts, Pullbacks, and Proper Model Categories. The axioms imply that the class of fibrations (respectively, acyclic fibrations) in a
model category is completely characterized by the right lifting property with
respect to acyclic cofibrations (respectively, cofibrations). Similarly, the class
of cofibrations (respectively, acyclic cofibrations) is completely characterized
by the left lifting property with respect to acyclic fibrations (respectively,
fibrations).

These observations imply further that the class of fibrations (respectively, acyclic fibrations) is stable under pullbacks, the class of cofibrations

(respectively, acyclic cofibrations) is stable under pushouts. The model category $\mathcal{A}$ is called *left* (respectively, *right*) *proper* if we have further:

P1 (*left properness axiom*): The class of weak-equivalences is stable under pushouts along cofibrations.

P2 (*right properness axiom*): The class of weak-equivalences is stable under pullbacks along fibrations.

The category $\mathcal{A}$ is *proper* if both axioms are satisfied. In this book, we only use axiom P1 of left properness.

Not all categories are left (or right) proper. But:

11.1.4 Proposition (see [27, Proposition 13.1.2]). *The following assertion holds in every model category $\mathcal{A}$:*

P1'. *The pushout of a weak-equivalence along a cofibration*

$$
\begin{array}{ccc}
A & \rightarrowtail & C \\
\sim\downarrow & & \vdots \\
B & \dashrightarrow & D
\end{array}
$$

gives a weak-equivalence $C \xrightarrow{\sim} D$ provided that A and B are cofibrant in $\mathcal{A}$.

Proof. See [27, Proposition 13.1.2]. ☐

The dual weak right properness property holds in every model category too.

Basic examples of model categories include the category of dg-modules over a ring, the category of topological spaces, and the category of simplicial sets. Before recalling the definition of these basic model categories, we still review the notion of a *cofibrantly generated model category* which applies to these examples. The categories of spectra in stable homotopy give other important instances of cofibrantly generated model categories. This example is studied briefly in §§11.6.8-11.6.11 as an instance of a symmetric monoidal model category with regular tensor powers.

For our purpose, the importance of cofibrantly generated model categories comes from the definition of new model categories by adjunction from a cofibrantly generated model structure. In this book, we apply the construction of adjoint model structures to the category of Σ_*-objects (see §11.4), the categories of right modules over operads (see §14), the category of operads itself (see §12.2), and categories of algebras over operads (see §12.3).

The core idea of a cofibrantly generated model category is to realize the factorizations of axiom M5 by successive cell attachments of reference cofibrations (respectively, acyclic cofibrations) which are also used to determine the class of acyclic fibrations (respectively, fibrations).

11.1.5 Cofibrantly Generated Model Categories. Formally, a *cofibrantly generated model category* consists of a model category $\mathcal{A}$ equipped with a *set of generating cofibrations* $\mathcal{I}$ and a set of *generating acyclic cofibrations* $\mathcal{J}$, for which the small object argument holds (see short recollections next and [27, 28] for further details), and so that:

G1. The fibrations are characterized by the right lifting property with respect to acyclic generating cofibrations $j \in \mathcal{J}$.

G2. The acyclic fibrations are characterized by the right lifting property with respect to generating cofibrations $i \in \mathcal{I}$.

Since the class of cofibrations (respectively, acyclic cofibrations) is also characterized by the left lifting property with respect to acyclic fibrations (respectively, fibrations), the structure of a cofibrantly generated model category is completely determined by its class of weak-equivalences, the set of generating cofibrations $\mathcal{I}$ and the set of generating acyclic cofibrations $\mathcal{J}$.

11.1.6 Relative Cell Complexes. Let $\mathcal{K}$ be any set of morphisms in a category $\mathcal{A}$ with colimits. We call $\mathcal{K}$-*cell attachment* any morphism $j : K \to L$ obtained by a pushout

$$
\begin{array}{ccc}
\bigoplus_\alpha C_\alpha & \longrightarrow & K \\
\downarrow & & \vdots\, j \\
\bigoplus_\alpha D_\alpha & \cdots\cdots\cdots\!\!\!\longrightarrow & L
\end{array}
$$

in which the right hand side consists of a sum of morphisms of $\mathcal{K}$. We call *relative $\mathcal{K}$-cell complex* any morphism $j : K \to L$ obtained by a (possibly transfinite) composite

$$
K = L_0 \to \cdots \to L_{\lambda-1} \xrightarrow{j_\lambda} L_\lambda \to \cdots \to \operatorname*{colim}_{\lambda < \mu} L_\lambda = L,
$$

over an ordinal μ, of $\mathcal{K}$-cell attachments $j_\lambda : L_{\lambda-1} \to L_\lambda$. We refer to the ordinal μ as the length of the relative $\mathcal{K}$-cell complex $j : K \to L$. We call $\mathcal{K}$-*cell complex* any object L so that the initial morphism $0 \to L$ is equipped with the structure of a relative $\mathcal{K}$-cell complex.

11.1.7 The Small Object Argument and Cellular Approximations. The small object argument (whenever it holds) associates to any morphism f a natural factorization $f = pi$ so that i is a relative $\mathcal{K}$-complex and p has the right lifting property with respect to morphisms of $\mathcal{K}$ (we refer to [27, §10.5] and [28, §2.1.2] for a detailed account of this construction). The small object argument holds for a set of morphisms $\mathcal{K}$ if the domain A of every morphism of $\mathcal{K}$ is small with respect to relative $\mathcal{K}$-cell complexes of length $\mu \geq \omega$, for some fixed ordinal ω: any morphism $f : A \to L$, where $L = \operatorname*{colim}_{\lambda<\mu} L_\lambda$ is a relative $\mathcal{K}$-cell complexes of length $\mu \geq \omega$, admits a factorization

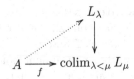

for some $\lambda < \mu$.

11.1.8 Relative Cell Complexes and (Acyclic) Cofibrations in Cofibrantly Generated Model Categories.
In the context of a cofibrantly generated model category, the small object argument returns:

- a factorization $f = pi$ such that i is a relative $\mathcal{I}$-cell complex and p has the right lifting property with respect to generating cofibrations,
- and a factorization $f = qj$ such that j is a relative $\mathcal{J}$-cell complex and q has the right lifting property with respect to generating acyclic cofibrations.

Every relative $\mathcal{I}$-cell (respectively, $\mathcal{J}$-cell) complex forms a cofibration (respectively, acyclic cofibration) because relative $\mathcal{I}$-cell (respectively, $\mathcal{J}$-cell) complexes inherits the left lifting properties of generating cofibrations (respectively, acyclic cofibrations). Thus we obtain the factorizations required by axiom M5 of model categories

If we assume that f is a cofibration, then we can deduce from the left lifting property that f forms a retract of the relative $\mathcal{I}$-cell complex i which occurs in the factorization $f = pi$. Similarly, if f is an acyclic cofibration, then we obtain that f forms a retract of the relative $\mathcal{J}$-cell complex j which occurs in the factorization $f = qj$. Hence the cofibrations (respectively, acyclic cofibrations) in a cofibrantly generated model category are the retracts of relative $\mathcal{I}$-cell (respectively, $\mathcal{J}$-cell) complexes.

In usual examples of model categories, we call *cofibrant cell objects* the cell complexes built from a natural set of generating cofibrations of the model category, and we use a similar convention for relative cell complexes.

11.1.9 Example: dg-Modules.
The category of dg-modules $\mathrm{dg}\,\Bbbk\,\mathrm{Mod}$ is equipped with a proper model structure so that a morphism $f : A \to B$ is a weak-equivalence if it induces an isomorphism in homology, a fibration if it is degreewise surjective, a cofibration if it has the left lifting property with respect to acyclic fibrations (see [28] for proofs).

Let E^d be the dg-module freely spanned by an element $e = e_d$ in degree d and an element $b = b_{d-1}$ in degree $d - 1$ so that $\delta(e_d) = b_{d-1}$. Let B^{d-1} be the submodule of E^d spanned by b_{d-1}. The embeddings $i : B^{d-1} \to E^d$ define a set of generating cofibrations of the category of dg-modules, and the null morphisms $j : 0 \to E^d$ define a set of generating acyclic cofibrations, so that $\mathrm{dg}\,\Bbbk\,\mathrm{Mod}$ forms a cofibrantly generated model category (we refer to [28] for details).

The structure of cofibrations in dg-modules is studied more thoroughly in §11.2.

11.1.10 Example: Topological Spaces. The category of topological spaces Top is equipped with a proper model structure so that (see [28]):

- A map $f : X \to Y$ is a weak-equivalence if $\pi_0(f) : \pi_0(X) \to \pi_0(Y)$ is bijective and $\pi_n(f) : \pi_n(X, *) \to \pi_n(Y, *)$ is a group isomorphism for all $n > 0$ and for every choice of base points;
- The fibrations are the Serre fibrations, the maps which have the homotopy lifting property with respect to cell complexes;
- The cofibrations are the maps which have the left lifting property with respect to acyclic fibrations.

Let Δ^n be the standard model of the n-simplex in the category of simplicial sets, defined by the morphism sets $\Delta^n = \mathrm{Mor}_\Delta(-, n)$ in the simplicial category. Form the subobjects $\partial \Delta^n = \bigcup_{i=1}^{n} d_i(\Delta^n) \subset \Delta^n$ and $\Lambda_k^n = \bigcup_{i \neq k} d_i(\Delta^n) \subset \Delta^n$, $0 \leq k \leq n$. The category of topological spaces has a set of generating cofibrations defined by the geometric realization of the embeddings $i : \partial \Delta^n \to \Delta^n$, a set of generating acyclic cofibrations defined by the geometric realization of the embeddings $j : \Lambda_k^n \to \Delta^n$, so that Top forms a cofibrantly generated category.

11.1.11 Example: Simplicial Sets. The category of simplicial sets $\mathcal{S}$ is equipped with a proper model structure so that a map $f : X \to Y$ is a weak-equivalence if its geometric realization $|f| : |X| \to |Y|$ is a weak-equivalence of topological spaces, a fibration if f is a fibration in Kan's sense, a cofibration if f is dimensionwise injective (as usual, we refer to [28]).

The category of simplicial sets $\mathcal{S}$ is also cofibrantly generated by the embeddings $i : \partial \Delta^n \to \Delta^n$ as generating cofibrations and the embeddings $j : \Lambda_k^n \to \Delta^n$ as generating acyclic cofibrations. The definition of a Kan fibration is actually equivalent to a map $f : X \to Y$ which has the right lifting property with respect to the embeddings $j : \Lambda_k^n \to \Delta^n$.

11.1.12 Model Structures and Adjunctions. Our motivation to use cofibrantly generated model categories is to derive new model structures by adjunction from a well-defined model category.

Say that a functor $U : \mathcal{A} \to \mathcal{X}$ between model categories *creates* weak-equivalence if the weak-equivalences of $\mathcal{A}$ are exactly the morphisms f such that $U(f)$ forms a weak-equivalence in $\mathcal{X}$, and adopt similar conventions with respect to other classes of morphisms. In many usual examples, we have a natural adjunction relation $F : \mathcal{X} \rightleftarrows \mathcal{A} : U$, where $\mathcal{X}$ is a reference model category, and a model structure on $\mathcal{A}$ is specified by assuming that the functor $U : \mathcal{A} \to \mathcal{X}$ creates weak-equivalences and fibrations. The class of cofibrations of $\mathcal{A}$ is determined by the left lifting property with respect to acyclic fibrations. The difficulty is to check the axioms of model categories to conclude that this definition returns an actual model structure on $\mathcal{A}$.

In the context of cofibrantly generated model categories, the verifications can be reduced to simple conditions:

11.1.13 Theorem (see [27, Theorem 11.3.2]). *Suppose we have an adjunction* $F : \mathcal{X} \rightleftarrows \mathcal{A} : U$, *where* $\mathcal{A}$ *is any category with limits and colimits and* $\mathcal{X}$ *is a cofibrantly generated model category. Let* $\mathcal{I}$, *respectively* $\mathcal{J}$, *be the set of generating (acyclic) cofibrations of* $\mathcal{X}$ *and set* $F\mathcal{I} = \{F(i),\ i \in \mathcal{I}\}$, *respectively* $F\mathcal{J} = \{F(j),\ j \in \mathcal{J}\}$. *Under assumptions (1-2) below, the category* $\mathcal{A}$ *inherits a cofibrantly generated model structure with* $F\mathcal{I}$ *(respectively,* $F\mathcal{J}$*) as generating (acyclic) cofibrations and so that the functor* $U : A \to \mathcal{X}$ *creates weak-equivalences. The functor* $U : A \to \mathcal{X}$ *creates the class of fibrations as well.*

(1) The small object argument holds for $F\mathcal{I}$ *and* $F\mathcal{J}$.
(2) Any relative $F\mathcal{J}$*-cell complex* $g : A \to B$ *forms a weak-equivalence in* $\mathcal{A}$.

In §12.1 we extend this theorem to semi-model structures. For this reason, we review the proof of this classical statement, which can also be found in [27].

Proof. The statement supposes that the weak-equivalences of $\mathcal{A}$ are the morphisms f such that $U(f)$ is a weak-equivalence in $\mathcal{X}$. Define the class of fibrations by the right lifting property with respect to morphisms $F(j)$, $j \in \mathcal{J}$, the class of cofibrations by the left lifting property with respect to morphisms which are both fibrations and weak-equivalences. By adjunction, we obtain that p forms a fibration in $\mathcal{A}$ if and only if $U(p)$ has the right lifting property with respect to morphisms $j \in \mathcal{J}$ and hence forms a fibration in $\mathcal{X}$. Thus the functor $U : \mathcal{A} \to \mathcal{X}$ creates both fibrations and weak-equivalences. As a byproduct we obtain that the functor $U : \mathcal{A} \to \mathcal{X}$ creates the class of acyclic fibrations as well.

The category $\mathcal{A}$ satisfies axiom M1 by assumption and inherits trivially axioms M2-M3 from the model category $\mathcal{X}$. Axiom M4.i is tautologically satisfied. We use the small object argument to check the factorization axioms M5.i-ii and we prove the other lifting axiom M4.ii afterwards.

The small object argument applied to $F\mathcal{I}$ returns a factorization $f = pi$, where i is a relative $F\mathcal{I}$-complex and p has the right lifting property with respect to morphisms of $F\mathcal{I}$. By adjunction, we obtain that $U(p)$ has the right lifting property with respect to morphisms of $\mathcal{I}$, and hence forms an acyclic fibration in $\mathcal{X}$, from which we conclude that p forms an acyclic fibration. Thus the small object argument applied to $F\mathcal{I}$ gives the factorization $f = pi$ required by axiom M5.i of model categories.

The small object argument applied to $F\mathcal{J}$ returns a factorization $f = qj$, where j is a relative $F\mathcal{J}$-complex and q has the right lifting property with respect to morphisms of $F\mathcal{J}$. The morphism q is a fibration by definition of fibrations in $\mathcal{A}$. Relative $F\mathcal{J}$-complexes are weak-equivalences by assumption and inherit the left lifting property with respect to fibrations from the morphisms of $F\mathcal{J}$. Thus we obtain that any relative $F\mathcal{J}$-complex forms an acyclic cofibration. Thus the small object argument applied to $F\mathcal{J}$ gives the factorization $f = qj$ required by axiom M5.ii of model categories. Note further that the acyclic cofibration which occurs in this factorization is ensured to have the left lifting property with respect to fibrations.

If we assume further that f is an acyclic cofibration, then the fibration q is also acyclic by the two-out-of-three axiom. By axiom M4.i, which is tautologically satisfied in $\mathcal{A}$, we have a morphism s so that $sf = j$ and $ps = \text{id}$, from which we deduce that f is a retract of j. Since retracts inherit left lifting properties, we obtain that f has the left lifting property with respect to fibrations as well. This argument proves that the lifting axiom M4.ii is also satisfied in $\mathcal{A}$ and achieves the proof of the theorem. □

In our constructions, we use the following proposition to check the conditions of theorem 11.1.13:

11.1.14 Proposition. *The conditions of theorem 11.1.13 are satisfied under the sufficient assumptions that:*

(1) The functor $U : \mathcal{A} \to \mathcal{X}$ preserves colimits over non-empty ordinals;
(2) For any pushout

$$
\begin{array}{ccc}
F(C) & \longrightarrow & A \\
{\scriptstyle F(i)}\downarrow & & \downarrow {\scriptstyle f} \\
F(D) & \dashrightarrow & B
\end{array}
\quad ,
$$

the morphism $U(f)$ forms a cofibration, respectively an acyclic cofibration, in $\mathcal{X}$ if i is so.

In this situation, the functor $U : \mathcal{A} \to \mathcal{X}$ preserves cofibrations in addition to create weak-equivalences and fibrations.

Proof. Let $j : A \to B$ be a relative $F\mathcal{I}$-cell (respectively, $F\mathcal{J}$-cell) complex in $\mathcal{A}$. The assumptions imply that the morphism $U(j)$ splits into a colimit

$$
U(A){=}U(B_0)\to\ldots\to U(B_{\lambda-1}) \xrightarrow{U(j_\lambda)} U(B_\lambda)\to\cdots \to \text{colim}_{\lambda<\mu}U(B_\lambda){=}U(B)
$$

so that $U(j_\lambda)$ forms a cofibration (respectively, an acyclic cofibration) in $\mathcal{X}$. Condition (2) of theorem 11.1.13 follows immediately from this decomposition.

To check condition (1), we construct a sequence of relative $\mathcal{I}$-cell (respectively, $\mathcal{J}$-cell) complexes so that we have retracts:

$$
\begin{array}{ccccccc}
L_0 & \dashrightarrow \cdots \dashrightarrow & L_{\lambda-1} & \dashrightarrow & L_\lambda & \dashrightarrow & \cdots \\
{\scriptstyle =}\Big\updownarrow{\scriptstyle =} & & {\scriptstyle s_{\lambda-1}}\Big\updownarrow{\scriptstyle q_{\lambda-1}} & & {\scriptstyle s_\lambda}\Big\updownarrow{\scriptstyle q_\lambda} & & \\
U(B_0) & \longrightarrow \cdots \longrightarrow & U(B_{\lambda-1}) & \longrightarrow & U(B_\lambda) & \longrightarrow & \cdots
\end{array}
$$

Suppose we have achieved the construction of the retracts $U(B_\lambda) \underset{s_\lambda}{\overset{q_\lambda}{\rightleftarrows}} L_\lambda$ till an ordinal λ. Pick a factorization

$$L_\lambda \xrightarrow{\ k_\lambda\ } L_{\lambda+1}$$

$$\downarrow q_\lambda \qquad\qquad \downarrow q_{\lambda+1}$$

$$U(B_\lambda) \xrightarrow[U(j_\lambda)]{} U(B_{\lambda+1}),$$

where k_λ is a relative $\mathcal{I}$-cell (respectively, $\mathcal{J}$-cell) complex and $q_{\lambda+1}$ is an acyclic fibration (respectively, a fibration) in $\mathcal{X}$. Observe that $q_{\lambda+1} \cdot k_\lambda \cdot s_\lambda = U(j_\lambda) \cdot q_\lambda \cdot s_\lambda = U(j_\lambda)$ so that the solid frame commutes in the diagram:

$$U(B_\lambda) \xrightarrow{\ s_\lambda\ } L_\lambda \xrightarrow{\ k_\lambda\ } L_{\lambda+1}$$

$$\downarrow U(j_\lambda) \qquad\qquad\qquad \downarrow q_{\lambda+1}$$

$$U(B_{\lambda+1}) \xrightarrow{\quad =\quad } U(B_{\lambda+1}).$$

Use the lifting axiom in $\mathcal{X}$ to pick a fill-in morphism $s_{\lambda+1}$, which satisfies $q_{\lambda+1} \cdot s_{\lambda+1} = \mathrm{id}$ and $s_{\lambda+1} \cdot U(j_\lambda) = k_\lambda \cdot s_\lambda$ by construction. The definition of $s_{\lambda+1}$ achieves the construction of the retracts till the ordinal $\lambda + 1$ and the construction can be carried on by induction.

Suppose we have a morphism $K \to \mathrm{colim}_{\lambda<\mu} U(B_\lambda)$, where $K \in \mathcal{X}$ is small with respect to relative $\mathcal{I}$-cell (respectively, $\mathcal{J}$-cell) complexes of length $\mu \geq \omega$. The sequence of relative $\mathcal{I}$-cell (respectively, $\mathcal{J}$-cell) complexes

$$L_0 \to \cdots \to L_{\lambda-1} \xrightarrow{\ k_\lambda\ } L_\lambda \to \cdots \to \mathrm{colim}_{\lambda<\mu} L_\lambda$$

can be refined into a relative $\mathcal{I}$-cell complex of length $\geq \mu$. From this observation, we deduce that the composite

$$K \to \mathrm{colim}_{\lambda<\mu} U(B_\lambda) \to \mathrm{colim}_{\lambda<\mu} L_\lambda$$

factors through some stage L_λ of the colimit. By using the retraction, we obtain readily that the initial morphism $K \to \mathrm{colim}_{\lambda<\mu} U(B_\lambda)$ factors through some $U(B_\lambda)$.

By adjunction, we obtain immediately that any morphism $F(K) \to \mathrm{colim}_{\lambda<\mu} B_\lambda$, where K is small with respect to relative $\mathcal{I}$-cell (respectively, $\mathcal{J}$-cell) complexes of length μ, admits a factorization:

$$B_\lambda$$
$$\nearrow \qquad \downarrow$$
$$F(K) \xrightarrow{\ f\ } \mathrm{colim}_{\lambda<\mu} B_\mu$$

for some λ. As a conclusion, we obtain that $F(K)$ is small with respect to relative $F\mathcal{I}$-cell (respectively, $F\,\mathcal{J}$-cell) complexes of length μ. Thus the small object argument holds for $F\mathcal{I}$ (respectively, $F\mathcal{J}$) and this achieves the proof of condition (1) of theorem 11.1.13.

Our first observation implies that the functor $U : \mathcal{A} \to \mathcal{X}$ maps relative $F\mathcal{I}$-cell complexes to cofibrations. As a byproduct, we obtain readily that $U : \mathcal{A} \to \mathcal{X}$ preserves cofibrations since cofibrations are retracts of relative $F\mathcal{I}$-cell complexes. □

11.1.15 Recollections on Quillen's Adjunctions. Now we suppose given an adjunction $F : \mathcal{X} \rightleftarrows \mathcal{A} : U$ between well-defined model categories $\mathcal{A}$ and $\mathcal{X}$. In accordance with usual conventions, we say that the functors (F, U) define a *Quillen adjunction* if any one of the following equivalent condition holds (we refer to [27, §8.5] and [28, §1.3]):

A1. The functor F preserves cofibrations and acyclic cofibrations.
A2. The functor U preserves fibrations and acyclic fibrations.
A3. The functor F preserves cofibrations and U preserves fibrations.

The functors (F, U) define a *Quillen equivalence* if we have further:

E1. For every cofibrant object $X \in \mathcal{X}$, the composite

$$X \xrightarrow{\eta X} UF(X) \to U(B)$$

forms a weak-equivalence in $\mathcal{X}$, where ηX refers to the adjunction unit and B is any fibrant replacement of $F(X)$.

E2. For every fibrant object $A \in \mathcal{A}$, the composite

$$F(Y) \to FU(A) \xrightarrow{\epsilon A} A$$

forms a weak-equivalence in $\mathcal{A}$, where ϵA refers to the adjunction augmentation and Y is any cofibrant replacement of $U(A)$.

The notion of a Quillen adjunction gives usual conditions to define adjoint derived functors on homotopy categories. The notion of a Quillen equivalence gives a usual condition to obtain adjoint equivalences of homotopy categories. In this book, we define examples of Quillen adjunctions (respectively, equivalences), but we do not address applications to the construction of derived functors for which we refer to the literature.

We use essentially:

(a) In a Quillen adjunction, the left adjoint $F : \mathcal{X} \to \mathcal{A}$ preserves weak-equivalences between cofibrant objects. This assertion is a consequence of the so-called Brown's lemma, recalled next in a generalized context.

(b) In a Quillen equivalence, the adjunction relation maps a weak-equivalence $\phi : X \xrightarrow{\sim} U(B)$ so that X is cofibrant and B is fibrant to a weak-equivalence $\phi_\sharp : F(X) \xrightarrow{\sim} B$.

11.1.16 Enlarged Classes of Cofibrations. In certain constructions, we have functors $F : \mathcal{A} \to \mathcal{X}$ which preserve more (acyclic) cofibrations than the genuine (acyclic) cofibrations of $\mathcal{A}$. To formalize the assumptions needed in applications, we introduce the notion of an *enlarged class of cofibrations*.

In general, we are given a class of $\mathcal{B}$-*cofibrations*, to refer to some class of morphisms in $\mathcal{B}$, and we define the class of *acyclic $\mathcal{B}$-cofibrations* by the morphisms which are both $\mathcal{B}$-cofibrations and weak-equivalences in $\mathcal{A}$. As usual, we say also that an object $X \in \mathcal{A}$ is $\mathcal{B}$-cofibrant if the initial morphism $0 \to X$ is a $\mathcal{B}$-cofibration.

The class of $\mathcal{B}$-cofibrations defines an enlarged class of cofibrations in $\mathcal{A}$ if the following axioms hold:

C0. The class of $\mathcal{B}$-cofibrations includes the cofibrations of $\mathcal{A}$.

C1. Any morphism $i : C \to D$ which splits into a sum

$$\bigoplus_\alpha i_\alpha : \bigoplus_\alpha C_\alpha \to \bigoplus_\alpha D_\alpha,$$

where the morphisms i_α are $\mathcal{B}$-cofibrations (respectively, acyclic $\mathcal{B}$-cofibrations) forms itself a $\mathcal{B}$-cofibration (respectively, an acyclic $\mathcal{B}$-cofibration).

C2. For any pushout

$$
\begin{array}{ccc}
C & \longrightarrow & S \\
\downarrow{\scriptstyle i} & & \vdots{\scriptstyle j} \\
D & \cdots\cdots\!\!\!\!\longrightarrow & D \oplus_C S,
\end{array}
$$

where $i : C \to D$ is a $\mathcal{B}$-cofibration (respectively, an acyclic $\mathcal{B}$-cofibration), the morphism j forms itself a $\mathcal{B}$-cofibration (respectively, an acyclic $\mathcal{B}$-cofibration).

C3. Any morphism $i : C \to D$ which splits into (possibly transfinite) composite of $\mathcal{B}$-cofibrations (respectively, acyclic $\mathcal{B}$-cofibrations)

$$C = D_0 \to \cdots \to D_{\lambda-1} \xrightarrow{j_\lambda} D_\lambda \to \cdots \to \operatorname*{colim}_\lambda D_\lambda = D,$$

forms itself a $\mathcal{B}$-cofibration (respectively, an acyclic $\mathcal{B}$-cofibration).

C4. If $i : C \to D$ is a retract of a $\mathcal{B}$-cofibration (respectively, of an acyclic $\mathcal{B}$-cofibration), then i is also a $\mathcal{B}$-cofibration (respectively, an acyclic $\mathcal{B}$-cofibration).

11.1.17 Enlarged Class of Cofibrations Created by Functors.

Throughout this book, we use model categories $\mathcal{A}$ equipped with an obvious forgetful functor $U : \mathcal{A} \to \mathcal{B}$, where $\mathcal{B}$ is another model category. In our constructions, the model structure of $\mathcal{A}$ is usually defined by applying theorem 11.1.13 and proposition 11.1.14. Note that the conditions of proposition 11.1.14 imply that the forgetful functor $U : \mathcal{A} \to \mathcal{B}$ preserves cofibrations and acyclic cofibrations, in addition to create weak-equivalences and fibrations in $\mathcal{B}$.

In this situation, we call $\mathcal{B}$-cofibrations the class of cofibrations created by the forgetful functor $U : \mathcal{A} \to \mathcal{B}$. If the forgetful functor $U : \mathcal{A} \to \mathcal{B}$ preserves colimits in addition to cofibrations and acyclic cofibrations, then the class of $\mathcal{B}$-cofibrations defines an enlarged class of cofibrations in $\mathcal{B}$.

In the sequel, we may use classes of $\mathcal{B}$-cofibrations created by forgetful functors $U : \mathcal{A} \to \mathcal{B}$ in cases where the $\mathcal{B}$-cofibrations do not form an enlarged class of cofibrations. In principle, we adopt the convention of §11.1.16 that an object $A \in \mathcal{A}$ is $\mathcal{B}$-cofibrant if the initial morphism $0 \to A$ is a $\mathcal{B}$-cofibration. In certain cases, the functor $U : \mathcal{A} \to \mathcal{B}$ does not preserves initial objects, but $U(0)$ forms a cofibrant object since we assume that $U : \mathcal{A} \to \mathcal{B}$ preserves cofibrations. Consequently, the object $U(A)$ is cofibrant if A is $\mathcal{B}$-cofibrant, but the converse implication does not hold.

The following statement gives the generalization of Brown's lemma to enlarged classes of cofibrations:

11.1.18 Proposition (Brown's lemma). *Let $F : \mathcal{A} \to \mathcal{X}$ be a functor, where $\mathcal{A}$ is a model category and $\mathcal{X}$ is a category equipped with a class of weak-equivalences that satisfies the two-out-of-three axiom. Suppose we have an enlarged class of $\mathcal{B}$-cofibrations in $\mathcal{A}$. If F maps acyclic $\mathcal{B}$-cofibrations between $\mathcal{B}$-cofibrant objects to weak-equivalences, then F maps all weak-equivalences between $\mathcal{B}$-cofibrant objects to weak-equivalences.*

Proof. The generalization is proved along the same lines as the standard Brown's lemma, for which we refer to [27, §7.7] and [28, Lemma 1.1.12]. □

11.2 Examples of Model Categories in the dg-Context

In the sequel, we illustrate our constructions by examples in dg-modules. For this reason, we study this instance of a cofibrantly generated model category more thoroughly.

The first purpose of the next paragraphs is to give a manageable representation of cofibrant objects in dg-modules, starting from the definition of a relative cell complex. For simplicity, we use the convention of §11.1.8 to call *cofibrant cell dg-modules* the cell complexes in dg-modules built from the generating cofibrations of §11.1.9, and similarly as regards relative cell complexes.

Then we study the model category of modules over an associative dg-algebra as an example of a model structure defined by adjunction from a basic model category. We also prove that the usual extension and restriction of structures of modules over dg-algebras give examples of Quillen adjunctions. We extend these classical results to modules over operads in the next part.

11.2.1 Twisting Cochains and Quasi-Free dg-Modules. In general, we assume that a dg-module C is equipped with an internal differential $\delta : C \to C$, fixed once and for all. But in certain constructions, the natural differential of a given dg-module C is *twisted* by a cochain $\partial \in Hom_{\mathrm{dg\,k\,Mod}}(C, C)$ of degree -1 to produce a new dg-module, which has the same underlying graded module as C, but whose differential is given by the sum $\delta + \partial : C \to C$. By convention, the pair (C, ∂), where C is a dg-module and ∂ is a *twisting cochain*, will refer to this new dg-module structure, obtained by the addition of the twisting cochain ∂ to the internal differential of C.

Note that the map $\delta + \partial$ satisfies the equation of a differential $(\delta + \partial)^2 = 0$ if and only if the twisting cochain ∂ verifies the equation $\delta(\partial) + \partial^2 = 0$ in $Hom_{\mathrm{dg\,k\,Mod}}(C, C)$.

As an example, recall that a dg-module D is *quasi-free* if it is free as a graded $\Bbbk$-module as long as we forget differentials. Clearly, this definition implies that a quasi-free dg-module D is equivalent to a twisted dg-module $D = (C, \partial)$ so that $C = \bigoplus_\alpha \Bbbk\, e_{d_\alpha}$ is a free graded module equipped with a trivial differential.

We use twisted dg-modules to determine the structure of cofibrations in dg-modules. Our results follow from the following easy observation:

11.2.2 Observation. *The cell attachments of generating cofibrations in dg-modules*

$$\begin{array}{ccc} \bigoplus_\alpha B^{d_\alpha - 1} & \xrightarrow{\ f\ } & K \\ {\scriptstyle i_{d_\alpha}}\Big\downarrow & & \Big\downarrow{\scriptstyle j} \\ \bigoplus_\alpha E^{d_\alpha} & \dashrightarrow & L \end{array}$$

are equivalent to twisted direct sums $L = (K \oplus E, \partial)$, so that E is a free graded $\Bbbk$-module $E = \bigoplus_\alpha \Bbbk\, e_{d_\alpha}$ (equipped with a trivial differential) and the twisting cochain ∂ is reduced to a component $\partial : E \to K$.

The twisting cochain is determined from the attaching map $f : \bigoplus_\alpha B^{d_\alpha - 1} \to K$ by the relation $\partial(e_{d_\alpha}) = f(b_{d_\alpha - 1})$.

This observation gives immediately:

11.2.3 Proposition. *In dg-modules, the relative cofibrant cell objects are equivalent to composites*

$$K = L_0 \to \cdots \to L_{\lambda-1} \xrightarrow{\ j_\lambda\ } L_\lambda \to \cdots \to \mathop{\mathrm{colim}}_\lambda L_\lambda = L,$$

whose terms L_λ are defined by twisted direct sums of the form of observation 11.2.2

$$L_\lambda = (L_{\lambda-1} \oplus E_\lambda, \partial).$$

As a corollary:

11.2.4 Proposition. *A cofibrant cell dg-module is equivalent to a quasi-free dg-module $L = (\bigoplus_\alpha \Bbbk\, e_{d_\alpha}, \partial)$, where the free graded $\Bbbk$-module $K = \bigoplus_\alpha \Bbbk\, e_{d_\alpha}$ is equipped with a basis filtration $K_\lambda = \bigoplus_{\alpha < \lambda} \Bbbk\, e_{d_\alpha}$ so that $\partial(K_\lambda) \subset K_{\lambda-1}$.*

In the sequel, we use the representation of cofibrations and cofibrant objects in dg-modules yielded by proposition 11.2.3 and proposition 11.2.4. Different characterizations of cofibrations in dg-modules occur in the literature (see for instance [28, §2.3]).

11.2.5 Modules over a dg-Algebra. The categories of modules over an associative dg-algebra give a first example of an application of theorem 11.1.13 and proposition 11.1.14. The model categories of modules over dg-algebras are also used in §13 and in §17.3, where we study the (co)homology of algebras over operads. The result has an easy generalization in the setting of monoidal model categories (whose axioms are recalled in the next section).

Let R be an associative algebra in dg-modules. By definition, a left module over R consists of a dg-module E together with a left R-action defined by a morphism of dg-modules $\lambda : R \otimes E \to E$ that satisfies the standard unit and associativity relations with respect to the unit and the product of R. The definition of a right R-module is symmetric. For these usual module categories, we adopt the standard notation $R\operatorname{Mod}$ and $\operatorname{Mod}R$.

The category of left R-modules comes equipped with an obvious forgetful functor $U : R\operatorname{Mod} \to \operatorname{dg}\Bbbk\operatorname{Mod}$. The forgetful functor has a left adjoint $R\otimes - : \operatorname{dg}\Bbbk\operatorname{Mod} \to R\operatorname{Mod}$ which maps a dg-module C to the tensor product $R \otimes C$ equipped the natural left R-action

$$R \otimes (R \otimes C) = R \otimes R \otimes C \xrightarrow{\mu \otimes C} R \otimes C,$$

in which $\mu : R \otimes R \to R$ refers to the product of R. The forgetful functor creates all limits and all colimits in the category of left R-modules. As usual, we can use the natural morphism $R\otimes\lim_{i\in I}(E_i) \to \lim_{i\in I}(R\otimes E_i)$ to provide any limit of left R-modules E_i with an R-module structure. On the other hand, the existence of a natural isomorphism $R \otimes \operatorname{colim}_{i\in I}(E_i) \xleftarrow{\simeq} \operatorname{colim}_{i\in I}(R\otimes E_i)$ implies that a colimit of left R-modules E_i inherits an R-module structure. The conclusion follows. Symmetric observations hold for the category of right R-modules $\operatorname{Mod}R$.

Recall that the notion of a left (respectively, right) module over an operad comes from a generalization of the definition of a left (respectively, right) module over an associative algebra, where the tensor product is replaced by the composition product of Σ_*-objects. But left modules over operads do not satisfy usual properties of left modules over algebras, because the composition product does not preserves all colimits on the right. (As an example, we observe in §3.3 that the forgetful functor from left modules over operads to Σ_*-objects does not preserve all colimits, unlike the forgetful functor on left modules over algebras.)

The application of proposition 11.1.14 to the adjunction

$$R \otimes - : \operatorname{dg}\Bbbk\operatorname{Mod} \rightleftarrows R\operatorname{Mod} : U$$

returns:

11.2.6 Proposition. *Suppose R is cofibrant as a dg-module.*

The category of left R-modules inherits a cofibrantly generated model structure so that the forgetful functor $U : R\,\mathrm{Mod} \to \mathrm{dg}\,\Bbbk\,\mathrm{Mod}$ creates weak-equivalences and where the generating cofibrations (respectively, acyclic cofibrations) are the morphisms of free left R-modules $R \otimes i : R \otimes C \to R \otimes D$ induced by generating cofibrations (respectively, acyclic cofibrations) of dg-modules.

The forgetful functor $U : R\,\mathrm{Mod} \to \mathrm{dg}\,\Bbbk\,\mathrm{Mod}$ creates fibrations as well and preserves cofibrations.

Symmetric assertions hold for the category of right R-modules.

If R is not cofibrant as a dg-module, then the conditions of proposition 11.1.14 are not satisfied. Nevertheless, the assumptions of theorem 11.1.13 can be checked directly without using the assumption of proposition 11.2.6. Thus the category of left (respectively, right) R-modules has still a cofibrantly generated model structure if R is not cofibrant as a dg-module, but the forgetful functor $U : R\,\mathrm{Mod} \to \mathrm{dg}\,\Bbbk\,\mathrm{Mod}$ does not preserve cofibrations.

Proof. Condition (1) of proposition 11.1.14 is satisfied since the forgetful functor $U : R\,\mathrm{Mod} \to \mathrm{dg}\,\Bbbk\,\mathrm{Mod}$ creates all colimits. Condition (2) is implied by the next observation, because the forgetful functor $U : R\,\mathrm{Mod} \to \mathrm{dg}\,\Bbbk\,\mathrm{Mod}$ creates all colimits (including pushouts) and cofibrations (respectively, acyclic cofibrations) are stable under pushouts in dg-modules. $\square$

11.2.7 Fact. *Let K be any cofibrant dg-module. If i is a cofibration (respectively, an acyclic cofibration) in the category of dg-modules, then so is the tensor product $K \otimes i : K \otimes C \to K \otimes D$.*

This fact holds in the context of monoidal model categories as a consequence of the pushout-product axiom, recalled next. But, in the context of dg-modules, we can also use the explicit form of cofibrations yielded by proposition 11.2.3 to check fact 11.2.7 directly.

11.2.8 Twisting Cochains and Quasi-Free Modules over dg-Algebras. The constructions and results of §§11.2.1-11.2.4 have obvious generalizations in the context of modules over dg-algebras. For a left R-module E, we assume that a twisting cochain $\partial : E \to E$ commutes with the R-action to ensure that the map $\delta + \partial : E \to E$ defines a differential of R-modules. Naturally, we say that a left R-module E is quasi-free if we have $E = (R \otimes C, \partial)$, where $\partial : R \otimes C \to R \otimes C$ is a twisting cochain of left R-modules. In this case, the commutation with the R-action implies that the twisting cochain $\partial : R \otimes C \to R \otimes C$ is determined by a homogeneous map $\partial : C \to R \otimes C$.

The extension of proposition 11.2.9 to modules over dg-algebras reads:

11.2.9 Proposition. *A cofibrant cell object in left R-modules is equivalent to a quasi-free R-module $L = (R \otimes K, \partial)$, where K is a free graded $\Bbbk$-module*

$K = \bigoplus_{\alpha} \Bbbk\, e_{d_\alpha}$ *equipped with a basis filtration* $K_\lambda = \bigoplus_{\alpha < \lambda} \Bbbk\, e_{d_\alpha}$ *so that* $\partial(K_\lambda) \in R \otimes K_{\lambda-1}$.

The cofibrant cell objects in right R-modules have a symmetric represen-tation. □

Again, a cofibrant cell object in left R-modules (respectively, right R-modules) refers by convention to a cell complex built from the generating cofibrations of $R\,\mathrm{Mod}$ (respectively, $\mathrm{Mod}\,R$).

Recall that any morphism of associative dg-algebras $\psi : R \to S$ gives adjoint extension and restriction functors on module categories:

$$\psi_! : R\,\mathrm{Mod} \rightleftarrows S\,\mathrm{Mod} : \psi^*.$$

The left R-module $\psi^* N$ obtained by restriction of structures from a left S-modules N is defined by the dg-module $\psi^* N = N$ underlying N on which the dg-algebra R acts through the morphism $\psi : R \to S$. The extension functor is given by the relative tensor product $\psi_! M = S \otimes_R M$.

The extension and restriction of left (respectively, right) modules over operads are generalizations of this construction.

The extension and restriction functors give examples of Quillen adjunction:

11.2.10 Proposition. *The extension and restriction functors*

$$\psi_! : R\,\mathrm{Mod} \rightleftarrows S\,\mathrm{Mod} : \psi^*$$

form a Quillen pair of adjoint functors, a Quillen equivalence if $\psi : R \to S$ is also a weak-equivalence.

Symmetric results hold for extensions and restrictions of right modules over dg-algebras. □

We sketch only the proof of this standard proposition. We assume again that R and S are cofibrant as dg-modules, but the proposition holds without this assumption.

In §16 we prove an analogous statement in the context of left (respectively, right) modules over operads and we give more detailed arguments in this case. The plan of the proof is the same. The proof of the Quillen equivalence is simply more difficult in the case of left modules over operads, because the forgetful functor does not preserve colimits.

Proof. Fibrations and acyclic fibrations are created by forgetful functors in categories of left module over dg-algebras. Since the restriction functor re-duces to the identity if we forget algebra actions, we obtain immediately that ψ^* preserves fibrations and acyclic fibrations. Hence we conclude readily that the pair $(\psi_!, \psi^*)$ forms a Quillen adjunction.

If the morphism ψ is a weak-equivalence in dg-modules, then the adjunc-tion unit

$$E \xrightarrow{\eta(E)} \psi^* \psi_!(E) = S \otimes_R E$$

defines a weak-equivalence for any cofibrant R-module E. This assertion is immediate for a free R-module $E = R \otimes C$, because we have $S \otimes_R (R \otimes C) = S \otimes C$ and $\eta(R \otimes C)$ is identified with the morphism $\psi \otimes C : R \otimes C \to S \otimes C$. The assertion can easily be generalized to cofibrant cell R-modules (by using an induction on the cell decomposition), because the functors $U : R\,\mathrm{Mod} \to \mathrm{dg}\,\Bbbk\,\mathrm{Mod}$ and $U(\psi^* \psi_! -) : R\,\mathrm{Mod} \to \mathrm{dg}\,\Bbbk\,\mathrm{Mod}$ map cell attachments in R-module to pushouts in dg-modules. Use that any cofibrant R-module is a retract of a cofibrant cell R-module to conclude.

Let F be any left S-module. Let $E \xrightarrow{\sim} \psi^* F$ be a cofibrant replacement of $\psi^* F$ in the category of left R-modules. Use the commutative diagram

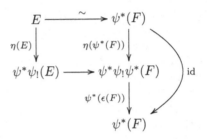

to prove that the composite

$$\psi_!(E) \to \psi_! \psi^*(F) \xrightarrow{\epsilon(F)} F$$

is a weak-equivalence if $\eta(E) : E \to \psi^* \psi_!(E)$ is so (see §16.1.5).

From these verifications, we conclude that the pair $(\psi_!, \psi^*)$ forms a Quillen equivalence if ψ is a weak-equivalence.   □

11.3 Symmetric Monoidal Model Categories over a Base

In the context of operads, we use model categories equipped with a symmetric monoidal structure. With a view to applications in homotopy theory, we have to put appropriate conditions on the tensor product and to introduce a suitable notion of a symmetric monoidal model category. Good axioms are formalized in [28, §§4.1-4.2]. The purpose of this section is to review these axioms in the relative context of symmetric monoidal categories over a base and to specify which axioms are really necessary for our needs.

11.3.1 The Pushout-Product. The *pushout-product* gives a way to assemble tensor products of (acyclic) cofibrations in symmetric monoidal model categories. The definition of the pushout-product makes sense for any bifunctor $T : \mathcal{A} \times \mathcal{B} \to \mathcal{X}$ where $\mathcal{X}$ is a category with colimits. The pushout-product of morphisms $f : A \to B$ and $g : C \to D$ is just the morphism

$$(f_*, g_*) : T(A, D) \bigoplus_{T(A,C)} T(B, C) \to T(B, D)$$

which arises from the commutative diagram

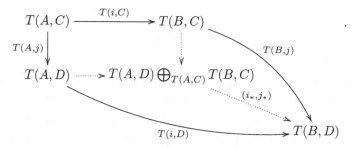

We apply this construction to the tensor product of a symmetric monoidal category, to the internal and external tensor products of a symmetric monoidal category over a base category, and next, to the bifunctors $(M, X) \mapsto S(M, X)$ and $(M, A) \mapsto S_R(M, A)$. We also use a dual construction in the context of enriched categories.

The word "product" in the expression "pushout-product" refers to a tensor product, to which the construction of the pushout-product is usually applied, but we keep using this expression "pushout-product" for any bifunctor $T : \mathcal{A} \times \mathcal{B} \to \mathcal{X}$.

In applications, a main task consists in proving that pushout-products of morphisms (i, j) form (acyclic) cofibrations. The next lemma simplifies the verification of this property:

11.3.2 Lemma. *Let $f : F \to G$ be a natural transformation of functors $F, G : \mathcal{A} \to \mathcal{X}$ where $\mathcal{X}$ is a model category. Let $\mathcal{K}$ be a set of morphisms in $\mathcal{A}$. Suppose that F, G preserves colimits. If the pushout-product*

$$(i_*, f_*) : G(A) \bigoplus_{F(A)} F(B) \to G(B)$$

forms a cofibration (respectively, an acyclic cofibration) for every $i \in \mathcal{K}$, then so does every pushout-product (i_*, f_*) such that i is a retract of a relative $\mathcal{K}$-complex. □

This lemma can be applied to the functors $F(-) = T(-, C)$ and $G(-) = T(-, D)$, and to the natural transformation $f = T(-, j)$, for any morphism $j : C \to D$, provided that $T : \mathcal{A} \times \mathcal{B} \to \mathcal{X}$ is a bifunctor which preserves colimits in the first variable. The argument is classical in the context of symmetric monoidal categories (we refer to [28, §4.2]).

Proof. The lemma follows from an immediate generalization of the arguments of [28, §4.2]. □

11.3.3 Symmetric Monoidal Model Categories. The next axioms are introduced in [28, §4.2] for symmetric monoidal categories equipped with a model structure:

MM0 (*unit axiom*): The unit object 1 is cofibrant in $\mathcal{C}$.
MM1 (*pushout-product axiom*): The natural morphism

$$(i_*, j_*) : A \otimes D \bigoplus_{A \otimes C} B \otimes C \to B \otimes D$$

induced by cofibrations $i : A \rightarrowtail B$ and $j : C \rightarrowtail D$ forms a cofibration, respectively an acyclic cofibration if i or j is also acyclic.

In certain arguments, we use MM0 and MM1. Therefore we take the convention that a *symmetric monoidal model category* is a symmetric monoidal category equipped with a model structure such that both axioms MM0-MM1 hold. The base symmetric monoidal category $\mathcal{C}$ is supposed to satisfy these requirements whenever it is equipped with a model structure.

The model categories of dg-modules, topological spaces, and simplicial sets fit our requirements. ¶ But certain stable model categories of spectra (examined next) do not satisfy the unit axiom MM0. However some of our results can be applied (with care) in the context of spectra. These possible generalizations are deferred to the note apparatus.

Naturally, we say that a symmetric monoidal category $\mathcal{E}$ over a base symmetric monoidal model category $\mathcal{C}$ forms a symmetric monoidal model category over $\mathcal{C}$ if axiom MM1 holds for the internal tensor product of $\mathcal{E}$ and for the external tensor product $\otimes : \mathcal{C} \times \mathcal{E} \to \mathcal{E}$. Again, we assume that axiom MM0 holds in $\mathcal{E}$. These axioms imply that the canonical functor $\eta : \mathcal{C} \to \mathcal{E}$ preserves (acyclic) cofibrations since we have by definition $\eta(C) = C \otimes 1$.

Of course, any symmetric monoidal model category forms a symmetric monoidal model category over itself. In the sequel, we prove that the category Σ_*-objects in a symmetric monoidal model category inherits the structure of a symmetric monoidal category over the base category, and similarly as regards modules over operads.

¶ Standard categories of spectra (for instance the symmetric spectra of [30]) form naturally a symmetric monoidal category over simplicial sets. In applications of operads, categories of spectra are often used as such. The canonical functor from simplicial sets to spectra is identified with the functor $\Sigma^\infty(-)_+$ which maps a simplicial set K to the suspension spectrum of K with a base point added. This functor $\Sigma^\infty(-)_+$ does not preserves cofibrations as axiom MM0 fails in spectra, but the pushout-product axiom MM1 also holds for the external tensor product of spectra over simplicial sets.

11.3.4 Enriched Categories. The pushout-product axiom applied to the external tensor product $\otimes : \mathcal{C} \times \mathcal{E} \to \mathcal{E}$ implies that the adjoint hom-bifunctor $\mathrm{Hom}_\mathcal{E}(-, -) : \mathcal{E}^{op} \times \mathcal{E} \to \mathcal{C}$ satisfies an analogue of axiom SM7 of simplicial model categories. Namely:

MM1'. The morphism

$$(i^*, p_*) : Hom_{\mathcal{E}}(B, X) \to Hom_{\mathcal{E}}(A, X) \times_{Hom_{\mathcal{E}}(A,Y)} Hom_{\mathcal{E}}(B, Y)$$

induced by a cofibration $i : A \rightarrowtail B$ and a fibration $p : X \twoheadrightarrow Y$ forms a fibration in $\mathcal{C}$, an acyclic fibration if i is also an acyclic cofibration or p is also an acyclic fibration.

In fact, we have an equivalence MM1 $\Leftrightarrow$ MM1'. To check this assertion, use that the lifting axioms M4.i-ii characterize cofibrations, acyclic cofibrations, fibrations, and acyclic fibrations and apply the adjunction relation to obtain an equivalence of lifting problems:

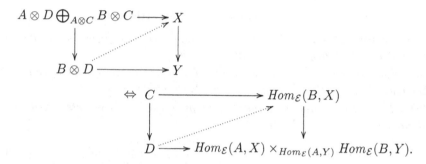

See [28, Lemma 4.2.2] for details.

11.4 The Model Category of Σ_*-Objects

In this book, we use the standard cofibrantly generated model structure of the category of Σ_*-objects in which a morphism $f : M \to N$ is a weak-equivalence (respectively a fibration) if the underlying collection of morphisms $f : M(n) \to N(n)$ consists of weak-equivalences (respectively fibrations) in $\mathcal{C}$. This model structure is an instance of the cofibrantly generated model structures defined in [27, §11.6] for categories of diagrams.

The model structure of Σ_*-objects can be deduced from the discussion of §§11.1.12-11.1.14. Consider the category $\mathcal{C}^{\mathbb{N}}$ formed by collections of objects $\{K(n) \in \mathcal{C}\}_{n \in \mathbb{N}}$ together with the obvious model structure of a product of model categories: a collection of morphisms $\{f : K(n) \to L(n)\}_{n \in \mathbb{N}}$ defines a weak-equivalence (respectively, a cofibration, a fibration) in $\mathcal{C}^{\mathbb{N}}$ if every component $f : K(n) \to L(n)$ forms a weak-equivalence (respectively, a cofibration, a fibration) in $\mathcal{C}$. This model category has a set generating (acyclic) cofibrations inherited componentwise from the base category $\mathcal{C}$. The forgetful functor $U : \mathcal{M} \to \mathcal{C}^{\mathbb{N}}$ has a left-adjoint $F : \mathcal{C}^{\mathbb{N}} \to \mathcal{M}$ which maps a collection K to the Σ_*-object such that $(\Sigma_* \otimes K)(n) = \Sigma_n \otimes K(n)$. The assumptions of proposition 11.1.14 are satisfied and we obtain readily:

Proposition 11.4.A. *The category of Σ_*-objects $\mathcal{M}$ inherits the structure of a cofibrantly generated model category from the base category $\mathcal{C}$ so that the forgetful functor $U : \mathcal{M} \to \mathcal{C}^{\mathbb{N}}$ creates weak-equivalences and fibrations in $\mathcal{M}$. The generating (acyclic) cofibrations of $\mathcal{M}$ are given by tensor products*

$$i \otimes F_r : C \otimes F_r \to D \otimes F_r,$$

where $i : C \to D$ ranges over generating (acyclic) cofibrations of the base category $\mathcal{C}$ and F_r ranges over the generators, defined in §2.1.12, of the category of Σ_*-objects.

If the base model category $\mathcal{C}$ is left (respectively, right) proper, then so is the model category of Σ_*-objects. □

The main purpose of this section is to prove further:

Proposition 11.4.B. *The category of Σ_*-object $\mathcal{M}$ forms a symmetric monoidal model category over $\mathcal{C}$ in our sense:*

(a) *If the unit object $1 \in \mathcal{C}$ is cofibrant in $\mathcal{C}$, then so is the equivalent constant Σ_*-object $1 \in \mathcal{M}$. Hence the unit axiom MM0 holds in $\mathcal{M}$.*

(b) *The internal and external tensor products of Σ_*-objects satisfy the pushout-product axiom MM1.*

¶ **Remark.** Of course, if $\mathcal{C}$ satisfies only the pushout-product axiom MM1, then $\mathcal{M}$ satisfies only the pushout-product axiom MM1 and forms a symmetric monoidal model category in the weak sense.

Before checking propositions 11.4.A-11.4.B, we recall terminologies about enlarged classes of cofibrations in the context of collections and Σ_*-objects. Usually, we say that a morphism in a model category $\mathcal{A}$ is a $\mathcal{B}$-cofibration if its image under an obvious forgetful functor $U : \mathcal{A} \to \mathcal{B}$ forms a cofibration in $\mathcal{A}$ (see §11.1.17). Similarly, we say that an object $X \in \mathcal{A}$ is $\mathcal{B}$-cofibrant if the initial morphism $0 \to X$ forms a $\mathcal{B}$-cofibration.

In the case where $\mathcal{B}$ is the category $\mathcal{M}$ of Σ_*-objects, the usage is to call Σ_*-*cofibrations* the $\mathcal{M}$-cofibrations and Σ_*-*cofibrant* objects the $\mathcal{M}$-cofibrant objects. In the case where $\mathcal{B}$ is the category $\mathcal{C}^{\mathbb{N}}$ of collections of $\mathcal{C}$-objects, we call for simplicity $\mathcal{C}$-*cofibrations* the $\mathcal{C}^{\mathbb{N}}$-cofibrations and $\mathcal{C}$-*cofibrant* objects the $\mathcal{C}^{\mathbb{N}}$-cofibrant objects. Accordingly, a morphism of Σ_*-objects $f : M \to N$ defines a $\mathcal{C}$-cofibration if every morphism $f : M(n) \to N(n)$, $n \in \mathbb{N}$, defines a cofibration in $\mathcal{C}$.

Observe that:

11.4.1 Proposition. *The $\mathcal{C}$-cofibrations in the category of Σ_*-objects satisfy the axioms of an enlarged class of cofibrations.*

Proof. Immediate because colimits in the category of Σ_*-objects are created in $\mathcal{C}^{\mathbb{N}}$, as well as weak-equivalences. Observe further that the generating (acyclic) cofibrations of $\mathcal{M}$ are (acyclic) $\mathcal{C}$-cofibrations to prove that all (acyclic) cofibrations of $\mathcal{M}$ are (acyclic) $\mathcal{C}$-cofibrations. □

The next statements are devoted to the verification of proposition 11.4.B.

11.4.2 Lemma. *The functor $C \mapsto C \otimes F_r$, from C to $\mathcal{M}$, preserve (acyclic) cofibrations.*

Proof. By standard categorical arguments (see [28, §4.2.5]), the assertion in the general case of an (acyclic) cofibration $i : C \to D$ follows from the case of generating (acyclic) cofibrations, asserted by proposition 11.4.A.   □

11.4.3 Lemma. *The pushout-product axiom MM1 holds for the internal tensor product of Σ_*-objects: the natural morphism*

$$(i_*, j_*) : K \otimes N \bigoplus_{K \otimes M} L \otimes M \to L \otimes N$$

induced by cofibrations $i : K \rightarrowtail L$ and $j : M \rightarrowtail N$ of the category of Σ_*-objects forms a cofibration, an acyclic cofibration if i or j is also acyclic.

Proof. By lemma 11.3.2 (or [28, §4.2.5]), we can reduce the verification of the claim to the case of generating (acyclic) cofibrations. Thus we consider (acyclic) cofibrations of the form

$$A \otimes F_r \xrightarrow{i \otimes F_r} B \otimes F_r \quad \text{and} \quad C \otimes F_s \xrightarrow{j \otimes F_s} D \otimes F_s,$$

where $i : A \to B$, respectively $j : C \to D$, is a generating (acyclic) cofibration of C.

Recall that $F_r = I^{\otimes r}$. Accordingly, we have $F_r \otimes F_s \simeq F_{r+s}$. For all $C, D \in \mathcal{C}$, we obtain

$$(C \otimes F_r) \otimes (D \otimes F_s) \simeq (C \otimes D) \otimes (F_r \otimes F_s) \simeq (C \otimes D) \otimes F_{r+s}$$

and, as a byproduct, the pushout of the morphisms

$$(A \otimes F_r) \otimes (C \otimes F_s)$$

$$(A \otimes F_r) \otimes (j \otimes F_s) \swarrow \qquad \searrow (i \otimes F_r) \otimes (C \otimes F_s)$$

$$(A \otimes F_r) \otimes (D \otimes F_s) \qquad\qquad (B \otimes F_r) \otimes (C \otimes F_s)$$

is isomorphic to $(A \otimes D \bigoplus_{A \otimes C} B \otimes C) \otimes F_{r+s}$. Accordingly, to prove our claim, we check the class of the morphism

$$(i_*, j_*) \otimes F_{r+s} : (A \otimes D \bigoplus_{A \otimes C} B \otimes C) \otimes F_{r+s} \to (B \otimes D) \otimes F_{r+s}$$

induced by generating (acyclic) cofibrations $i : A \rightarrowtail B$ and $j : C \rightarrowtail D$ of C. At this stage, we can use axiom MM1 in the base category and lemma 11.4.2 to conclude.   □

11.4.4 Lemma. *The canonical functor $\eta : C \to M$ which associates a constant Σ_*-object to any $C \in C$ preserve cofibrations and acyclic cofibrations.*

In particular, if the monoid unit 1 forms a cofibrant object in C, then so does the equivalent constant Σ_*-object in M.

Proof. This lemma is an immediate corollary of lemma 11.4.2 since we have identities $\eta C = C \otimes 1 = C \otimes F_0$.   □

11.4.5 Lemma. *The cofibration axiom MM1 holds for the external tensor product of Σ_*-objects: the natural morphism*

$$(i_*, j_*) : C \otimes N \bigoplus_{C \otimes M} D \otimes N \to D \otimes N$$

induced by a cofibration $i : C \rightarrowtail D$ in C and a cofibration $j : M \rightarrowtail N$ in M forms a cofibration in M, and an acyclic cofibration if i or j is also acyclic.

Proof. This lemma is an immediate consequence of lemma 11.4.3 and lemma 11.4.4 since we have the identity $C \otimes M = \eta(C) \otimes M$, which connects the external tensor product to the internal tensor product in any symmetric monoidal category over C.   □

This lemma achieves the proof of proposition 11.4.B.   □

The next assertions are used repeatedly in §14:

11.4.6 Lemma.

(a) If M is a cofibrant object in M, then so is $M^{\otimes r}$, for all $r > 0$.

(b) If M is C-cofibrant, then so is $M^{\otimes r}$, for all $r > 0$.

Proof. Assertion (a) is a direct consequence of the axioms of a symmetric monoidal model category for Σ_*-objects.

By §2.1.8, we have

$$M^{\otimes r}(n) = \bigoplus_{n_1 + \cdots + n_r = n} \Sigma_n \otimes_{\Sigma_{n_1} \times \cdots \times \Sigma_{n_r}} M(n_1) \otimes \cdots \otimes M(n_r)$$

Accordingly, the object $M^{\otimes r}(n)$ splits as a sum in the category C

$$M^{\otimes r}(n) \simeq \bigoplus_{n_1 + \cdots + n_r = n} \Sigma_n / \Sigma_{n_1} \times \cdots \times \Sigma_{n_r} \otimes M(n_1) \otimes \cdots \otimes M(n_r)$$

$$\simeq \bigoplus_{\substack{n_1 + \cdots + n_r = n \\ w \in \Sigma_n / \Sigma_{n_1} \times \cdots \times \Sigma_{n_r}}} M(n_1) \otimes \cdots \otimes M(n_r).$$

Assertion (b) follows immediately.   □

11.5 The Pushout-Product Property for Symmetric Tensors

In this section, we study the homotopy invariance of the bifunctor $(M, X) \to$ $S(M, X)$, where M ranges over the category of Σ_*-objects $\mathcal{M}$ and X ranges over any symmetric monoidal category $\mathcal{E}$ over the base category $\mathcal{C}$. The result, stated in proposition 11.5.3, is classical. But, for our needs, we prove an analogue of the pushout-product axiom of symmetric monoidal category, from which the usual homotopy invariance properties follow. Namely:

11.5.1 Lemma. *Suppose* $i : M \rightarrowtail N$ *is a cofibration in* $\mathcal{M}$ *and* $j : X \rightarrowtail Y$ *is a cofibration in* $\mathcal{E}$. *Suppose that the object* X *is also cofibrant in* $\mathcal{E}$. *Then the pushout-product*

$$(i_*, j_*) : S(M, Y) \bigoplus_{S(M,X)} S(N, X) \to S(N, Y)$$

forms a cofibration. If i *or* j *is also acyclic, then the pushout-product* (i_*, j_*) *forms an acyclic cofibration.*

Note the additional assumption about the object $X \in \mathcal{E}$. This statement is also proved in [22] in the case $\mathcal{E} = \mathcal{M}$ (for the composition product of Σ_*-objects).

Proof. Since the functor $(M, X) \mapsto S(M, X)$ preserves colimits in M, we can apply lemma 11.3.2 to reduce the claim to the case where $i : M \to N$ is a generating (acyclic) cofibration:

$$u \otimes F_r : C \otimes F_r \to D \otimes F_r.$$

For modules of the form $M = C \otimes F_r$, we have $S(M, X) = S(C \otimes F_r, X) = C \otimes X^{\otimes r}$. As a byproduct, we obtain that the morphism

$$(i_*, j_*) : S(C \otimes F_r, Y) \bigoplus_{S(C \otimes F_r, X)} S(D \otimes F_r, X) \to S(D \otimes F_r, Y)$$

is identified with the pushout-product

$$(u_*, j_*^{\otimes r}) : C \otimes Y^{\otimes r} \bigoplus_{C \otimes X^{\otimes r}} D \otimes X^{\otimes r} \to D \otimes Y^{\otimes r}$$

of $u : C \to D$ and $j^{\otimes r} : X^{\otimes r} \to Y^{\otimes r}$. Hence our claim reduces to an immediate consequence of axioms of symmetric monoidal model categories for the category $\mathcal{E}$. Note that $j^{\otimes r} : X^{\otimes r} \to Y^{\otimes r}$ forms a cofibration because X is supposed to be cofibrant. □

Lemma 11.5.1 gives as a corollary:

11.5.2 Lemma.

(a) If M is a cofibrant object in $\mathcal{M}$, then the functor $X \mapsto S(M, X)$ preserves cofibrations and acyclic cofibrations with a cofibrant domain.

(b) If X is a cofibrant object in $\mathcal{E}$, then the functor $M \mapsto S(M, X)$ preserves cofibrations and acyclic cofibrations.   □

And the classical Brown lemma implies:

11.5.3 Proposition.

(a) The morphism

$$S(M, f) : S(M, X) \to S(M, Y)$$

induced by a weak-equivalence $f : X \xrightarrow{\sim} Y$ between cofibrant objects $X, Y \in \mathcal{E}$ is a weak-equivalence as long as $M \in \mathcal{M}$ is cofibrant.

(b) The morphism

$$S(f, X) : S(M, X) \to S(N, X)$$

induced by a weak-equivalence $f : M \xrightarrow{\sim} N$ between cofibrant objects $M, N \in \mathcal{M}$ is a weak-equivalence as long as $X \in \mathcal{E}$ is cofibrant.   □

This proposition can also be deduced from the pointwise adjunction of §2.3:

$$\mathrm{Mor}_{\mathcal{E}}(S(M, X), Y) \simeq \mathrm{Mor}_{\mathcal{M}}(M, End_{X,Y}).$$

Assertion (b) of proposition 11.5.3 has a converse:

11.5.4 Proposition. *Assume $\mathcal{E} = \mathcal{C}$.*

Let $f : M \to N$ be a morphism between cofibrant objects $M, N \in \mathcal{M}$. If the morphism $S(f, X) : S(M, X) \to S(N, X)$ induced by f forms a weak-equivalence in $\mathcal{C}$ for every cofibrant object $X \in \mathcal{C}$, then f is a weak-equivalence as well.

If $\mathcal{C}$ is pointed, then this implication holds without the assumption that $M, N \in \mathcal{M}$ are cofibrant.

Recall that a category $\mathcal{C}$ is pointed if its initial object $0 \in \mathcal{C}$ is also a final object (and hence defines a null object).

Proof. In general in a model category $\mathcal{A}$, a sum of morphisms $f \oplus g : A_0 \oplus B_0 \to A_1 \oplus B_1$, between cofibrant objects A_0, A_1, B_0, B_1 defines a weak-equivalence if and only if f and g are both weak-equivalences. To check this claim, use simply that the morphism induced by $f \oplus g$ on homotopy classes is identified with the cartesian product

$$f^* \times g^* : [A_1, X] \times [B_1, X] \to [A_0, X] \times [B_0, X]$$

as long as X is a fibrant object of $\mathcal{C}$.

In the context of a pointed model category, the existence of a null object implies that the canonical morphisms of a sum $A \xrightarrow{i} A \oplus B \xleftarrow{j} B$ admit natural retractions $A \xleftarrow{r} A \oplus B \xrightarrow{s} B$. As a corollary, we obtain that f and g are retracts of $f \oplus g$. Hence, in the context of a pointed model category, our observation holds without the assumption that A_0, A_1, B_0, B_1 are cofibrant.

In light of these observations, the proposition is a consequence of the functorial splitting of proposition 2.3.9 for objects of the form $X = 1^{\oplus r}$.   □

11.6 ¶ Symmetric Monoidal Categories with Regular Tensor Powers

For certain (possibly reduced) categories $\mathcal{E}$, the assertions of proposition 11.5.4 hold not only for cofibrant Σ_*-objects, but for $\mathcal{C}$-cofibrant Σ_*-objects as well.

The purpose of this section is to make explicit a sufficient condition, the *axiom of regular tensor powers*, which implies this property. To simplify we assume $\mathcal{E} = \mathcal{C}$. The axiom of regular tensor powers is used to improve applications of lemma 11.5.1. These refinements are addressed in remarks and proofs are usually omitted.

11.6.1 Iterated Pushout-Products. To state the axiom, we use iterated pushout-products of morphisms. These higher pushout-products are also used in §18.2 to study pushout-products of functors on algebras over operads. For the moment, we give only the definition of the $n$fold pushout-product of a morphism $f : X \to Y$.

Let $T_0 = X, T_1 = Y$. The tensor products $T_{\epsilon_1} \otimes \cdots \otimes T_{\epsilon_n}$ and the morphisms

$$T_{\epsilon_1} \otimes \cdots \otimes T_0 \otimes \cdots \otimes T_{\epsilon_n} \xrightarrow{T_{\epsilon_1} \otimes \cdots \otimes i \otimes \cdots \otimes T_{\epsilon_n}} T_{\epsilon_1} \otimes \cdots \otimes T_1 \otimes \cdots \otimes T_{\epsilon_n}$$

are vertices and edges of an n-dimensional cubical diagram. The tensor power $T_n(Y/X) = Y^{\otimes n} = T_1 \otimes \cdots \otimes T_1$ is put at the terminal vertex of the cube. The n-fold pushout-product of f is the canonical morphism $\lambda(f) : L_n(Y/X) \to T_n(Y/X)$ associated to the colimit

$$L_n(Y/X) = \operatorname*{colim}_{(\epsilon_1,\ldots,\epsilon_n)<(1,\ldots,1)} T_{\epsilon_1} \otimes \cdots \otimes T_{\epsilon_n}.$$

11.6.2 The Axiom of Regular Tensor Powers. The symmetric group Σ_n acts on $L_n(Y/X)$ and $T_n(Y/X)$. Moreover the n-fold pushout product

$$\lambda(f) : L_n(Y/X) \to T_n(Y/X)$$

is a Σ_n-equivariant morphism. For a Σ_*-object M, we set

$$T_n M(Y/X) = M(n) \otimes_{\Sigma_n} T_n(Y/X) = M(n) \otimes_{\Sigma_n} Y^{\otimes n},$$

$$L_n M(Y/X) = M(n) \otimes_{\Sigma_n} L_n(Y/X)$$

and we consider the morphism

$$\lambda M(f) : L_n M(Y/X) \to T_n M(Y/X)$$

induced by $\lambda(f) : L_n(Y/X) \to T_n(Y/X)$.

We say that $\mathcal{E}$ has *regular tensor powers* with respect to an enlarged class of $\mathcal{B}$-cofibrations if the following axiom holds:

R1. The pushout-product

$$(i_*, \lambda M(f)_*) : T_n M(Y/X) \bigoplus_{L_n M(Y/X)} L_n N(Y/X) \to T_n N(Y/X)$$

forms a $\mathcal{B}$-cofibration if i is a $\mathcal{B}$-cofibration and j is a $\mathcal{B}$-cofibration, an acyclic $\mathcal{B}$-cofibration if i or j is also acyclic.

Of course, a morphism of Σ_*-objects $i : M \to N$ is a $\mathcal{B}$-cofibration if its components $i : M(n) \to N(n)$ are $\mathcal{B}$-cofibrations.

The axiom of regular tensor powers implies:

11.6.3 Lemma. *Let $i : M \to N$ be a morphism of Σ_*-objects in $\mathcal{E}$. Let $j : X \to Y$ be a morphism in $\mathcal{E}$.*

If $\mathcal{E}$ has regular tensor powers with respect to an enlarged class of $\mathcal{B}$-cofibrations, then the pushout-product

$$(i_*, j_*) : S(M, Y) \bigoplus_{S(M,X)} S(N, X) \to S(N, Y)$$

forms a $\mathcal{B}$-cofibration whenever i is a $\mathcal{B}$-cofibration and j is a $\mathcal{B}$-cofibration with a $\mathcal{B}$-cofibrant domain, an acyclic $\mathcal{B}$-cofibration if i or j is also acyclic.

Proof (sketch). We give only the idea of the proof of the implication "Axiom R1 $\Rightarrow$ Lemma 11.6.3". An analogous implication is proved in §19 for the bifunctor $(M, A) \mapsto S_R(M, A)$ associated to an operad and the arguments are similar, but less complicated in the context of lemma 11.6.3.

The idea is to extend the pushout-product property to the shifted functors $M[X] = S[M, X]$ defined in §4.1.4. The morphism $\lambda M(f) = L_n M(Y/X) \to T_n M(Y/X)$ has an obvious generalization in the context of shifted functors since the shifted functor $M[X] = S[M, X]$ is nothing but the functor $S(M[\cdot], X)$ associated to a shifted Σ_*-object $M[\cdot]$.

For a $\mathcal{B}$-cofibrant object X, the morphism

$$S[i, X] : S[M, X] \to S[N, X]$$

forms a $\mathcal{B}$-cofibration (respectively, an acyclic $\mathcal{B}$-cofibration) in Σ_*-objects if i is so. This assertion is an immediate consequence of axiom R1 since

$L_n M[X/0] = 0$ and $S[M, X] = \bigoplus_{n=0}^{\infty} T_n M[X/0]$ for the initial morphism $0 \to X$.

For a Σ_*-object M and a morphism $j : X \to Y$, observe that the morphism $S[M, j] : S[M, X] \to S[M, Y]$ has a natural decomposition

$$S[M, X] = S[M, Y]_0 \to \ldots$$

$$\cdots \to S[M, Y]_{n-1} \xrightarrow{j_n} S[M, Y]_n \to \ldots$$

$$\cdots \to \operatorname*{colim}_{n} S[M, Y]_n = S[M, Y]$$

so that each morphism fits a pushout

$$
\begin{array}{ccc}
S[L_n M[Y/X], X] & \longrightarrow & S[M, Y]_{n-1} \\
{\scriptstyle S[\lambda M[j], X]} \downarrow & & \downarrow {\scriptstyle j_n} \\
S[T_n M[Y/X], X] & \longrightarrow & S[M, Y]_n
\end{array}
$$

(take $R = I$ in the construction of §18.2).

Study the pushout-products

$$(i_*, j_{n*}) : S[M, Y]_n \bigoplus_{S[M,Y]_{n-1}} S[N, Y]_{n-1} \to S[N, Y]_n$$

and use patching techniques to prove that the pushout-product

$$(i_*, j_*) : S[M, Y] \bigoplus_{S[M,X]} S[N, X] \to S[N, Y]$$

forms a $\mathcal{B}$-cofibration (respectively, an acyclic $\mathcal{B}$-cofibration) under the assumption of the lemma (see the arguments of §19.2).   □

The improved pushout-product lemma implies:

11.6.4 Proposition. *Assume $\mathcal{E}$ has regular tensor powers with respect to a class of $\mathcal{B}$-cofibrations.*

(a) The morphism

$$S(M, f) : S(M, X) \to S(M, Y)$$

induced by a weak-equivalence $f : X \xrightarrow{\sim} Y$ between $\mathcal{B}$-cofibrant objects $X, Y \in \mathcal{E}$ is a weak-equivalence as long as $M \in \mathcal{M}$ is $\mathcal{B}$-cofibrant.

(b) The morphism

$$S(f, X) : S(M, X) \to S(N, X)$$

induced by a weak-equivalence $f : M \xrightarrow{\sim} N$ is a weak-equivalence as long as $X \in \mathcal{E}$ is $\mathcal{B}$-cofibrant and the Σ_*-objects M, N are $\mathcal{B}$-cofibrant.

Proof. Similar to proposition 11.5.3.   □

In usual cases, we use the next observation to prove that a category $\mathcal{E}$ has regular tensor powers:

11.6.5 Observation. *Suppose the class of $\mathcal{B}$-cofibrations is created by a forgetful functor $U : \mathcal{E} \to \mathcal{B}$ which preserves colimits, cofibrations, and weak-equivalences.*

The category $\mathcal{E}$ has regular tensor powers with respect to $\mathcal{B}$-cofibrations if, for every $n \in \mathbb{N}$, the nfold pushout-product $\lambda(i) : L_n(Y/X) \to T_n(Y/X)$ defines a cofibration (respectively, an acyclic cofibration) in the projective model category of Σ_n-objects in $\mathcal{B}$ whenever i is a $\mathcal{B}$-cofibration (respectively, an acyclic $\mathcal{B}$-cofibration) in $\mathcal{E}$.

The examples of reduced categories with regular tensor powers introduced in this book arise from:

11.6.6 Proposition. *The category of connected Σ_*-objects $\mathcal{M}^0$ has regular tensor powers with respect to $\mathcal{C}$-cofibrations.*

Proof. Let $M \in \mathcal{M}^0$. In [14, §1.3.7-1.3.9] we observe that the object $M^{\otimes r}(n)$ expands as a sum of tensor products $M^{\otimes r}(n) = \bigoplus_{(I_1,\dots,I_r)} M(I_1) \otimes \cdots \otimes M(I_r)$ where $(I_1,\dots,I_r)$ ranges over partitions $I_1 \amalg \cdots \amalg I_r = \{1,\dots,n\}$ so that $I_k \neq \emptyset$, for every $k \in \{1,\dots,r\}$. The symmetric group Σ_r operates on $M^{\otimes r}(n)$ by permuting the factors of the tensor product $N(I_1) \otimes \cdots \otimes N(I_r)$ and the summands of the expansion. The assumption $I_k \neq \emptyset$ implies that Σ_r acts freely on partitions $(I_1,\dots,I_r)$. As a corollary, we have an isomorphism of Σ_r-objects

$$M^{\otimes r}(n) \simeq \Sigma_r \otimes \left\{ \bigoplus_{(I_1,\dots,I_r)'} M(I_1) \otimes \cdots \otimes M(I_r) \right\}$$

where the sum ranges of representative of the coset of partitions $(I_1,\dots,I_r)$ under the action of Σ_r (see *loc. cit.* for details).

From this assertion, we deduce readily that the category of connected Σ_*-objects $\mathcal{M}^0$ satisfies the assumption of observation 11.6.5. Hence we conclude that $\mathcal{M}^0$ has regular tensor powers with respect to $\mathcal{C}$-cofibrations.   □

11.6.7 Remark: The Example of Simplicial Modules. Proposition 11.5.3 can also be improved in the context of simplicial modules*.

According to [9], the natural extension of any functor $F : \Bbbk \operatorname{Mod} \to \Bbbk \operatorname{Mod}$ to the category of simplicial $\Bbbk$-modules maps a weak-equivalence between cofibrant objects to a weak-equivalence. From this result, it is easy to deduce that assertion (a) of proposition 11.5.3 holds for every Σ_*-object in simplicial $\Bbbk$-modules M and not only for cofibrant Σ_*-objects.

---

* Observe however that the category of simplicial sets does not satisfies the axiom of regular tensor powers.

11.6.8 The Example of Model Categories of Spectra. Models of stable homotopy (S-*modules* [10], *symmetric spectra* [30], *orthogonal spectra* [48]) give examples of symmetric monoidal model categories for which proposition 11.5.3 can be improved (see for instance lemma 15.5 in [42]). Unfortunately, this refinement supposes to deal with a symmetric monoidal model structure such that the monoidal unit is not a cofibrant object. Nevertheless, some of our results can partly be applied to spectra and we address these refinements in remarks. Proofs are always omitted.

In our applications, we use the *positive stable flat model category of symmetric spectra*. The stable flat cofibrations of symmetric spectra are introduced in [30] (where they are called S-cofibrations). The axioms of the stable flat model category are verified in [55]. The reader is referred to these articles and to the book [59] for the background of symmetric spectra. In the sequel, we also refer to [23, 24] for applications of the stable flat model category of symmetric spectra in the context of operads.

We adopt the usual notation Sp^Σ to refer to the category of symmetric spectra. We have:

11.6.9 Proposition. *Let $i : M \to N$ be a morphism of Σ_*-object in symmetric spectra. Let $f : X \to Y$ be a morphism of symmetric spectra. If the morphisms $i : M(n) \to N(n)$ define positive flat cofibrations in Sp^Σ, the morphism $f : X \to Y$ is a positive flat cofibration and X is cofibrant with respect to the positive flat model structure, then the pushout-product*

$$(i_*, \lambda M(f)_*) : T_n M(Y/X) \bigoplus_{L_n M(Y/X)} L_n N(Y/X) \to T_n N(Y/X)$$

forms a positive flat cofibration as well. If i is also a positive stable weak-equivalence, then so is $(i_*, \lambda M(f)_*)$. □

The arguments of proposition 11.6.6 can easily be adapted to obtain this proposition (flat cofibrations are related to the $\mathcal{C}$-cofibrations of Σ_*-objects). This proposition is stated for the needs of remarks and we omit to give more than this idea of the proof.

We only have one half of the axiom of regular tensor powers. But this partial result is sufficient for applications.

First, proposition 11.6.9 is sufficient to obtain the homotopy invariance of the bifunctor $(M, E) \mapsto S(M, E)$ on the left-hand side:

11.6.10 Proposition. *Let $f : M \xrightarrow{\sim} N$ be a positive stable weak-equivalence between Σ_*-objects in spectra such that every $M(n)$ (respectively, $N(n)$) forms a cofibrant object in Sp^Σ with respect to the positive flat model structure. Let $X \in \mathrm{Sp}^\Sigma$. Suppose X is also cofibrant with respect to the positive flat model structure. Then the morphism*

$$S(f, X) : S(M, X) \to S(N, X)$$

forms a positive stable equivalence in Sp^Σ.

Proof. Adapt the arguments of proposition 11.6.4.   □

But, our results are also sufficient to obtain the homotopy invariance of the bifunctor $(M, X) \mapsto S(M, X)$ on the right-hand side:

11.6.11 Proposition. *Let M be a Σ_*-object in spectra such that every $M(n)$ forms a cofibrant object in Sp^Σ with respect to the positive flat model structure. Let $f : X \xrightarrow{\sim} Y$ be a positive stable equivalence of spectra such that X, Y are cofibrant with respect to the positive flat model structure. The morphism*

$$S(M, f) : S(M, X) \to S(M, Y)$$

forms a positive stable equivalence.

Proof. Pick a cofibrant replacement $0 \rightarrowtail P \xrightarrow{\sim} M$ in the model category of Σ_*-objects. We have a commutative diagram

$$
\begin{array}{ccc}
S(P, X) & \xrightarrow{S(P,f)} & S(P, Y) \\
{\scriptstyle\sim}\downarrow & & \downarrow{\scriptstyle\sim} \\
S(M, X) & \xrightarrow[S(M,f)]{} & S(M, Y).
\end{array}
$$

in which vertical morphisms are stable equivalences by proposition 11.6.10. The morphism $S(P, f)$ forms a stable equivalence as well by proposition 11.5.3. The conclusion follows from the two-out-of-three axiom.   □

11.6.12 Symmetric Spectra as a Symmetric Monoidal Category over Simplicial Sets. The category of symmetric spectra forms naturally a symmetric monoidal category over simplicial sets and is often used as such in applications of the theory of operads. The canonical functor from simplicial sets to symmetric spectra is identified with a functor $\Sigma^\infty(-)_+ : \mathcal{S} \to \mathrm{Sp}^\Sigma$ which maps a simplicial set K to the suspension spectrum of K with a base point added. The external tensor product $\otimes : \mathcal{S} \times \mathrm{Sp}^\Sigma \to \mathrm{Sp}^\Sigma$ satisfies the pushout-product axiom MM1, but the functor $\Sigma^\infty(-)_+ : \mathcal{S} \to \mathrm{Sp}^\Sigma$ does not preserves cofibrations if Sp^Σ is equipped with the positive stable flat model structure. The proof of proposition 11.6.10 and 11.6.11 can be adapted to Σ_*-objects in simplicial sets (which are no more cofibrant in Sp^Σ). In particular, we obtain that the symmetric functors $X \mapsto (X^{\wedge n})_{\Sigma_n}$ preserves weak-equivalences (see lemma 15.5 in [10]).

Chapter 12
The Homotopy of Algebras over Operads

Introduction

In this chapter we apply the adjoint construction of model structures to the category of operads and to categories algebras over operads. For this purpose, we use the adjunction $F : \mathcal{M} \rightleftarrows \mathcal{O} : U$ between operads and Σ_*-objects, respectively the adjunction $P(-) : \mathcal{E} \rightleftarrows {}_P\mathcal{E} : U$ between P-algebras and their underlying category $\mathcal{E}$.

The construction of these adjoint model structures is studied in [26] in the dg-context and in [4, 58] in a more general setting. The difficulty is to check condition (2) in proposition 11.1.14. Indeed, in many usual cases, this condition is not satisfied unless we restrict ourself to cellular objects. For this reason, we have to use *semi-model categories*, structures introduced in [29] to enlarge the applications of theorem 11.1.13. The rough idea is to restrict the lifting and factorization axioms of model categories to morphisms with a cofibrant domain. By [58], the category of operads inherits such a semi-model structure, and so do the categories of algebras over a Σ_*-cofibrant operads.

The main purpose of this chapter is to review the definition of these semi-model categories. First of all, in §12.1, we recall the definition of a semi-model category, borrowed from [29], and we review the construction of adjoint model structures in this setting. In §12.2, we survey briefly the definition of semi-model structures on categories of operads. In §12.3, we address the definition of semi-model structures on categories of algebras over operads.

In this book, the semi-model category of operads only occurs in examples of applications of the main results. For this reason, we only sketch the proof of the axioms for this semi-model category. But we give comprehensive proofs of the axioms of semi-model categories for categories of algebras over operads, because we use this structure in the next part. The main verification is a particular case of statements about modules over operads used in the next part of the book and deferred to an appendix.

B. Fresse, *Modules over Operads and Functors*, Lecture Notes in Mathematics 1967, 185
DOI: 10.1007/978-3-540-89056-0_12, © Springer-Verlag Berlin Heidelberg 2009

In §12.4, we study the semi-model categories of algebras over a cofibrant operad, for which the lifting and factorization axioms of semi-model categories hold in wider situations. In §12.5, we survey results about the homotopy of extension and restriction functors $\phi_! : {}_P\mathcal{E} \rightleftarrows {}_Q\mathcal{E} : \phi^*$ associated to an operad morphism $\phi : P \to Q$.

The statements of §§12.4-12.5 are proved in the next part of the book as applications of our results on the homotopy of modules over operads.

12.1 Semi-Model Categories

The rough idea of semi-model categories is to assume all axioms of model categories, including the lifting axiom M4 and the factorization axiom M5, but only for morphisms $f : X \to Y$ whose domain X is a cofibrant object. This restriction allows us to relax condition (2) of proposition 11.1.14 in the definition of model categories by adjunction and to enlarge the applications of this construction.

12.1.1 The Axioms of Semi-Model Categories. Explicitly, the structure of a *semi-model category* consists of a category $\mathcal{A}$ equipped with classes of weak-equivalences, cofibrations and fibrations so that axioms M1, M2, M3 of model categories hold, but where the lifting axiom M4 and the factorization axiom M5 are replaced by the weaker requirements:

M4'. i. The fibrations have the right lifting property with respect to the acyclic cofibrations $i : A \to B$ whose domain A is cofibrant.
ii. The acyclic fibrations have the right lifting property with respect to the cofibrations $i : A \to B$ whose domain A is cofibrant.

M5'. i. Any morphism $f : A \to B$ such that A is cofibrant has a factorization $f = pi$, where i is a cofibration and p is an acyclic fibration.
ii. Any morphism $f : A \to B$ such that A is cofibrant has a factorization $f = qj$, where j is an acyclic cofibration and q is a fibration.

Besides, a semi-model category is assumed to satisfy:

M0' (*initial object axiom*): The initial object of $\mathcal{A}$ is cofibrant.

In the context of semi-model categories, the lifting axiom M4' and the factorization axiom M5' are not sufficient to imply that the initial object is cofibrant. Therefore we add this assertion as an axiom.

In a semi-model category, the class of (acyclic) cofibrations is not fully characterized by the left lifting axioms M4', and similarly as regards the class of (acyclic) fibrations. As a byproduct, the class of (acyclic) cofibrations is not stable under the composition of morphisms, and similarly as regards the class of acyclic fibrations. The axioms imply only that a (possibly transfinite) composite of (acyclic) cofibrations with a cofibrant domain forms still an (acyclic) cofibration.

Similarly, not all (acyclic) cofibrations are stable under pushouts, not all (acyclic) fibrations are stable under pullbacks. The axioms imply only that (acyclic) cofibrations are stable under pushouts over cofibrant domains.

On the other hand, since usual semi-model categories are defined by adjunction from a cofibrantly generated model category (see next), the class of (acyclic) fibrations is stable under composites and pullbacks in applications. Besides, these properties are used to generalize the construction of the homotopy category of model categories. For these reasons, the next assertions are taken as additional axioms of semi-model categories:

M6' (*fibration axioms*):

i. The class of (acyclic) fibrations is stable under (possibly transfinite) composites.
ii. The class of (acyclic) fibrations is stable under pullbacks.

But we do not use these properties in this book.

The result of proposition 11.1.4 can be generalized in the context of semi-model categories:

12.1.2 Proposition. *The following assertion holds in every semi-model category $\mathcal{A}$:*

P1'. The pushout of a weak-equivalence along a cofibration

gives a weak-equivalence $C \xrightarrow{\sim} D$ provided that A and B are cofibrant in $\mathcal{A}$.

Proof. Careful inspection of the proof of proposition 11.1.4 in [27, Proposition 13.1.2]. □

The properness axiom P1 does not make sense in the context of semi-model categories because only cofibrations with a cofibrant domain are characterized by the axioms.

12.1.3 Cofibrantly Generated Semi-Model Categories. The notion of a cofibrantly generated model category has a natural generalization in the context of semi-model categories. Again, a cofibrantly generated semi-model category consists of a semi-model category $\mathcal{A}$ equipped with a set of generating cofibrations $\mathcal{I}$, respectively a set of generating acyclic cofibrations $\mathcal{J}$, so that:

G1. The fibrations are characterized by the right lifting property with respect to acyclic generating cofibrations $j \in \mathcal{J}$.

G2. The acyclic fibrations are characterized by the right lifting property with respect to generating cofibrations $i \in \mathcal{I}$.

The small object argument is also supposed to hold for the set of generating cofibrations $\mathcal{I}$ (respectively, generating acyclic cofibrations $\mathcal{J}$) but we can relax the smallness assumption. Namely, we may assume:

S1'. The domain A of every generating cofibration (respectively, generating acyclic cofibration) is small with respect to relative $\mathcal{I}$-cell (respectively, $\mathcal{J}$-cell) complexes

$$K = L_0 \to \cdots \to L_{\lambda-1} \xrightarrow{j_\lambda} L_\lambda \to \cdots \to \operatorname*{colim}_{\lambda < \mu} L_\lambda = L$$

such that K is a cofibrant object.

In a cofibrantly generated semi-model category, the axioms imply only that:

K1'. The relative $\mathcal{I}$-cell (respectively, $\mathcal{J}$-cell) complexes with a cofibrant domain are cofibrations (respectively, acyclic cofibrations).
K2'. The cofibrations (respectively, acyclic cofibrations) with a cofibrant domain are retracts of relative $\mathcal{I}$-cell (respectively, $\mathcal{J}$-cell) complexes.

Our motivation to use semi-model categories comes from the following proposition which weaken the conditions of proposition 11.1.14 to define semi-model structures by adjunction:

12.1.4 Theorem. *Suppose we have an adjunction $F : \mathcal{X} \rightleftarrows \mathcal{A} : U$, where $\mathcal{A}$ is any category with limits and colimits and $\mathcal{X}$ is a cofibrantly generated model category. Let $\mathcal{I}$, respectively $\mathcal{J}$, be the set of generating (acyclic) cofibrations of $\mathcal{X}$ and set $F\mathcal{I} = \{F(i), i \in \mathcal{I}\}$, respectively $F\mathcal{J} = \{F(j), j \in \mathcal{J}\}$. Consider also the set $F\mathcal{X}_c = \{F(i), i \text{ cofibration in } \mathcal{X}\}$.*

Under assumptions (1-3) below, the category $\mathcal{A}$ inherits a cofibrantly generated semi-model structure with $F\mathcal{I}$ (respectively, $F\mathcal{J}$) as generating (acyclic) cofibrations and so that the functor $U : A \to \mathcal{X}$ creates weak-equivalences.

(1) The functor $U : \mathcal{A} \to \mathcal{X}$ preserves colimits over non-empty ordinals.
(2) For any pushout

$$
\begin{array}{ccc}
F(K) & \longrightarrow & A \\
{\scriptstyle F(i)} \downarrow & & \downarrow {\scriptstyle f} \\
F(L) & \dashrightarrow & B
\end{array}
$$

such that A is an $\mathcal{X}$-cofibrant $F\mathcal{X}_c$-cell complex, the morphism $U(f)$ forms a cofibration (respectively an acyclic cofibration) in $\mathcal{X}$ whenever i is a cofibration (respectively an acyclic cofibration) with a cofibrant domain.
(3) The object $UF(0)$ is cofibrant.

The functor $U : \mathcal{A} \to \mathcal{X}$ creates the class of fibrations too and preserves cofibrations with a cofibrant domain.

In accordance with the conventions of §11.1.17, we say that an object $A \in \mathcal{A}$ is $\mathcal{X}$-cofibrant if the functor $U : \mathcal{A} \to \mathcal{X}$ maps the initial morphism $F(0) \to A$ to a cofibration.

Proof. This theorem follows from a careful inspection of the arguments of theorem 11.1.13 and proposition 11.1.14. Use the next lemma to apply condition (2) in the case where the domain of generating (acyclic) cofibrations is not cofibrant.   □

12.1.5 Lemma. *Suppose we have a pushout*

$$
\begin{array}{ccc}
F(K) & \longrightarrow & A \\
{\scriptstyle F(i)}\downarrow & & \downarrow{\scriptstyle f} \\
F(L) & \dashrightarrow & B
\end{array}
$$

such that $U(A)$ is cofibrant and $i : K \to L$ is a cofibration (respectively, an acyclic cofibration), but where K is not necessarily cofibrant. Then we can form a new pushout

$$
\begin{array}{ccc}
F(M) & \longrightarrow & A \\
{\scriptstyle F(j)}\downarrow & & \downarrow{\scriptstyle f} \\
F(N) & \dashrightarrow & B
\end{array}
$$

such that $j : M \to N$ is still a cofibration (respectively, an acyclic cofibration), but where M is now cofibrant.

Proof. Set $M = U(A)$ and consider the pushout

$$
\begin{array}{ccc}
K & \longrightarrow & U(A) =: M \\
{\scriptstyle i}\downarrow & & \downarrow{\scriptstyle j} \\
L & \dashrightarrow & L \oplus_K U(A) =: N
\end{array}
$$

where $K \to U(A)$ is the adjoint morphism of $F(K) \to A$. By straightforward categorical constructions, we can form a new pushout

$$
\begin{array}{ccc}
F(U(A)) & \longrightarrow & A \\
{\scriptstyle F(j)}\downarrow & & \downarrow{\scriptstyle f} \\
F(L \oplus_K U(A)) & \longrightarrow & B
\end{array}
$$

in which j is substituted to i.

The object $M = U(A)$ is cofibrant by assumption. The morphism j forms a cofibration (respectively, an acyclic cofibration) if i is so. Hence all our requirements are satisfied.   □

For our needs, we record that Brown's lemma is also valid in the context of semi-model categories:

12.1.6 Proposition (Brown's lemma). *Let $F : A \to X$ be a functor, where A is a semi-model category and X is a category equipped with a class of weak-equivalences that satisfies the two-out-of-three axiom. If F maps acyclic cofibrations between cofibrant objects to weak-equivalences, then F maps all weak-equivalences between cofibrant objects to weak-equivalences.*

Proof. The proposition follows from a straightforward generalization of the proof of the standard Brown's lemma.   □

¶ The dual version of this statement, in which cofibrations are replaced by fibrations, holds only under a weaker form:

12.1.7 ¶ Proposition (Brown's lemma). *Let $U : A \to X$ be a functor, where A is a semi-model category and X is a category equipped with a class of weak-equivalences that satisfies the two-out-of-three axiom. If F maps acyclic fibrations between fibrant objects to weak-equivalences, then F maps to weak-equivalences the weak-equivalences $f : X \to Y$ so that X is both cofibrant and fibrant and Y is fibrant.*   □

The proof of this statement uses axiom M6'.

12.1.8 Quillen Adjunctions Between Semi-Model Categories. Some care is necessary to generalize the notion of a Quillen adjunction in the context of semi-model categories: as the lifting axiom M4' is not sufficient to characterize the class of (acyclic) cofibrations and the class of (acyclic) fibrations, the usual equivalent conditions of the definition of a Quillen adjunction are no more equivalent.

Therefore, we say that adjoint functors $F : A \rightleftarrows X : U$ between semi-model categories A and X define a Quillen adjunction if every one of the following conditions hold:

A1'. The functor F preserves cofibrations and acyclic cofibrations between cofibrant objects.

A2'. Same as A2: The functor U preserves fibrations and acyclic fibrations.

Observe however that A2' implies (but is not equivalent to) A1'. Thus, in applications, we only check condition A2'. These properties imply that the pair (F, U) yields an adjunction between homotopy categories, as in the context of model categories.

Say that the functors (F, U) define a Quillen equivalence if we have further:

E1'. For every cofibrant object $X \in \mathcal{X}$, the composite

$$X \xrightarrow{\eta X} UF(X) \xrightarrow{U(i)} U(B)$$

forms a weak-equivalence in $\mathcal{X}$, where ηX refers to the adjunction unit and i arises from a factorization $F(X) \xrightarrow{\sim} B \twoheadrightarrow *$ of the terminal morphism $F(X) \to *$.

E2'. Same as E2: For every fibrant object $A \in \mathcal{A}$, the composite

$$F(Y) \to FU(A) \xrightarrow{\epsilon A} A$$

forms a weak-equivalence in $\mathcal{A}$, where ϵA refers to the adjunction augmentation and Y is any cofibrant replacement of $U(A)$.

The derived functors of a Quillen equivalence of semi-model categories define adjoint equivalences of homotopy categories, as in the context of model categories.

12.1.9 Relative Semi-Model Structures. In certain semi-model categories, the lifting and factorization axioms hold under weaker assumptions on the domain of morphisms. These properties are formalized in a relative notion of a semi-model structure. Next we explain that operads and algebras over cofibrant operads inherits such improved lifting and factorization axioms. But we do not really use these improved semi-model structures that we recall for the sake of completeness only.

Suppose we have an adjunction $F : \mathcal{X} \rightleftarrows \mathcal{A} : U$, where $\mathcal{A}$ is a category with limits and colimits and $\mathcal{X}$ is a model category. Suppose $\mathcal{A}$ is equipped with a semi-model structure so that:

– The functor $U : \mathcal{A} \to \mathcal{X}$ creates weak-equivalences, creates fibrations, and maps the cofibrations $i : A \to B$ so that A is $\mathcal{X}$-cofibrant to cofibrations.

Again, we say that an object $A \in \mathcal{A}$ is $\mathcal{X}$-cofibrant if the functor $U : \mathcal{A} \to \mathcal{X}$ maps the initial morphism $F(0) \to A$ to a cofibration.

In this situation, it makes sense to require the following lifting and factorization axioms:

M4". i. The fibrations have the right lifting property with respect to the acyclic cofibrations $i : A \to B$ such that A is $\mathcal{X}$-cofibrant.

ii. The acyclic fibrations have the right lifting property with respect to the cofibrations $i : A \to B$ such that A is $\mathcal{X}$-cofibrant.

M5". i. Any morphism $f : A \to B$ has a factorization $f = pi$, where i is a cofibration and p is an acyclic fibration, provided that A is $\mathcal{X}$-cofibrant.

ii. Any morphism $f : A \to B$ has a factorization $f = qj$, where j is an acyclic cofibration and q is a fibration, provided that A is $\mathcal{X}$-cofibrant.

If these properties are satisfied, then we say that $\mathcal{A}$ forms a *semi-model category over* $\mathcal{X}$.

The initial object of $\mathcal{A}$ is supposed to be cofibrant by axiom M0' of semi-model categories. As a byproduct, the assumption on the functor $U : \mathcal{A} \to \mathcal{X}$ implies that $U(A)$ is cofibrant if A is cofibrant in $\mathcal{A}$. Accordingly, the lifting and factorization axioms M4''-M5'' are stronger than the lifting and factorization axioms M4'-M5' of semi-model categories.

Suppose now we have an adjunction $F : \mathcal{X} \rightleftarrows \mathcal{A} : U$, where $\mathcal{A}$ is any category with limits and colimits and $\mathcal{X}$ is a cofibrantly generated model category. Suppose we have:

(1) Same as assumption (1) of theorem 12.1.4: "*The functor* $U : \mathcal{A} \to \mathcal{X}$ *preserves colimits over non-empty ordinals.*"
(2) Drop the condition that A is an $F\mathcal{X}_c$-cell complex in assumption (2) of theorem 12.1.4: "*For any pushout*

$$
\begin{array}{ccc}
F(K) & \longrightarrow & A \\
{\scriptstyle F(i)}\downarrow & & \downarrow{\scriptstyle f} \\
F(L) & \dashrightarrow & B
\end{array}
$$

such that A is $\mathcal{X}$-cofibrant, the morphism $U(f)$ forms a cofibration (respectively an acyclic cofibration) in $\mathcal{X}$ whenever i is a cofibration (respectively an acyclic cofibration) with a cofibrant domain."
(3) Same as assumption (3) of theorem 12.1.4: "*The object $UF(0)$ is cofibrant.*"

Then $\mathcal{A}$ inherits an adjoint semi-model structure from $\mathcal{X}$ since the requirements of theorem 12.1.4 are fulfilled.

But we have better:

12.1.10 Proposition. *Under these assumptions (1-3) the category $\mathcal{A}$ forms a semi-model category over $\mathcal{X}$.*

Proof. This proposition, like theorem 12.1.4, follows from a careful inspection of the arguments of theorem 11.1.13 and proposition 11.1.14. $\square$

12.1.11 Relative Properness Axioms. In the case of a semi-model category $\mathcal{A}$ over a model category $\mathcal{X}$, it makes sense to improve the properness property of proposition 12.1.2 to cofibrations $i : A \to B$ such that A is $\mathcal{X}$-cofibrant. If the next axiom holds, then we say that $\mathcal{A}$ forms a *(left) proper semi-model category over $\mathcal{X}$*:

P1". The pushout of a weak-equivalence along a cofibration

gives a weak-equivalence $C \xrightarrow{\sim} D$ provided that A and B are $\mathcal{X}$-cofibrant.

12.2 The Semi-Model Category of Operads

In this section, we survey briefly the application of model structures to categories of operads. For this purpose, we assume that the base category $\mathcal{C}$ is equipped with a model structure and forms a cofibrantly generated symmetric monoidal category. Recall that $\mathcal{C}$ is supposed to satisfy the pushout product axiom MM1, as well as the unit axiom MM0.

In this book, the semi-model structure of the category of operads is only used in §17.4, where we study applications of the homotopy theory of modules over operads to categories of algebras over cofibrant dg-operads. Usually, we only deal with the underlying model category of Σ_*-objects and we only use Σ_*-cofibrations of operads and Σ_*-cofibrant operads. Recall that, according to our convention, a Σ_*-cofibration refers to a morphism of operads $\phi : P \to Q$ which forms a cofibration in the underlying category of Σ_*-objects and an operad P is Σ_*-cofibrant if the unit morphism $\eta : I \to P$ forms a Σ_*-cofibration.

For our needs, we only recall the statement of the result, for which we refer to [4, 26, 58], and we make explicit the structure of cofibrant operads in dg-modules.

Proposition 12.1.10 can be applied to the adjunction

$$\mathrm{F} : \mathcal{M} \rightleftarrows \mathcal{O} : U$$

between operads and Σ_*-objects and returns the following statement:

Theorem 12.2.A (see [26, 58]). *The category of operads $\mathcal{O}$ forms a semi-model category over the category of Σ_*-objects, so that:*

- *The forgetful functor $U : \mathcal{O} \to \mathcal{M}$ creates weak-equivalences, creates fibrations, and maps the cofibrations of operads $i : P \to Q$ such that P is Σ_*-cofibrant to cofibrations.*
- *The morphisms of free operads $\mathrm{F}(i) : \mathrm{F}(M) \to \mathrm{F}(N)$, where $i : M \to N$ ranges over generating (acyclic) cofibrations of Σ_*-objects, form generating (acyclic) cofibrations of the category of operads.*
- *The lifting axiom M4" holds for the (acyclic) cofibrations of operads $i : P \to Q$ such that P is Σ_*-cofibrant.*

  – *The factorization axiom M5" holds for the operad morphisms $\phi : \mathrm{P} \to \mathrm{Q}$ such that P is Σ_*-cofibrant.*

$\square$

    Thus an operad morphism $\phi : \mathrm{P} \to \mathrm{Q}$ forms a weak-equivalence, respectively a fibration, if its components $\phi : \mathrm{P}(n) \to \mathrm{Q}(n)$ are weak-equivalences, respectively fibrations, in the base model category $\mathcal{C}$.

    One has to study the structure of free operads and coproducts to check that assumptions (1-3) of proposition 12.1.10. are fulfilled. This task is achieved in [26] in the context of dg-modules and in [58] in a wider context to return the result of theorem 12.2.A.

    By [58], we also have:

Theorem 12.2.B (see [58]). *The semi-model category of operads satisfies the axiom of relative properness:*

P1". The pushout of a weak-equivalence along a cofibration

gives a weak-equivalence $\mathrm{R} \xrightarrow{\sim} \mathrm{S}$ provided that P and Q are Σ_*-cofibrant.

    The authors of [4] observe that the proof of theorem 12.2.A can be simplified in certain usual situations to give a better result:

Theorem 12.2.C (see [4]). *Under assumptions (1-3) below, the adjunction*

$$F : \mathcal{M} \rightleftarrows \mathcal{O} : U$$

creates a full model structure on the category of non-unitary operads $\mathcal{O}_0$.

(1) There is a fixed ordinal μ, such that the domains of generating (acyclic) cofibrations are small with respect to all colimits

$$C = D_0 \to \cdots \to D_{\lambda-1} \xrightarrow{j_\lambda} D_\lambda \to \cdots \to \operatorname*{colim}_{\lambda<\mu} D_\lambda = D.$$

(2) There is a functor of symmetric monoidal model categories $R : \mathcal{C} \to \mathcal{C}$ which associates a fibrant replacement to any object $C \in \mathcal{C}$.

(3) The category $\mathcal{C}$ is equipped with a commutative Hopf interval (we refer to loc. cit. for the definition of this notion). $\square$

    Recall that an operad P is non-unitary if we have $\mathrm{P}(0) = 0$.

    The assumptions hold for the category of dg-modules, for the category of simplicial sets, but assumption (1) fails for the category of topological spaces (see [28, §2.4]).

The Hopf interval is used to associate a canonical path object to any operad. An argument of Quillen permits to turn round the study of relative cell complexes of operads to prove directly condition (2) of theorem 12.1.4 by using the existence of such path objects (see *loc. cit.* for details).

In the remainder of this section, we take $\mathcal{C} = \mathrm{dg\,k\,Mod}$ and we study the semi-model category of operads in dg-modules. Our purpose is to review the explicit structure of *cofibrant cell dg-operads* obtained in [26]. The result is used in applications of §17.4.

In summary, we check that cofibrant cell dg-operads are equivalent to certain quasi-free objects in operads, just like we prove in §11.2 that cofibrant cell dg-modules are equivalent to quasi-free objects equipped with an appropriate filtration.

12.2.1 Quasi-Free Operads in dg-Modules.

To begin with, we recall the definition of a twisting cochain and of a quasi-free object in the context of operads. For more background, we refer to [14, 17, 26].

First, a *twisting cochain of operads* consists of a collection of twisting cochains of dg-modules $\partial \in Hom_{\mathcal{C}}(\mathrm{P}(n), \mathrm{P}(n))$ that commute with the action of symmetric groups and satisfy the derivation relation

$$\partial(p \circ_e q) = \partial(p) \circ_e q + \pm p \circ_e \partial(q)$$

with respect to operadic composites. These assumptions ensure that the collection of twisted dg-modules $(\mathrm{P}(n), \partial)$ forms still a dg-operad with respect to the operad structure of P.

An operad P is *quasi-free (as an operad)* if we have $\mathrm{P} = (\mathrm{F}(M), \partial)$, for a certain twisting cochain of operads $\partial : \mathrm{F}(M) \to \mathrm{F}(M)$, where $\mathrm{F}(M)$ is a free operad. Note that $\partial : \mathrm{F}(M) \to \mathrm{P}(M)$ is determined by its restriction to the generating Σ_*-object $M \subset \mathrm{F}(M)$ since we have the derivation relation

$$\begin{aligned}
\partial((\cdots((\xi_1 \circ_{e_2} \xi_2) \circ_{e_3} \cdots) \circ_{e_r} \xi_r) &= (\cdots((\partial\xi_1 \circ_{e_2} \xi_2) \circ_{e_3} \cdots) \circ_{e_r} \xi_r \\
&+ (\cdots((\xi_1 \circ_{e_2} \partial\xi_2) \circ_{e_3} \cdots) \circ_{e_r} \xi_r \\
&+ \cdots + (\cdots((\xi_1 \circ_{e_2} \xi_2) \circ_{e_3} \cdots) \circ_{e_r} \partial\xi_r
\end{aligned}$$

for any formal composite $(\cdots((\xi_1 \circ_{e_2} \xi_2) \circ_{e_3} \cdots) \circ_{e_r} \xi_r \in \mathrm{F}(M)$, where $\xi_1, \ldots, \xi_r \in M$.

By construction, the generating cofibrations of dg-operads are the morphism of free operads $\mathrm{F}(i \otimes F_r) : \mathrm{F}(C \otimes F_r) \to \mathrm{F}(D \otimes F_r)$, where $i : C \to D$ ranges over the generating cofibrations of dg-modules and F_r is a free Σ_*-object. Recall that the generating cofibrations of dg-modules are inclusions $i : B^{d-1} \to E^d$ where E^d is spanned by an element e_d of degree d, by an element b_{d-1} of degree $d-1$, together with the differential $\delta(e_d) = b_{d-1}$, and B^{d-1} is the submodule of E^d spanned by b_{d-1}.

We use the convention of §11.1.8 to call *cofibrant cell operads* the cell complexes in operads built from generating cofibrations. We prove that cofibrant cell operads are quasi-free operads equipped with a suitable filtration. To obtain this result, we examine the structure of cell attachments:

12.2.2 Lemma. *For a quasi-free operad* $P = (F(M), \partial)$, *a cell attachment of generating cofibrations*

$$
\begin{array}{ccc}
F(\bigoplus_\alpha B^{d_\alpha - 1} \otimes F_{r_\alpha}) & \xrightarrow{\;f\;} & P \\
{\scriptstyle F(i_{d_\alpha} \otimes F_{r_\alpha})}\Big\downarrow & & \Big\downarrow{\scriptstyle j} \\
F(\bigoplus_\alpha E^{d_\alpha} \otimes F_{r_\alpha}) & \cdots\cdots\rightarrow & Q
\end{array}
$$

returns a quasi-free operad such that $Q = (F(M \oplus E), \partial)$, *where* E *is a free* Σ_*-*object in graded* $\Bbbk$-*modules* $E = \bigoplus_\alpha \Bbbk\, e_{d_\alpha} \otimes F_{r_\alpha}$ *(together with a trivial differential).*

The twisting cochain $\partial : F(M \oplus E) \to F(M \oplus E)$ *is given by the twisting cochain of* P *on the summand* $M \subset F(M \oplus E)$ *and is determined on the summand* $E = \bigoplus_\alpha \Bbbk\, e_{d_\alpha} \otimes F_{r_\alpha} \subset F(M \oplus E)$ *by the relation* $\partial(e_{d_\alpha}) = f(b_{d_\alpha - 1})$, *where* $f : \bigoplus_\alpha \Bbbk\, b_{d_\alpha - 1} \otimes F_{r_\alpha} \to P$ *represents the attaching map.*

Proof. Straightforward verification.   □

By induction, we obtain immediately:

12.2.3 Proposition. *A cofibrant cell operad is equivalent to a quasi-free operad* $P = (F(L), \partial)$ *where* L *is a free* Σ_*-*object in graded* $\Bbbk$-*modules* $L = \bigoplus_\alpha \Bbbk\, e_{d_\alpha} \otimes F_{r_\alpha}$ *equipped with a basis filtration* $L_\lambda = \bigoplus_{\alpha < \lambda} \Bbbk\, e_{d_\alpha} \otimes F_{r_\alpha}$ *such that* $\partial(L_\lambda) \subset F(L_{\lambda - 1})$.   □

12.3 The Semi-Model Categories of Algebras over Operads

The purpose of this section is to define the semi-model structure of the category of algebras over an operad. For this aim, we apply theorem 12.1.4 to the adjunction

$$ P(-) : \mathcal{E} \rightleftarrows {}_P\mathcal{E} : U $$

between the category of P-algebras and the underlying category $\mathcal{E}$. As usual, we assume that $\mathcal{E}$ is any symmetric monoidal category over the base category $\mathcal{C}$ in which the operad is defined. For the needs of this section, we assume as well that $\mathcal{E}$ is equipped with a model structure and forms a cofibrantly generated symmetric monoidal category over $\mathcal{C}$. Recall that $\mathcal{E}$ is supposed to satisfy the pushout product axiom MM1, as well as the unit axiom MM0, like the base category $\mathcal{C}$.

The main result reads:

Theorem 12.3.A. *If* P *is a* Σ_*-*cofibrant operad, then the category of* P-*algebras inherits a cofibrantly generated semi-model structure so that the forgetful functor* U : $_P\mathcal{E}$ → $\mathcal{E}$ *creates weak-equivalences and fibrations. The generating (acyclic) cofibrations are the morphisms of free* P-*algebras* $P(i) : P(K) \to P(L)$ *such that* $i : K \to L$ *is a generating (acyclic) cofibrations of the underlying category* $\mathcal{E}$.

¶ In the context where the category $\mathcal{E}$ has regular tensor powers, we obtain further:

¶ **Theorem.** *If* $\mathcal{E}$ *is a (reduced) symmetric monoidal category with regular tensor powers, then the definition of theorem 12.3.A returns a semi-model structure as long as the operad* P *is* C-*cofibrant.*

This theorem gives as a corollary:

¶ **Proposition.** *Let* P *be a reduced operad. The category of* P-*algebras in connected* Σ_*-*objects* $_P\mathcal{M}^0$ *forms a semi-model category as long as the operad* P *is* C-*cofibrant.*   □

¶ The positive stable model category of symmetric spectra Sp^Σ does not satisfy axiom MM0, but this difficulty can be turned round. Moreover, a better result holds: according to [23], the category of P-algebras in the positive stable flat model category of symmetric spectra inherits a full model structure, for every operad P in Sp^Σ (not necessarily cofibrant in any sense). Note however that the forgetful functor $U : {}_P\mathrm{Sp}^\Sigma \to \mathrm{Sp}^\Sigma$ does not preserves cofibrations in general.

In many usual situations, the operad P is the image of an operad in simplicial sets under the functor $\Sigma^\infty(-)_+ : \mathcal{S} \to \mathrm{Sp}^\Sigma$. In our sense, we use the category of simplicial sets $\mathcal{C} = \mathcal{S}$ as a base model category and the category of spectra $\mathcal{E} = \mathrm{Sp}^\Sigma$ as a symmetric monoidal category over $\mathcal{S}$. The forgetful functor $U : {}_P\mathrm{Sp}^\Sigma \to \mathrm{Sp}^\Sigma$ seems to preserve cofibrations for an operad in simplicial sets though P does no form an Sp^Σ-cofibrant object in spectra (see for instance the case of the commutative operad in [55]). For a cofibrant P-algebra in spectra A, this property implies that the initial morphism $\eta : P(0) \to A$ is a cofibration, but A does not form a cofibrant object in the underlying category of spectra unless we assume $P(0) = \mathrm{pt}$.

The proof of theorem 12.3.A (and theorem 12.3) is outlined in the next paragraph. The technical verifications are achieved in the appendix, §20.1.

Under the assumption of theorem 12.3.A, the pushout-product property of proposition 11.5.1 implies that the functor $S(P) : \mathcal{E} \to \mathcal{E}$ preserves cofibrations, respectively acyclic cofibrations, with a cofibrant domain. In §2.4, we observe that the functor $S(P) : \mathcal{E} \to \mathcal{E}$ preserves all filtered colimits as well. Thus condition (1) of theorem 12.1.4 is easily seen to be satisfied (and similarly in the context of theorem 12.3). The difficulty is to check condition (2):

12.3.1 Lemma. *Under the assumption of theorem 12.3.A, for any pushout*

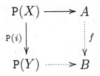

such that A is an $\mathcal{E}$-cofibrant $P(\mathcal{E}_c)$-cell complex, the morphism f forms a cofibration (respectively, an acyclic cofibration) in the underlying category $\mathcal{E}$ if $i : X \to Y$ is so.

The same result holds in the context of theorem 12.3. The technical verification of this lemma is postponed to §20.1.

As usual, we call $\mathcal{E}$-cofibrations the morphisms of P-algebras $i : A \to B$ which form a cofibration in the underlying category $\mathcal{E}$, we call $\mathcal{E}$-cofibrant objects the P-algebras A such that the initial morphism $\eta : P(0) \to A$ is an $\mathcal{E}$-cofibration.

The verification of lemma 12.3.1 includes a proof that:

12.3.2 Proposition. *The cofibrations of P-algebras $i : A \to B$ such that A is a cofibrant P-algebra are $\mathcal{E}$-cofibrations. Any cofibrant P-algebra A is $\mathcal{E}$-cofibrant.*

The definition of the semi-model structure in theorem 12.3.A is natural with respect to the underlying category $\mathcal{E}$ in the following sense:

12.3.3 Proposition. *Let P be any Σ_*-cofibrant operad. Let $\rho_! : \mathcal{D} \rightleftarrows \mathcal{E} : \rho^*$ be a Quillen adjunction of symmetric monoidal model categories over $\mathcal{C}$. The functors*

$$\rho_! : {}_P\mathcal{D} \rightleftarrows {}_P\mathcal{E} : \rho^*$$

induced by $\rho_!$ and ρ^* define a Quillen adjunction of semi-model categories.

Proof. Fibrations and acyclic fibrations are created by forgetful functors in the semi-model categories of P-algebras. For this reason we obtain immediately that ρ^* preserves fibrations and acyclic fibrations. Since the functor $\rho_!$ maps (acyclic) cofibrations to (acyclic) cofibrations and preserves free objects by proposition 3.2.14, we obtain that $\rho_!$ maps generating (acyclic) cofibrations of ${}_P\mathcal{D}$ to (acyclic) cofibrations in ${}_P\mathcal{E}$. Since the functor $\rho_!$ preserves colimits and retracts, we obtain further that $\rho_!$ maps all (acyclic) cofibrations of ${}_P\mathcal{D}$ to (acyclic) cofibrations in ${}_P\mathcal{E}$. □

We have further:

12.3.4 Proposition. *If $\rho_! : \mathcal{D} \rightleftarrows \mathcal{E} : \rho^*$ is a Quillen equivalence, then $\rho_! : {}_P\mathcal{D} \rightleftarrows {}_P\mathcal{E} : \rho^*$ defines a Quillen equivalence as well.*

Proof. Suppose A is a cofibrant object in ${}_P\mathcal{D}$. By proposition 12.3.2, the morphism $\eta : P(0) \to A$ forms a cofibration in $\mathcal{D}$. Since P is supposed to be

Σ_*-cofibrant in $\mathcal{D}$, the P-algebra A forms a cofibrant object in $\mathcal{D}$ as well. Since the forgetful functor $U : {}_P\mathcal{E} \to \mathcal{E}$ creates fibrations, any fibrant replacement of $\rho_! A$ in ${}_P\mathcal{E}$ defines a fibrant replacement of $\rho_! A$ in the underlying category. From these observations, we conclude that the composite

$$A \xrightarrow{\eta A} \rho^* \rho_! A \to \rho^* B,$$

where ηA refers to the adjunction unit and B is any fibrant replacement of $\rho_! A$ in ${}_P\mathcal{E}$, forms a weak-equivalence in $\mathcal{D}$ and hence forms a weak-equivalence of P-algebras in $\mathcal{D}$.

Suppose B is a fibrant object in ${}_P\mathcal{E}$. Pick a cofibrant replacement $P(0) \rightarrowtail A \xrightarrow{\sim} \rho^* B$ of $\rho^* B$ in ${}_P\mathcal{D}$. Use again that the forgetful functor $U : {}_P\mathcal{E} \to \mathcal{E}$ creates fibrations and that the forgetful functor $U : {}_P\mathcal{D} \to \mathcal{D}$ preserves cofibrations to conclude that the composite

$$\rho_! A \to \rho_! \rho^* B \xrightarrow{\epsilon} B,$$

where ϵB refers to the adjunction augmentation, forms a weak-equivalence in $\mathcal{E}$ and hence forms a weak-equivalence of P-algebras in $\mathcal{E}$. $\square$

12.3.5 ¶ Remark. If $\rho_! : \mathcal{D}^0 \rightleftarrows \mathcal{E}^0 : \rho^*$ is a Quillen adjunction, respectively a Quillen equivalence, between (reduced) categories with regular tensor powers, then propositions 12.3.3-12.3.4 hold as long as P is a $\mathcal{C}$-cofibrant (non-unitary) operad.

12.3.6 Quasi-Free Algebras over Operads in dg-Modules. In the remainder of this section, we take $\mathcal{C} = \mathrm{dg}\,\Bbbk\,\mathrm{Mod}$ and we study the structure of cofibrant algebras over a Σ_*-cofibrant dg-operad P. To simplify, we also take $\mathcal{E} = \mathrm{dg}\,\Bbbk\,\mathrm{Mod}$ but we can use the principle of generalized point-tensors to extend our results to P-algebras in Σ_*-objects and to P-algebras in right modules over operads.

As usual, we prove that *cofibrant cell* P-*algebras* in dg-modules are equivalent to quasi-free P-algebras equipped with an appropriate filtration. The plan of our constructions parallels the case of operads, addressed in §§12.2.1-12.2.3.

First, we review the definition of a twisting cochain and of a quasi-free object in the category of P-algebras.

A twisting cochain of dg-modules $\partial \in Hom_{\mathcal{C}}(A, A)$ defines a *twisting cochain of* P-*algebras* if $\partial : A \to A$ satisfies the derivation relation

$$\partial(p(a_1, \ldots, a_n)) = \sum_{i=1}^{n} \pm p(a_1, \ldots, \partial(a_i), \ldots, a_n),$$

for every $p \in P(n)$, $a_1, \ldots, a_n \in A$. This assumption ensures that the twisted dg-module (A, ∂) inherits the structure of a P-algebra in dg-modules.

A P-algebra A is *quasi-free* if we have $A = (\mathrm{P}(C), \partial)$ for a certain twisting cochain of P-algebras $\partial : \mathrm{P}(C) \to \mathrm{P}(C)$. Note that $\partial : \mathrm{P}(C) \to \mathrm{P}(C)$ is determined by its restriction to the generating dg-module $C \subset \mathrm{P}(C)$ of the free P-algebra $\mathrm{P}(C)$ since we have the relation

$$\partial(p(x_1, \ldots, x_n)) = \sum_{i=1}^{n} \pm p(x_1, \ldots, \partial(x_i), \ldots, x_n),$$

for every element $p(x_1, \ldots, x_n) \in \mathrm{P}(C)$.

Recall that the generating cofibrations of P-algebras in dg-modules are the morphism of free P-algebras $\mathrm{P}(i) : \mathrm{P}(C) \to \mathrm{P}(D)$ induced by generating cofibrations of dg-modules. Recall that the generating cofibrations of dg-modules are inclusions $i : B^{d-1} \to E^d$ where E^d is spanned by an element e_d of degree d, by an element b_{d-1} of degree $d-1$, together with the differential $\delta(e_d) = b_{d-1}$, and B^{d-1} is the submodule of E^d spanned by b_{d-1}.

We use the convention of §11.1.8 to call *cofibrant cell P-algebras* the cell complexes in P-algebras obtained by successive attachments of generating cofibrations of the category of P-algebras. We prove that cofibrant cell P-algebras are quasi-free P-algebras equipped with a suitable filtration. To obtain this result, we examine the structure of cell attachments on quasi-free P-algebras:

12.3.7 Lemma. *For a quasi-free P-algebra $A = (\mathrm{P}(C), \partial)$, a cell attachment of generating cofibrations*

$$
\begin{array}{ccc}
\mathrm{P}(\bigoplus_\alpha B^{d_\alpha - 1}) & \xrightarrow{\ f\ } & A \\
{\scriptstyle \mathrm{P}((i_{d_\alpha}))} \downarrow & & \downarrow {\scriptstyle j} \\
\mathrm{P}(\bigoplus_\alpha E^{d_\alpha}) & \dashrightarrow & B
\end{array}
$$

returns a quasi-free P-algebra such that $B = (\mathrm{P}(C \oplus E), \partial)$, where E is a free graded $\Bbbk$-module $E = \bigoplus_\alpha \Bbbk\, e_{d_\alpha}$ (equipped with a trivial differential).

The twisting cochain $\partial : \mathrm{P}(C \oplus E) \to \mathrm{P}(C \oplus E)$ is given by the twisting cochain of A on the summand $C \subset \mathrm{P}(C \oplus E)$ and is determined by the relation $\partial(e_{d_\alpha}) = f(b_{d_\alpha - 1})$ on the summand $E = \bigoplus_\alpha \Bbbk\, e_{d_\alpha} \subset \mathrm{P}(C \oplus E)$, where $f : \bigoplus_\alpha \Bbbk\, b_{d_\alpha - 1} \to A$ represents the attaching map.

Proof. Straightforward verification. □

By induction, we obtain immediately:

12.3.8 Proposition. *A cofibrant cell P-algebra is equivalent to a quasi-free P-algebra $A = (\mathrm{P}(C), \partial)$ where C is a free graded $\Bbbk$-module $C = \bigoplus_\alpha \Bbbk\, e_{d_\alpha}$ equipped with a basis filtration $C_\lambda = \bigoplus_{\alpha < \lambda} \Bbbk\, e_{d_\alpha}$ such that $\partial(C_\lambda) \subset \mathrm{P}(C_{\lambda-1})$.* □

12.4 Addendum: The Homotopy of Algebras over Cofibrant Operads

The result of theorem 12.3.A can be improved if we assume that P is a cofibrant operad:

Theorem 12.4.A (see [58]). *If P is a cofibrant operad, then the category of P-algebras in $\mathcal{E}$ forms a semi-model category over $\mathcal{E}$, so that:*

- *The forgetful functor $U : {}_P\mathcal{E} \to \mathcal{E}$ creates weak-equivalences, creates fibrations, and maps the cofibrations of P-algebras $i : A \to B$ such that A is $\mathcal{E}$-cofibrant to cofibrations.*
- *The lifting axiom M4" holds for the (acyclic) cofibrations of P-algebras $i : A \to B$ such that A is $\mathcal{E}$-cofibrant.*
- *The factorization axiom M5" holds for the morphisms of P-algebras $f : A \to B$ such that A is $\mathcal{E}$-cofibrant.*

Besides:

Theorem 12.4.B (see [58]). *The semi-model category of algebras over a cofibrant operad P satisfies the axiom of relative properness:*

P1". The pushout of a weak-equivalence along a cofibration

gives a weak-equivalence $C \xrightarrow{\sim} D$ provided that A and B are $\mathcal{E}$-cofibrant.

We refer to [58] for the proof of these theorems. We do not use these results, which are only mentioned for the sake of completeness.

12.5 The Homotopy of Extension and Restriction Functors – Objectives for the Next Part

By §3.3.5, any morphism of operads $\phi : P \to Q$ gives rise to adjoint extension and restriction functors:

$$\phi_! : {}_P\mathcal{E} \rightleftarrows {}_Q\mathcal{E} : \phi^*.$$

In §16, we use the homotopy of modules over operads to prove:

Theorem 12.5.A. *Suppose P (respectively, Q) is a Σ_*-cofibrant operad so that the category of P-algebras (respectively, Q-algebras) comes equipped with a semi-model structure.*

The extension and restriction functors

$$\phi_! : {}_P\mathcal{E} \rightleftarrows {}_Q\mathcal{E} : \phi^*.$$

define a Quillen adjunction, a Quillen equivalence if ϕ is a weak-equivalence.

¶ In the context of a (reduced) symmetric monoidal category with regular tensor powers, this statement holds whenever the operads P and Q are $\mathcal{C}$-cofibrant.

Property A2' (the right adjoint preserves fibrations and acyclic fibrations) of a Quillen adjunction is immediate to check, because fibrations and acyclic fibrations are created by forgetful functors and the restriction functor ϕ^* : ${}_Q\mathcal{E} \to {}_P\mathcal{E}$ reduces to the identity if we forget algebra structures.

In §16, we use the representation of the functors $\phi_! : {}_P\mathcal{E} \rightleftarrows {}_Q\mathcal{E} : \phi^*$ by modules over operads and the results of §15 to prove properties E1'-E2' of a Quillen equivalence when $\phi : P \to Q$ is a weak-equivalence of operads.

We refer to this chapter for full details on the proof of theorem 12.5.A.

Chapter 13
The (Co)homology of Algebras over Operads – Some Objectives for the Next Part

Introduction

The *cohomology* $H_P^*(A, E)$ *of an algebra A over an operad* P *with coefficients in a representation E* is defined by the homology of the derived functor of derivations $B \mapsto \mathrm{Der}_P(B, E)$ on the homotopy category of P-algebras over A. There is also a *homology theory* $H_*^P(A, E)$ defined by the homology of the derived functor of $B \mapsto E \otimes_{U_P(B)} \Omega_P^1(B)$, where $U_P(B)$ refers to the enveloping algebra of B, the coefficient E is a right $U_P(A)$-module, and $\Omega_P^1(B)$ is the module of Kähler differentials of B. The first purpose of this chapter, carried out in §13.1, is to survey the definition of these derived functors for algebras and operads in dg-modules.

There are universal coefficients spectral sequences which connect the cohomology $H_P^*(A, E)$, respectively the homology $H_*^P(A, E)$, to standard derived functors of homological algebra. The construction of these spectral sequences is reviewed in §13.2. The universal coefficients spectral sequences degenerate in the case of the associative operad A and the Lie operad L, but not in the case of the commutative operad C. This observation implies that the cohomology (respectively, homology) of associative and Lie algebras is determined by an Ext-functor (respectively, a Tor-functor). In the next part we use that the functor of Kähler differentials is represented by a right module over an operad, the theorems of §15, and the homotopy theory of modules over operads to give a new interpretation of this result.

The cotriple construction gives an alternative definition of a (co)homology theory for algebras over operads. In §13.3 we review the definition of the cotriple construction in the dg-context and we prove that the associated (co)homology theory agrees with the derived functor construction of §13.1.

The proofs of the main theorems of this chapter are only completed in §17 as applications of the homotopy theory of modules over operads. In this sense this chapter gives first objectives for the next part of the book. Nevertheless, the results of this chapter are not completely new: the (co)homology with

B. Fresse, *Modules over Operads and Functors*, Lecture Notes in Mathematics 1967,   203
DOI: 10.1007/978-3-540-89056-0_13, © Springer-Verlag Berlin Heidelberg 2009

coefficients for algebras in dg-modules is already defined in [26] by methods of homotopical algebra, the universal coefficient spectral sequences are defined in [1] for algebras in dg-modules (in the case of non-negatively graded dg-modules).

Throughout this chapter, we use the short notation $U_P(A) = U_P(A)(1)$ to refer to the enveloping algebra of P-algebras.

13.1 The Construction

The definition of the cohomology and homology of algebras over operads makes sense for operads and algebras in simplicial modules, or in dg-modules. In this section, we give a summary of the constructions in the dg-module setting. We take $C = \mathrm{dg}\,\Bbbk\,\mathrm{Mod}$ as a base category and $\mathcal{E} = \mathrm{dg}\,\Bbbk\,\mathrm{Mod}$ as an underlying category of algebras. The principle of generalized point-tensors can be used to extend the construction to algebras in Σ_*-objects, and to algebras in right modules over operads.

We fix an operad P. We require that P is Σ_*-cofibrant to ensure that P-algebras form a semi-model category.

13.1.1 Derivations and Cohomology. Recall that a map $\theta : A \to E$, where A is a P-algebra and E is a representation of A, forms a derivation if we have the derivation relation

$$\theta(p(a_1, \ldots, a_n)) = \sum_{i=1}^{n} p(a_1, \ldots, \theta(a_i), \ldots, a_n)$$

for all operations $p \in P(n)$ and all $a_1, \ldots, a_n \in A$. In the dg-context, we extend this definition to homogeneous maps $\theta : A \to E$ to obtain a dg-module of derivations.

The *cohomology of A with coefficients in E* is defined by:

$$H_P^*(A, E) = H^*(\mathrm{Der}_P(Q_A, E)),$$

where Q_A is any cofibrant replacement of A in the category of P-algebras. In this definition, we use the augmentation of the cofibrant replacement $Q_A \xrightarrow{\sim} A$ to equip E with the structure of a representation of Q_A.

More generally, for any P-algebra B over A, we can use the restriction of structures through the augmentation $\epsilon : B \to A$ to equip E with the structure of a representation of B. Accordingly, we have a well-defined functor from the category of P-algebras over A to the category of dg-modules defined by the map $\mathrm{Der}_P(-, E) : B \mapsto \mathrm{Der}_P(B, E)$.

The next proposition ensures that the definition of the cohomology makes sense:

13.1.2 Proposition (see [26]). *The functor $B \mapsto \mathrm{Der}_{\mathsf{P}}(B, E)$ preserves weak-equivalences between cofibrant objects in the category of P-algebras over A.*

The proof of this proposition is postponed to the end of this section, assuming a general result, proved in the next part, on the homotopy invariance of functors associated to right modules over operads.

13.1.3 Kähler Differentials and Homology. To define the homology of P-algebras, we use the module of Kähler differentials $\Omega_{\mathsf{P}}^1(A)$ and the coefficients consist of right modules over the enveloping algebra $U_{\mathsf{P}}(A)$. These objects are defined in §4 for operads and algebras in $\Bbbk$-modules. The principle of generalized point tensors can be used to extend the construction to the dg-context. Then the definition returns a dg-algebra $U_{\mathsf{P}}(A)$ and a dg-module $\Omega_{\mathsf{P}}^1(A)$ so that we have an isomorphism

$$\mathrm{Der}_{\mathsf{P}}(A, E) \simeq Hom_{U_{\mathsf{P}}(A)\,\mathrm{Mod}}(\Omega_{\mathsf{P}}^1(A), E)$$

in the category of dg-modules.

The *homology of A with coefficients in a right $U_{\mathsf{P}}(A)$-module E* is defined by:

$$H_*^{\mathsf{P}}(A, E) = H_*(E \otimes_{U_{\mathsf{P}}(Q_A)} \Omega_{\mathsf{P}}^1(Q_A)),$$

where, again, Q_A is any cofibrant replacement of A in the category of P-algebras. The next proposition ensures, as in the context of the cohomology, that this definition makes sense:

13.1.4 Proposition (see [26]). *The functor $B \mapsto E \otimes_{U_{\mathsf{P}}(B)} \Omega_{\mathsf{P}}^1(B)$ preserves weak-equivalences between cofibrant objects in the category of P-algebras over A.*

The proof of this lemma is also postponed to the end of this section.

13.1.5 Classical Examples. The definition of the cohomology $H_{\mathsf{P}}^*(A, E)$, respectively homology $H_*^{\mathsf{P}}(A, E)$, can be applied to the usual operads $\mathsf{P} = \mathsf{A}, \mathsf{L}, \mathsf{C}$, provided that the ground ring $\Bbbk$ has characteristic 0 in the case $\mathsf{P} = \mathsf{L}, \mathsf{C}$ (since these operads are not Σ_*-cofibrant in positive characteristic). The cohomology $H_{\mathsf{P}}^*(A, E)$, respectively homology $H_*^{\mathsf{P}}(A, E)$, defined by the theory of operads agrees with:

- the classical Hochschild cohomology, respectively homology, in the case of the associative operad $\mathsf{P} = \mathsf{A}$,
- the Chevalley-Eilenberg cohomology, respectively homology, in the case of the Lie operad $\mathsf{P} = \mathsf{L}$,
- the Harrison cohomology, respectively homology, in the commutative operad $\mathsf{P} = \mathsf{C}$.

These assertions are consequences of results of [17, 18] (see [46]). The idea is to use particular cofibrant replacements, the Koszul resolution, and to identify the dg-module $\mathrm{Der}_{\mathsf{P}}(Q_A, E)$, respectively $E \otimes_{U_{\mathsf{P}}(Q_A)} \Omega^1(Q_A)$, associated to this resolution with the usual Hochschild (respectively, Chevalley-Eilenberg, Harrison) complex.

The Hochschild cohomology and the Chevalley-Eilenberg cohomology (respectively, homology) can also be defined by an Ext-functor (respectively, a Tor-functor). This result has an interpretation in terms of a general universal coefficient spectral sequence, defined in the next section, which relates any cohomology (respectively, homology) associated to an operad to Ext-functors (respectively, Tor-functors).

13.1.6 ¶ Remark. If we use simplicial $\Bbbk$-modules rather than dg-modules, then the definition of the cohomology $H^*_\mathsf{P}(A, E)$, respectively homology $H^\mathsf{P}_*(A, E)$, makes sense for any simplicial operad P which is cofibrant in simplicial $\Bbbk$-modules, and not only for Σ_*-cofibrant operads. In particular, we can use P-algebras in simplicial $\Bbbk$-modules to associate a well-defined (co)homology theory to the Lie, respectively commutative, operad in positive characteristic. In the case of the Lie operad, this simplicial (co)homology agrees again with the Chevalley-Eilenberg (co)homology. In the case of the commutative operad, this simplicial (co)homology agrees with André-Quillen' (co)homology.

13.1.7 Universal Coefficients. The remainder of the section is devoted to the proof of proposition 13.1.2 and proposition 13.1.4, the homotopy invariance of the dg-modules $\mathrm{Der}_\mathsf{P}(Q_A, E)$ and $E \otimes_{U_\mathsf{P}(Q_A)} \Omega^1_\mathsf{P}(Q_A)$, for Q_A a cofibrant replacement of A.

For this aim, we use the representation

$$T^1_\mathsf{P}(Q_A) = U_\mathsf{P}(A) \otimes_{U_\mathsf{P}(Q_A)} \Omega^1_\mathsf{P}(Q_A)$$

which determines the homology of A with coefficients in the enveloping algebra $U_\mathsf{P}(A)$. The enveloping algebra $U_\mathsf{P}(A)$ gives universal coefficients for the homology of A since we have identities

$$\mathrm{Der}_\mathsf{P}(Q_A, E) = Hom_{U_\mathsf{P}(Q_A)\,\mathrm{Mod}}(\Omega^1_\mathsf{P}(Q_A), E)$$
$$= Hom_{U_\mathsf{P}(A)\,\mathrm{Mod}}(T^1_\mathsf{P}(Q_A), E)$$
$$\text{and} \quad E \otimes_{U_\mathsf{P}(Q_A)} \Omega^1_\mathsf{P}(Q_A) = E \otimes_{U_\mathsf{P}(A)} T^1_\mathsf{P}(Q_A).$$

The idea is to prove that the functor $Q_A \mapsto T^1_\mathsf{P}(Q_A)$ maps cofibrant P-algebras over A to cofibrant left $U_\mathsf{P}(A)$-modules and preserves weak-equivalences between cofibrant P-algebras. Then we apply the classical homotopy theory of modules over dg-algebras to conclude that the functors $Q_A \mapsto Hom_{U_\mathsf{P}(A)\,\mathrm{Mod}}(T^1_\mathsf{P}(Q_A), E)$ and $Q_A \mapsto E \otimes_{U_\mathsf{P}(A)} T^1_\mathsf{P}(Q_A)$ preserve weak-equivalences between cofibrant P-algebras over A.

In our verifications, we use repeatedly that the extension and restriction functors

$$\phi_! : R\,\mathrm{Mod} \rightleftarrows S\,\mathrm{Mod} : \phi^*$$

associated to a morphism of associative dg-algebras $\phi : R \to S$ form (by proposition §11.2.10) a Quillen pair and a Quillen equivalence if ϕ is also a weak-equivalence. Accordingly, the extension functor $\phi_! M = S \otimes_R M$ preserves cofibrant objects, weak-equivalences between cofibrant objects, and the adjoint morphism of a weak-equivalence $f : M \xrightarrow{\sim} N$, where M is a cofibrant right R-module, defines a weak-equivalence $f_\sharp : S \otimes_R M \xrightarrow{\sim} N$ if ϕ is also a weak-equivalence.

In our constructions, the cofibrancy requirements are derived from the following lemma, proved by a direct verification:

13.1.8 Lemma. *If Q a cofibrant cell P-algebra, then $\Omega^1_P(Q)$ forms a cofibrant cell $U_P(Q)$-module.*

Proof. For a free P-algebra $Q = P(C)$, we have an isomorphism of left $U_P(P(C))$-modules

$$U_P(P(C)) \otimes C \xrightarrow{\simeq} \Omega^1(Q)$$

which identifies the generating elements $x \in C$ to the formal differentials $dx \in \Omega^1(P(C))$ (see §§4.4.2-4.4.3). For a quasi-free P-algebra $Q = (P(C), \partial)$, we obtain that $\Omega^1(Q)$ forms a quasi-free left $U_P(Q)$-module

$$\Omega^1(Q) = (U_P(P(C)) \otimes C, \partial)$$

so that the differential of dx in $\Omega^1(Q)$ is deduced from the differential of x in $P(C)$ and the derivation relation $dp(x_1, \ldots, x_n) = \sum_i p(x_1, \ldots, dx_i, \ldots, x_n)$.

Suppose $Q = (P(C), \partial)$ forms a cofibrant cell P-algebra. Thus the generating module C is a free graded $\Bbbk$-module equipped with a basis filtration

$$0 = C_0 \subset \cdots \subset C_\lambda \subset \cdots \subset \operatorname*{colim}_\lambda C_\lambda = C$$

and the twisting cochain satisfies $\partial(C_\lambda) \subset P(C_{\lambda-1})$. For a differential $dx \in \Omega^1_P(Q)$, where $x \in C_\lambda$, we obtain the relation

$$\partial(dx) \in U_P(P(C_{\lambda-1})) \otimes C_{\lambda-1} \subset U_P(Q) \otimes C_{\lambda-1}.$$

Hence we conclude that $\Omega^1(Q) = (U_P(P(C)) \otimes C, \partial)$ forms a cofibrant cell $U_P(Q)$-module. □

Lemma 13.1.8 implies:

13.1.9 Lemma. *If Q_A is a cofibrant P-algebra, then $\Omega^1_P(Q_A)$ forms a cofibrant $U_P(Q_A)$-module.*

Proof. According to §12.3, any cofibrant P-algebra Q_A is a retract of a cofibrant cell P-algebra Q. The existence of morphisms of P-algebras $i : Q_A \leftrightarrows Q : r$ so that $ri = \mathrm{id}$ implies by an easy inspection that $\Omega^1_P(Q_A)$ forms a retract of $U_P(Q_A) \otimes_{U_P(Q)} \Omega^1_P(Q)$ in the category of left $U_P(Q_A)$-modules, where we use the augmentation $r : Q \to Q_A$ to determine the morphism

$U_\mathsf{P}(r) : U_\mathsf{P}(Q) \to U_\mathsf{P}(Q_A)$. According to recollections of §13.1.7, the extension and restriction functors

$$r_! : U_\mathsf{P}(Q) \operatorname{Mod} \rightleftarrows U_\mathsf{P}(Q_A) \operatorname{Mod} : r^*$$

define a Quillen pair so that $r_! = U_\mathsf{P}(Q_A) \otimes_{U_\mathsf{P}(Q)} -$ preserves cofibrant objects. Therefore, the object $U_\mathsf{P}(Q_A) \otimes_{U_\mathsf{P}(Q)} \Omega^1_\mathsf{P}(Q)$ is cofibrant in the category of left $U_\mathsf{P}(Q_A)$-modules, as well as $\Omega^1_\mathsf{P}(Q_A)$ since the class of cofibrations is stable under retracts.   □

And:

13.1.10 Lemma. *If Q_A is a cofibrant P-algebra over A, then $T^1_\mathsf{P}(Q_A) = U_\mathsf{P}(A) \otimes_{U_\mathsf{P}(Q_A)} \Omega^1_\mathsf{P}(Q_A)$ forms a cofibrant $U_\mathsf{P}(A)$-module.*

Proof. Immediate consequence of recollections of §13.1.7.   □

To prove the homotopy invariance of $T^1_\mathsf{P}(Q_A)$ we use the following lemma which is a consequence of the general theorems of the next part, about functors associated to modules over operads:

13.1.11 Lemma. *The morphism $U_\mathsf{P}(f) : U_\mathsf{P}(Q_A) \to U_\mathsf{P}(Q_B)$, respectively $\Omega^1_\mathsf{P}(f) : \Omega^1_\mathsf{P}(Q_A) \to \Omega^1_\mathsf{P}(Q_B)$, induced by a weak-equivalence of P-algebras $f : Q_A \xrightarrow{\sim} Q_B$ forms a weak-equivalence if Q_A and Q_B are cofibrant.*

Assuming this lemma, we have:

13.1.12 Lemma. *The morphism*

$$T^1_\mathsf{P}(f) : T^1_\mathsf{P}(Q_A) \to T^1_\mathsf{P}(Q_B)$$

induced by a weak-equivalence of P-algebras over A

$$f : Q_A \xrightarrow{\sim} Q_B$$

forms a weak-equivalence of left $U_\mathsf{P}(A)$-modules if Q_A and Q_B are cofibrant.

Proof. By lemma 13.1.9 and lemma 13.1.11, we have a weak-equivalence of left $U_\mathsf{P}(Q_A)$-modules

$$\Omega^1_\mathsf{P}(f) : \Omega^1_\mathsf{P}(Q_A) \to \Omega^1_\mathsf{P}(Q_B)$$

such that $\Omega^1_\mathsf{P}(Q_A)$ is cofibrant. By lemma 13.1.11, the morphism $U_\mathsf{P}(f) : U_\mathsf{P}(Q_A) \to U_\mathsf{P}(Q_B)$ defines a weak-equivalence of dg-algebras. Accordingly, the extension and restriction functors

$$U_\mathsf{P}(f)_! : U_\mathsf{P}(Q_A) \operatorname{Mod} \rightleftarrows U_\mathsf{P}(Q_B) \operatorname{Mod} : U_\mathsf{P}(f)^*$$

define a Quillen equivalence, from which we deduce that the morphism

$$\Omega^1_\mathsf{P}(f)_\sharp : U_\mathsf{P}(Q_B) \otimes_{U_\mathsf{P}(Q_A)} \Omega^1_\mathsf{P}(Q_A) \to \Omega^1_\mathsf{P}(Q_B),$$

adjoint to $\Omega_\mathrm{P}^1(f)$ forms a weak-equivalence between cofibrant left $U_\mathrm{P}(Q_B)$-modules. By Quillen adjunction, we conclude that the morphism

$$T_\mathrm{P}^1(f) = U_\mathrm{P}(A) \otimes_{U_\mathrm{P}(Q_B)} \otimes \Omega_\mathrm{P}^1(f)_\sharp$$

forms a weak-equivalence too.   □

The objects $T_\mathrm{P}^1(Q_A)$ and $T_\mathrm{P}^1(Q_B)$ are also cofibrant left $U_\mathrm{P}(A)$-modules by lemma 13.1.10. Therefore, according to the standard homotopy theory of modules over dg-algebras, we obtain:

13.1.13 Lemma. *Let $f : Q_A \xrightarrow{\sim} Q_B$ be a weak-equivalence of P-algebras over A. If Q_A and Q_B are cofibrant, then:*

(a) *The morphism*

$$f^* : Hom_{U_\mathrm{P}(A)\,\mathrm{Mod}}(T_\mathrm{P}^1(Q_B), E) \to Hom_{U_\mathrm{P}(A)\,\mathrm{Mod}}(T_\mathrm{P}^1(Q_B), E)$$

induced by f forms a weak-equivalence, for every left $U_\mathrm{P}(A)$-module E.

(b) *The morphism*

$$f_* : E \otimes_{U_\mathrm{P}(A)} T_\mathrm{P}^1(Q_A) \to E \otimes_{U_\mathrm{P}(A)} T_\mathrm{P}^1(Q_B)$$

induced by f forms a weak-equivalence, for every right $U_\mathrm{P}(A)$-module E.   □

Assuming lemma 13.1.11, this statement achieves the proof of proposition 13.1.2 and proposition 13.1.4.   □

13.2 Universal Coefficient Spectral Sequences

In general, the cohomology $H_\mathrm{P}^*(A, E)$, respectively the homology $H_*^\mathrm{P}(A, E)$, associated to an algebra A over an operad P can not be reduced to a derived functor on the category of left (respectively, right) $U_\mathrm{P}(A)$-modules. Nevertheless, we have universal coefficient spectral sequences which connect the cohomology $H_\mathrm{P}^*(A, E)$, respectively the homology $H_*^\mathrm{P}(A, E)$, to standard derived functors of homological algebra. The purpose of this section is to give the definition of these spectral sequences.

To simplify, we assume that P is an operad in $\Bbbk$-modules and A is a P-algebra in $\Bbbk$-modules. To define the homology of A, we use that A is equivalent to a dg-object concentrated in degree 0, and similarly as regards the operad P.

13.2.1 Construction. Fix a cofibrant replacement $\mathrm{P}(0) \rightarrowtail Q_A \xrightarrow{\sim} A$. The definition of the universal coefficients spectral sequences arises from the identities of §13.1.7:

$$\mathrm{Der}_{\mathsf{P}}(Q_A, E) = Hom_{U_{\mathsf{P}}(Q_A)\,\mathrm{Mod}}(\Omega^1_{\mathsf{P}}(Q_A), E) = Hom_{U_{\mathsf{P}}(A)\,\mathrm{Mod}}(T^1_{\mathsf{P}}(Q_A), E)$$

$$\text{and} \quad E \otimes_{U_{\mathsf{P}}(Q_A)} \Omega^1_{\mathsf{P}}(Q_A) = E \otimes_{U_{\mathsf{P}}(A)} T^1_{\mathsf{P}}(Q_A),$$

$$\text{where} \quad T^1_{\mathsf{P}}(Q_A) = U_{\mathsf{P}}(A) \otimes_{U_{\mathsf{P}}(Q_A)} \Omega^1_{\mathsf{P}}(Q_A).$$

The idea is to apply the standard hypercohomology spectral sequence

$$E_2^{st} = \mathrm{Ext}^s_{U_{\mathsf{P}}(A)}(H_t(T^1_{\mathsf{P}}(Q_A)), E),$$

respectively the standard hyperhomology spectral sequence

$$E_{st}^2 = \mathrm{Tor}^{U_{\mathsf{P}}(A)}_s(E, H_t(T^1_{\mathsf{P}}(Q_A))),$$

associated to this dg-hom, respectively tensor product, over the enveloping algebra $U_{\mathsf{P}}(A)$.

The object $T^1_{\mathsf{P}}(Q_A)$ forms a cofibrant left $U_{\mathsf{P}}(A)$-module by lemma 13.1.10. As a consequence, the hypercohomology spectral sequence abuts to the cohomology $H^*(Hom_{U_{\mathsf{P}}(A)\,\mathrm{Mod}}(T^1_{\mathsf{P}}(Q_A), E))$, the hyperhomology spectral sequence abuts to the homology $H_*(E \otimes_{U_{\mathsf{P}}(A)} T^1_{\mathsf{P}}(Q_A))$. By definition of $T^1_{\mathsf{P}}(Q_A)$, we also have $H_t(T^1_{\mathsf{P}}(Q_A)) = H^{\mathsf{P}}_t(A, U_{\mathsf{P}}(A))$. Thus we obtain finally:

13.2.2 Proposition (compare with [1, Propositions 5.3.2-5.3.3]). *We have universal coefficient spectral sequences*

$$E^2_{s,t} = \mathrm{Tor}^{U_{\mathsf{P}}(A)}_s(E, H^{\mathsf{P}}_t(A, U_{\mathsf{P}}(A))) \Rightarrow H^{\mathsf{P}}_{s+t}(A, E)$$

$$\text{and} \quad E_2^{s,t} = \mathrm{Ext}^s_{U_{\mathsf{P}}(A)}(H^{\mathsf{P}}_t(A, U_{\mathsf{P}}(A)), E) \Rightarrow H^{s-t}_{\mathsf{P}}(A, E). \quad \square$$

As a byproduct:

13.2.3 Proposition (compare with [1, Remark 5.3.4] and with [64]). *If the homology $H^{\mathsf{P}}_*(A, U_{\mathsf{P}}(A))$ vanishes in degree $* > 0$, then we have*

$$H^{\mathsf{P}}_*(A, E) = \mathrm{Tor}^{U_{\mathsf{P}}(A)}_*(E, \Omega^1_{\mathsf{P}}(A)) \quad \text{and} \quad H^*_{\mathsf{P}}(A, E) = \mathrm{Ext}^*_{U_{\mathsf{P}}(A)}(\Omega^1_{\mathsf{P}}(A), E).$$

Proof. The condition implies that the universal coefficients spectral sequences degenerate and the identities of the proposition follow. $\square$

Conversely, if the homology $H^{\mathsf{P}}_*(A, E)$ is given by a Tor-functor over the enveloping algebra $U_{\mathsf{P}}(A)$, then we have necessarily $H^{\mathsf{P}}_*(A, U_{\mathsf{P}}(A)) = 0$, for $* > 0$, since $\mathrm{Tor}^{U_{\mathsf{P}}(A)}_*(U_{\mathsf{P}}(A), F)$ vanishes in degree $* > 0$ for every left $U_{\mathsf{P}}(A)$-module F.

13.2.4 Classical Examples. The identities of proposition 13.2.3 are known to hold for the Hochschild (co)homology (case of the associative operad $\mathsf{P} = \mathsf{A}$) and for the Chevalley-Eilenberg (co)homology (case of the Lie operad $\mathsf{P} = \mathsf{L}$). The reader is referred to his favorite textbook. In §17.3, we prove that the assertion about Hochschild (co)homology is implied by the deeper property, checked in §10.3, that the module Ω^1_{A} forms a free right A-module, and similarly as regards the Chevalley (co)homology.

In the case $P = C$, we have $U_C(A) = A$. But the homology $H_*^C(A, A)$ does not vanish in general. This negative result is verified for the $\Bbbk$-algebra of an hypersurface (or a complete intersection) with isolated singularities. Thus the result of proposition 13.2.3 fails in the case $P = C$. The non-vanishing of $H_*^C(A, A)$ implies that Ω_C^1 does not form a projective right C-module and gives an obstruction for this property.

13.3 The Operadic Cotriple Construction

In this section, we review a usual construction of the theory of operads, the cotriple construction, which associates a kind of simplicial resolution to any algebra over an operad.

The original cotriple construction, defined in [3], is a simplicial object $B_\Delta(F, T, A)$ associated an algebra A over a monad T, with coefficients in a functor F equipped with a right action of the monad T. This construction is applied in [47] to the monad $T = S(P)$ associated to a topological operad P. If we take $F = S(P)$, then the construction returns a simplicial P-algebra $B_\Delta(P, P, A)$ whose realization is homotopy equivalent to A.

The definition of the simplicial object $B_\Delta(P, P, A)$ makes sense for any operad P in a symmetric monoidal category $\mathcal{C}$ and for any P-algebra A in a symmetric monoidal category $\mathcal{E}$ over $\mathcal{C}$. The construction returns a simplicial P-algebra $B_\Delta(P, P, A)$ together with an augmentation $\epsilon : B_\Delta(P, P, A) \to A$. In the sequel, we apply the construction $B_\Delta(P, P, A)$ to operads in the category of dg-modules $\mathcal{C} = \mathrm{dg}\,\Bbbk\,\mathrm{Mod}$ and we review the definition of $B_\Delta(P, P, A)$ in this context. Throughout this section, we assume to simplify that A is also a P-algebra in the category of dg-modules $\mathcal{E} = \mathrm{dg}\,\Bbbk\,\mathrm{Mod}$. but again we can extend the construction to P-algebras in Σ_*-objects $\mathcal{E} = \mathcal{M}$, respectively in right modules over operads $\mathcal{E} = \mathcal{M}_R$, by using the principle of generalized point-tensors. These generalizations are studied in §17.2.

In the context of simplicial sets or simplicial $\Bbbk$-modules, the construction $B_\Delta(P, P, A)$ forms a bisimplicial object and the diagonal of this bisimplicial object gives a cofibrant replacement of A in the category of P-algebras. In the dg-context, we take the normalized complex of $B_\Delta(P, P, A)$ to recover a P-algebra in the original category of dg-objects $\mathcal{E}$. The augmentation of $B_\Delta(P, P, A)$ induces a weak-equivalence $\epsilon : N_*(B_\Delta(P, P, A)) \to A$, but the object $N_*(B_\Delta(P, P, A))$ forms no more a cofibrant P-algebra. However, we shall observe that the simplicial object $B_\Delta(P, P, A)$ can be used to determine the cohomology $H_P^*(A, E)$ and the homology $H_*^P(A, E)$.

The normalized complex plays the role of geometric realization in the dg-context. Every cofibrantly generated model category has a good realization functor, but the author ignores if the realization of $B_\Delta(P, P, A)$ gives a P-algebra equivalent to A in general. This result is proved for topological operads in May's original monograph [47, §9], by using explicit homotopies, and for operads in spectra in [24].

Throughout this section, we use the notation $\mathcal{A}_\Delta$ to refer to the category of simplicial objects in any category $\mathcal{A}$. For the category of algebras over a discrete operad P, we have an identity $(_P\mathcal{E})_\Delta = {_P}(\mathcal{E}_\Delta)$, where we consider the category of P-algebras in $\mathcal{E}_\Delta$ on the right-hand side.

13.3.1 The Simplicial Cotriple Resolution. The simplicial object $B_\Delta(P, P, A)$ is given in dimension n by the composite

$$B_\Delta(P, P, A)_n = S(P)^{\circ n+1}(A) = \underset{0}{S(P)} \circ \underset{1}{S(P)} \circ \cdots \circ \underset{n}{S(P)}(A),$$

where we consider the functor $S(P) : \mathcal{E} \to \mathcal{E}$ associated to the operad P. The face $d_i : B_\Delta(P, P, A)_n \to B_\Delta(P, P, A)_{n-1}$, $i = 0, \ldots, n - 1$, is the morphism induced by the operad composition product

$$S(P) \circ S(P) = S(P \circ P) \xrightarrow{S(\mu)} S(P)$$

on factors $(i, i+1)$ of the composite $S(P)^{\circ n+1}(A)$. The face $d_n : B_\Delta(P, P, A)_n \to B_\Delta(P, P, A)_{n-1}$ is yielded by the evaluation product of the P-algebra

$$S(P)(A) = S(P, A) \xrightarrow{\lambda} A$$

on the last factor of $S(P)^{\circ n+1}(A)$. The degeneracy $s_j : B_\Delta(P, P, A)_n \to B_\Delta(P, P, A)_{n+1}$, $j = 0, \ldots, n$, is given by the insertion of an operad unit

$$\mathrm{Id} = S(I) \xrightarrow{S(\eta)} S(P)$$

between factors $(j, j + 1)$ of $S(P)^{\circ n+1}(A)$. The relations of faces and degeneracies are straightforward consequences of associativity and unit properties of operad composition products and operad actions.

The construction of this simplicial object $B_\Delta(P, P, A)$ makes sense for any operad P in a symmetric monoidal category $\mathcal{C}$, and for any P-algebra A in a symmetric monoidal category $\mathcal{E}$ over $\mathcal{C}$. Each object $B_\Delta(P, P, A)_n = S(P)^{\circ n+1}(A)$ forms a free P-algebra since we have an identity $P(-) = S(P, -)$. Observe further that the faces and degeneracies of $B_\Delta(P, P, A)$ are morphisms of P-algebras. Thus the construction $B_\Delta(P, P, A)$ returns a simplicial object in the category of P-algebras in $\mathcal{E}$.

The evaluation product $S(P)(A) = S(P, A) \xrightarrow{\lambda} A$ also gives a morphism of P-algebras $\epsilon : B_\Delta(P, P, A)_0 \to A$ so that $\epsilon d_0 = \epsilon d_1$. Equivalently, we have a morphism of simplicial P-algebras $\epsilon : B_\Delta(P, P, A) \to A$, where A is identified to a constant simplicial object in the category of P-algebras.

13.3.2 Normalized Complexes. In the context of dg-modules $\mathcal{E} = \mathcal{C} = \mathrm{dg}\Bbbk\mathrm{Mod}$, we apply a normalized complex functor to recover a genuine P-algebra in $\mathcal{E}$ from the simplicial P-algebra $B_\Delta(P, P, A)$. In §17.2, we observe that this construction can be extended to P-algebras in the category of Σ_*-objects $\mathcal{E} = \mathcal{M}$, or in categories of right modules over an operad $\mathcal{E} = \mathcal{M}_R$.

For the moment, we recall only the definition of the normalized complex functor in the context of dg-modules $\mathcal{E} = \mathrm{dg}\,\Bbbk\,\mathrm{Mod}$.

For any simplicial dg-module C, we form the quotient object:

$$N_m(C) = C_m / s_0 C_{m-1} + \cdots + s_{m-1} C_{m-1}.$$

Each object $N_m(C)$ has a grading $N_m(C) = \bigoplus_* N_m(C)_*$ given by the internal grading of the dg-module C_m and comes equipped with a differential $\delta : N_m(C)_* \to N_m(C)_{*-1}$ induced by the internal differential of C_m. The alternate sum of faces $\partial = \sum_{i=0}^{m} \pm d_i$ yields a differential $\partial : N_m(C)_* \to N_{m-1}(C)_*$ so that $\partial\delta + \delta\partial = 0$. Thus the bigraded collection $N_m(C)_*$ forms a bicomplex. Define the normalized complex $N_*(C)$ as the total complex of this bicomplex. There is a dual definition of a conormalized complex $N^*(D)$ in the context of cosimplicial dg-modules.

The normalized complex functor $N_*(-) : \mathcal{E}_\Delta \to \mathcal{E}$ does not define a functor of symmetric monoidal categories in the strong sense, but we have a natural weak-equivalence

$$\nabla : N_*(C) \otimes N_*(D) \xrightarrow{\sim} N_*(C \otimes D),$$

the Eilenberg-Mac Lane morphism, which preserves symmetric monoidal structures like the structure isomorphism of a functor of symmetric monoidal categories. If C is a (discrete) dg-module, then we have $N_*(C) = C$ and the Eilenberg-Mac Lane morphism reduces to an obvious isomorphism

$$\nabla : C \otimes N_*(D) \xrightarrow{\simeq} N_*(C \otimes D),$$

for every simplicial dg-module $D \in \mathcal{E}_\Delta$.

The normalized complex of any simplicial P-algebra in dg-modules B_Δ inherits a natural P-algebra structure defined by the evaluation morphism

$$\mathrm{P}(r) \otimes N_*(B_\Delta)^{\otimes r} \xrightarrow{\nabla} N_*(\mathrm{P}(r) \otimes B_\Delta^{\otimes r}) \xrightarrow{N_*(\lambda)} N_*(B_\Delta).$$

In particular, we obtain that the normalized complex of the simplicial bar construction $B_\Delta = B_\Delta(\mathrm{P}, \mathrm{P}, A)$ forms a P-algebra in dg-modules.

The augmentation morphism $\epsilon : B_\Delta(\mathrm{P}, \mathrm{P}, A) \to A$ induces a morphism of P-algebras in dg-modules

$$N_*(B_\Delta(\mathrm{P}, \mathrm{P}, A)) \xrightarrow{N_*(\epsilon)} N_*(A) = A.$$

We have further:

13.3.3 Lemma. *The morphism $\epsilon : N_*(B_\Delta(\mathrm{P}, \mathrm{P}, A)) \to A$ induced by the augmentation $\epsilon : B_\Delta(\mathrm{P}, \mathrm{P}, A) \to A$ forms a weak-equivalence in ${}_\mathrm{P}\mathcal{E}$.*

Proof. The augmentation $\epsilon : B_\Delta(\mathrm{P}, \mathrm{P}, A) \to A$ has a natural section $\eta : A \to B_\Delta(\mathrm{P}, \mathrm{P}, A)$ in the underlying category $\mathcal{E}$. This section is given by the morphism

$$A = S(I, A) \xrightarrow{S(\eta, A)} S(P, A) = B_{\Delta}(P, P, A)_0$$

induced by the operad unit $\eta : I \to P$. Since $\epsilon\eta = \mathrm{id}$, we have $N_*(\epsilon) \cdot N_*(\eta) = \mathrm{id}$ at the chain level.

The object $B_{\Delta}(P, P, A)$ comes also equipped with an extra degeneracy

$$s_{-1} : B_{\Delta}(P, P, A)_{n-1} \to B_{\Delta}(P, P, A)_n$$

given by the insertion of an operad unit

$$S(P \circ P^{\circ n-1}, A) \xrightarrow{S(\eta \circ P^{\circ n}, A)} S(P \circ P^{\circ n}, A)$$

at position 0 of the composite $S(P) \circ \cdots \circ S(P)(A) = S(P \circ P^{\circ n}, A)$. This extra degeneracy determines a morphisms

$$N_*(B_{\Delta}(P, P, A)) \otimes N_*(\Delta^1) \xrightarrow{h} N_*(B_{\Delta}(P, P, A))$$

which gives a chain homotopy $N_*(\eta) \cdot N_*(\epsilon) \simeq \mathrm{id}$. Hence we conclude that $N_*(\epsilon)$ forms a weak-equivalence and this achieves the proof of the lemma. $\square$

This lemma also holds for the generalization of the construction $N_*(B_{\Delta}(P, P, A))$ to P-algebras in Σ_*-objects, respectively in right modules over an operad.

Theoretically, we have to apply the functor $\Omega_P^1(-)$ to a cofibrant replacement of A to determine the cohomology $H_P^*(A, E)$ and the homology $H_*^P(A, E)$. The simplicial P-algebra $B_{\Delta}(P, P, A)$ is cofibrant in some natural sense (equip the category $\mathcal{E}_{\Delta}$ with the Reedy model structure), but the associated P-algebra in dg-modules $N_*(B_{\Delta}(P, P, A))$, for which we have a weak-equivalence $\epsilon : N_*(B_{\Delta}(P, P, A)) \xrightarrow{\sim} A$, does not form a cofibrant object in $_P\mathcal{E}$. Nevertheless:

13.3.4 Theorem (compare with [1, Theorem 7.7]). *We have*

$$H_P^*(A, E) = H^* N^* \big(\mathrm{Der}_P(B_{\Delta}(P, P, A), E) \big)$$
$$and \qquad H_*^P(A, E) = H_* N_* \big(E \otimes_{U_P(A)} \Omega_P^1(B_{\Delta}(P, P, A)) \big),$$

where on the right hand side we use a dimensionwise evaluation of functors to form a cosimplicial (respectively, simplicial) object in dg-modules and we take the cohomology (respectively, homology) of the associated conormalized (respectively, normalized) chain complex.

This theorem is proved in §17.3 as an application of the homotopy theory of functors associated to modules over operads.

Bibliographical Comments on Part III

First applications of model categories in algebra occur in Quillen's original monograph [50] for certain categories of simplicial algebras, including monoids, commutative algebras, ... (see also the article [52]).

The (semi-)model category of algebras over an operad is studied in several references in various contexts: the case of algebras in differential graded modules is addressed in [17, 26]; the case of algebras in simplicial sets and simplicial modules in [54]; the case of algebras in spectra in [21, 23, 36]. The articles [4] and [58] handle the definition of the (semi-)model structure in a general axiomatic setting. In the reference [58], the axioms of (semi-)model categories are proved by a technical study of pushouts in categories of algebras over operads. In [4], it is observed that verifications of [58] can be simplified for certain operads, including cofibrant operads, if the underlying category has a monoidal fibrant replacement functor and comes equipped with a good interval object (see *loc. cit.* for precise requirements).

The paper [22] includes another general construction of model structures for algebras and left modules over non-symmetric operads.

The definition of the (semi-)model category of operads occur in [26] for operads in dg-modules, and in [4, 58] in a more general setting. Note however that the homotopy theory of dg-operads was studied before by methods of differential graded algebra. In particular, the paper [45] gives a generalization to operads of the minimal models of rational homotopy. The Koszul duality of [18] has an interpretation in terms of these minimal models.

The notion of a semi-model category arises from [29] and gives the basis of the theory of [58]. Note however that our axioms differ slightly from these original references. The article [41] uses the semi-model structures of [29, 58], but most authors still prefer to deal with full model structures.

The definition of the homology with trivial coefficients occurs in [17, 18] for algebras over operads in dg-modules. The (co)homology with non-trivial coefficients is defined in [1]. In all these references, the authors deal with algebras in non-negatively graded dg-modules. The generalization of the (co)homology to algebras in unbounded dg-modules occurs in [26]. Universal coefficients spectral sequences are defined in [1] (for algebras in non-negatively graded dg-modules) by using the cotriple construction.

The work [20, 21, 62, 63] tackles applications of the cohomology of algebras over operads to realization problems in stable homotopy. In this setting, it is more appropriate to deal with simplicial objects and simplicial algebras rather than differential graded algebras, but this does not change much the definition of the (co)homology of algebras over operads.

The definition of the cotriple construction goes back to [3] and [47]. The idea to represent the cotriple construction of functors at the operad level occurs in [56, 57] and in [54]. The two-sided version of the cotriple construction is also studied in [14] in the differential graded context and is related to other complexes of the theory of operads, notably the Koszul complex of [18]. The recent preprint [24] focuses on applications of the operadic cotriple construction in the context of spectra.

Part IV
The Homotopy of Modules over Operads and Functors

Chapter 14
The Model Category of Right Modules over an Operad

Introduction

In this part, we aim to prove that right R-module structures give convenient models for the homotopy of associated functors.

First of all, we check in this chapter that the category of right R-modules inherits a convenient model structure. This verification is carried out in §14.1.

In §14.2, we make explicit the structure of cell complexes of right R-modules in the case where $\mathcal{C} = \mathrm{dg}\,\Bbbk\,\mathrm{Mod}$ is the category of dg-modules over a ring $\Bbbk$.

In §14.3, we apply the general results of §11.1 to the symmetric monoidal model category of right R-modules. To summarize, we obtain that the category of P-algebras in right R-modules, where P a Σ_*-cofibrant operad, is equipped with a natural semi model structure, inherited from right R-modules. In the case of a non-unitary operad R, we obtain that the category of P-algebras in connected right R-modules is equipped with a semi model structure as long as the operad P is $\mathcal{C}$-cofibrant.

14.1 The Symmetric Monoidal Model Category of Right Modules over an Operad

In this section, we check the definition of the symmetric monoidal model structure on the category of right modules over an operad R. For this purpose, we apply the general result of theorem 11.1.13 to the composite adjunction

$$\mathcal{C}^{\mathbb{N}} \underset{U}{\overset{\Sigma_* \otimes -}{\rightleftarrows}} \mathcal{M} \underset{U}{\overset{- \circ \mathrm{R}}{\rightleftarrows}} \mathcal{M}_{\mathrm{R}}$$

between the category of collections $\{K(n) \in \mathcal{C}\}_{n \in \mathbb{N}}$ and the category of right R-modules. Recall that $\mathcal{C}^{\mathbb{N}}$ is equipped with the obvious model structure of a product of model categories. The result reads:

B. Fresse, *Modules over Operads and Functors*, Lecture Notes in Mathematics 1967, 219
DOI: 10.1007/978-3-540-89056-0_14, © Springer-Verlag Berlin Heidelberg 2009

Proposition 14.1.A. *Let* R *be an operad. Assume that* R *is* C-*cofibrant.*

The category of right R-*modules* $\mathcal{M}_R$ *is equipped with the structure of a cofibrantly generated model category so that a morphism* $f : M \to N$ *is a weak-equivalence (respectively, a fibration) if the morphisms* $f : M(n) \to N(n)$ *define weak-equivalences (respectively, fibrations) in the base category* C. *The generating (acyclic) cofibrations of* $\mathcal{M}_R$ *are the tensor products*

$$i \otimes F_r \circ R : C \otimes F_r \circ R \to D \otimes F_r \circ R,$$

where $i : C \to D$ *ranges over the generating (acyclic) cofibrations of the base category* C *and* F_r *ranges over the generators, defined in* §2.1.12, *of the category of* Σ_*-*objects.*

Recall that an operad R is C-cofibrant if the underlying collection $\{R(n)\}_{n\in\mathbb{N}}$ consists of cofibrant objects in C.

In the case where C is the category of simplicial sets $C = \mathcal{S}$ and the category of simplicial $\Bbbk$-modules $C = \Bbbk\,\mathrm{Mod}_\Delta$, a different (but Quillen equivalent) model structure is used in [54, §3.3.8], namely the adjoint of the equivariant model structure of Σ_*-objects.

The forgetful functors

$$\mathcal{M}_R \xrightarrow{U} \mathcal{M} \xrightarrow{U} C^{\mathbb{N}}$$

creates all colimits. For this reason, condition (1) of proposition 11.1.14 is trivially satisfied and the proof of the sufficient conditions of proposition 11.1.14 reduces to:

Claim. *Under the assumption that* R *is* C-*cofibrant, the tensor product* $i \otimes F_r \circ R : C \otimes F_r \circ R \to D \otimes F_r \circ R$ *defines a* C-*cofibration (respectively, an acyclic* C-*cofibration) if* $i : C \to D$ *is a cofibration (respectively, an acyclic cofibration) in* C.

Proof. Lemma 11.4.6 implies that $F_r \circ R = R^{\otimes r}$ is C-cofibrant if the operad R is so. As a consequence, we obtain that the tensor products $- \otimes (F_r \circ R(n)) : C \to C$, $n \in \mathbb{N}$, preserve (acyclic) cofibrations and this proves our claim. ☐

This claim achieves the proof of proposition 14.1.A. ☐

Observe further:

Proposition 14.1.B. *The category of right* R-*modules* $\mathcal{M}_R$ *forms a symmetric monoidal model category over* C *in our sense:*

(a) *If the unit object* $1 \in C$ *is cofibrant in* C, *then so is the equivalent constant right* R-*module* $1 \in \mathcal{M}_R$ *so that axiom MM0 holds in* $\mathcal{M}_R$;

(b) *The internal and external tensor products of right* R-*modules satisfy the pushout-product axiom MM1.*

Proof. The verifications are similar to the case of Σ_*-objects handled in §11.4. Replace simply the objects $F_r = I^{\otimes r}$ by the right R-modules $F_r \circ R = I^{\otimes r} \circ R = R^{\otimes r}$. $\qquad\square$

In the remainder of this section, we prove additional properties of the model category of right R-modules.

14.1.1 Proposition.

(a) *If the operad* R *is* Σ_**-cofibrant, then the forgetful functor* $U : \mathcal{M}_R \to \mathcal{M}$ *preserves cofibrations.*

(b) *If the operad* R *is just* C*-cofibrant, then the image of a cofibration* $f : M \to N$ *under the forgetful functor* $U : \mathcal{M}_R \to \mathcal{M}$ *is only a* C*-cofibration.*

Note that proposition 11.1.14, which is applied to define the model structure of $\mathcal{M}_R$, gives assertion (b).

Proof. It is sufficient to check that the generating cofibrations

$$i \otimes F_r \circ R : C \otimes F_r \circ R \to D \otimes F_r \circ R$$

define cofibrations in $\mathcal{M}$, respectively C-cofibrations in $\mathcal{M}$. In lemma 11.4.6, we observe that $F_r \circ R = R^{\otimes r}$ forms a cofibrant object in $\mathcal{M}$ if R is Σ_*-cofibrant, respectively a C-cofibrant object in $\mathcal{M}$ if R is just C-cofibrant. The claim follows immediately. $\qquad\square$

Proposition 14.1.1 has the following easy consequences:

14.1.2 Proposition.

(a) *If the operad* R *is* Σ_**-cofibrant, then the* Σ_**-cofibrations form an enlarged class of cofibrations in* $\mathcal{M}_R$.

(b) *If the operad* R *is just* C*-cofibrant, then the* C*-cofibrations form an enlarged class of cofibrations in* $\mathcal{M}_R$.

Proof. Use that the forgetful functors

$$\mathcal{M}_R \xrightarrow{U} \mathcal{M} \xrightarrow{U} C^{\mathbb{N}}$$

creates colimits to check axioms C1-4 of an enlarged class of cofibrations. Proposition 14.1.1 gives axiom C0. $\qquad\square$

14.1.3 Proposition. *If* C *is a proper model category and the operad* R *is* C*-cofibrant, then* $\mathcal{M}_R$ *is proper as well.*

Proof. By definition of model structures, the forgetful functors

$$\mathcal{M}_R \to \mathcal{M} \to C^{\mathbb{N}}$$

create fibrations and weak-equivalences in $\mathcal{M}$ and in $\mathcal{M}_R$. In proposition 14.1.1, we observe that the forgetful functor $U : \mathcal{M}_R \to \mathcal{C}^N$ preserves cofibrations as well as long as R is $\mathcal{C}$-cofibrant. Recall that $U : \mathcal{M}_R \to \mathcal{C}^N$ creates colimits and limits.

The claim that $\mathcal{M}_R$ inherits a proper model structure if $\mathcal{C}$ forms a proper model category is an immediate consequence of these observations. $\square$

14.2 Relative Cell Complexes in Right Modules over Operads

In this section, we study the structure of cofibrant cell complexes of right R-modules in the case where $\mathcal{C} = \mathrm{dg}\, \Bbbk \,\mathrm{Mod}$. As usual, we use the convention of §11.1.8 to call cofibrant cell objects the cell complexes build from generating cofibrations.

We generalize the construction of §11.2 to give an explicit representation of cofibrant cell complexes of right R-modules. Since the construction are strictly parallel, we only make explicit the definition of a twisting cochain and of a quasi-free object in the context of right R-modules before giving the result.

The notion of a quasi-free module over an operad is used extensively in [14]. We refer to this article for a detailed study of this structure.

14.2.1 Twisting Cochains and Quasi-Free Modules over dg-Operads. A *twisting cochain of right* R-*modules* $\partial : M \to M$ consists of a collection of twisting cochains of dg-modules $\partial \in Hom_{\mathcal{C}}(M(n), M(n))$ that commutes with the action of symmetric groups and with the operad action. Equivalently, we assume that the collection $\partial \in Hom_{\mathcal{C}}(M(n), M(n))$ defines an element of $Hom_{\mathcal{M}_R}(M, M)$, the dg-hom of the category of right R-modules, and the equation of twisting cochains $\delta(\partial) + \partial^2 = 0$ holds in $Hom_{\mathcal{M}_R}(M, M)$.

A right R-module M is quasi-free if we have $M = (K \circ R, \partial)$, for a free right R-module $K \circ R$, where $\partial : K \circ R \to K \circ R$ is a twisting cochain of right R-modules. In the case of a quasi-free object, the commutation with the R-action implies that the twisting cochain $\partial : K \circ R \to K \circ R$ is determined by a homomorphism of Σ_*-objects $\partial : K \to K \circ R$.

The extension of propositions 11.2.9-11.2.9 to modules over operads reads:

14.2.2 Proposition. *A cofibrant cell object in right* R-*modules is equivalent to a quasi-free right* R-*module* $M = (K \circ R, \partial)$, *where* K *is a free* Σ_*-*object in graded* $\Bbbk$-*modules* $K = \bigoplus_\alpha \Bbbk\, e_{d_\alpha} \otimes F_{r_\alpha}$, *together with a trivial differential and a basis filtration* $K_\lambda = \bigoplus_{\alpha < \lambda} \Bbbk\, e_{d_\alpha} \otimes F_{r_\alpha}$ *so that* $\partial(K_\lambda) \in K_{\lambda-1} \circ R$. $\square$

14.3 Model Categories of Algebras in Right Modules over an Operad

In this section, we record applications of general results of §12.3 to algebras in right modules over an operad R. Since we prove that the category of right R-modules forms a symmetric monoidal model category over $\mathcal{C}$, proposition 12.3 returns:

14.3.1 Proposition. *Let R be any $\mathcal{C}$-cofibrant operad. If P is a Σ_*-cofibrant operad, then the P-algebras in right R-modules form a cofibrantly generated semi-model category so that the forgetful functor $U : {}_P\mathcal{M}_R \to \mathcal{M}_R$ creates weak-equivalences and fibrations.*

The generating (acyclic) cofibrations are the morphisms of free P-algebras

$$P(i \otimes F_r \circ R) : P(C \otimes F_r \circ R) \to P(D \otimes F_r \circ R),$$

where $i : C \to D$ ranges over the generating (acyclic) cofibrations of $\mathcal{C}$.

Recall that P-algebras in right R-modules are equivalent to P-R-bimodules.

¶ In §5.1.4, we observe that the category of connected right R-modules $\mathcal{M}_R^0$, where R is a non-unitary operad, forms a reduced symmetric monoidal category. For this reduced symmetric monoidal category, we obtain further:

14.3.2 ¶ Lemma. *The category of connected right R-modules $\mathcal{M}_R^0$ has regular tensor powers with respect to the class of $\mathcal{C}$-cofibrations in the category of Σ_*-objects and to the class of $\mathcal{C}$-cofibrations in the category of right R-modules.*

Note that the $\mathcal{C}$-cofibrations form an enlarged class of cofibrations in $\mathcal{M}_R^0$, like the Σ_*-cofibrations, just because the forgetful functors $\mathcal{M}_R \to \mathcal{M} \to \mathcal{C}^{\mathbb{N}}$ create colimits.

Proof. This property is inherited from the category of Σ_*-objects (for which we refer to proposition 11.6.6) because, on one hand, weak-equivalences of right R-modules are created in the category of Σ_*-objects and, on the other hand, the tensor product of right R-modules is created in the category of Σ_*-objects.   □

As a byproduct, we obtain:

14.3.3 ¶ Proposition. *Let R be a $\mathcal{C}$-cofibrant non-unitary operad. Let P be a non-unitary operad. The category of connected P-R-bimodules ${}_P\mathcal{M}_R^0$ forms a semi-model category as long as P is $\mathcal{C}$-cofibrant.*   □

Chapter 15
Modules and Homotopy Invariance of Functors

Introduction

In this chapter, we study the homotopy invariance of the functors $S_R(M) : {}_R\mathcal{E} \to \mathcal{E}$ associated to right modules over an operad R. In summary, we aim to prove that:

- A weak-equivalence of R-algebras $f : A \xrightarrow{\sim} B$ induces a weak-equivalence $S_R(M, f) : S_R(M, A) \xrightarrow{\sim} S_R(N, B)$ under reasonable assumptions on M, A and B,
- A weak-equivalence of right R-modules $f : M \xrightarrow{\sim} N$ induces a pointwise weak-equivalence of functors $S_R(f, A) : S_R(M, A) \xrightarrow{\sim} S_R(N, A)$, under reasonable assumptions on M, N and A.

In §15.1, we assume that the right R-modules are cofibrant. In this context, the homotopy invariance properties hold under very mild assumptions on the R-algebras.

In §15.2, we address the case where M and N are only cofibrant in the category of Σ_*-objects. In this context, we have to assume that the R-algebras are cofibrant to obtain our homotopy invariance properties.

In §15.3, we study the homotopy invariance of the relative composition product of modules over operads. Results can be deduced from theorems of §15.1-15.2, applied to R-algebras in Σ_*-objects. We study the homotopy invariance of the relative composition product for connected R-algebras in Σ_*-objects. We prove that the general result can be improved in this context.

B. Fresse, *Modules over Operads and Functors*, Lecture Notes in Mathematics 1967, 225
DOI: 10.1007/978-3-540-89056-0_15, © Springer-Verlag Berlin Heidelberg 2009

15.1 Cofibrant Modules and Homotopy Invariance of Functors

The purpose of this section is to prove:

Theorem 15.1.A. *Let* R *be a* C*-cofibrant operad.*

(a) The morphism

$$S_R(f, A) : S_R(M, A) \to S_R(N, A)$$

induced by a weak-equivalence of cofibrant right R*-modules* $f : M \xrightarrow{\sim} N$ *is a weak-equivalence for every* R*-algebra* $A \in {}_R\mathcal{E}$ *which is* $\mathcal{E}$*-cofibrant.*

(b) If M *is a cofibrant right* R*-module, then the morphism*

$$S_R(M, f) : S_R(M, A) \to S_R(M, B)$$

induced by a weak-equivalence of R*-algebras* $f : A \xrightarrow{\sim} B$ *is a weak-equivalence as long as* A, B *are* $\mathcal{E}$*-cofibrant.*

Recall that an R-algebra A is $\mathcal{E}$-cofibrant if the initial morphism $\eta : R(0) \to A$ is a cofibration in $\mathcal{E}$. Since the functor $\eta : \mathcal{C} \to \mathcal{E}$ is supposed to preserve cofibrations and the operad R is cofibrant in the underlying category, we obtain that any $\mathcal{E}$-cofibrant R-algebra A forms cofibrant object in $\mathcal{E}$.

This observation is used in the proof of theorem 15.1.A. Besides, we use that the forgetful functor $U : {}_R\mathcal{E} \to \mathcal{E}$ preserves cofibrations.

¶ Remark. Theorem 15.1.A holds in the context symmetric spectra $\mathcal{E} = \mathrm{Sp}^\Sigma$ and for an operad in simplicial sets R provided that $R(0) = \emptyset$ and we can ensure that the forgetful functor $U : {}_R\mathcal{E} \to \mathcal{E}$ preserves cofibrations. The assumption $R(0) = \emptyset$ is necessary for an operad in simplicial sets since the functor $\eta : \mathcal{C} \to \mathcal{E}$ does not preserve cofibrations in the case $\mathcal{E} = \mathrm{Sp}^\Sigma$ and $\mathcal{C} = \mathcal{S}$ (see §11.6.12).

The remainder of this section is devoted to the proof of theorem 15.1.A. Our argument is based on the following statement:

15.1.1 Lemma. *The pushout-product*

$$(i_*, f_*) : S_R(M, B) \bigoplus_{S_R(M,A)} S_R(N, A) \to S_R(N, B)$$

induced by $i : M \to N$*, morphism of right* R*-modules, and* $f : A \to B$*, morphism of* R*-algebras, is a cofibration as long as* i *is a cofibration in* $\mathcal{M}_R$*, the morphism* f *forms an* $\mathcal{E}$*-cofibration, and* A *is* $\mathcal{E}$*-cofibrant. The pushout-product* (i_*, f_*) *is an acyclic cofibration if* i *or* f *is also acyclic.*

Proof. By lemma 11.3.2, we can reduce the claim to the case where $i : M \to N$ is a generating (acyclic) cofibration

$$j \otimes F_r \circ \mathrm{R} : C \otimes F_r \circ \mathrm{R} \to D \otimes F_r \circ \mathrm{R}$$

since the functor $(M, A) \mapsto \mathrm{S}_\mathrm{R}(M, A)$ preserves colimits in M.

Consider the more general case of a morphism of free right R-modules

$$k \circ \mathrm{R} : K \circ \mathrm{R} \to L \circ \mathrm{R}$$

induced by a cofibration (an acyclic cofibration) of Σ_*-objects $k : K \to L$. In the case of a right R-module of the form $M = K \circ \mathrm{R}$, we have

$$\mathrm{S}_\mathrm{R}(M, A) = \mathrm{S}_\mathrm{R}(K \circ \mathrm{R}, A) = \mathrm{S}(K, A)$$

by theorem 7.1.1. Therefore we obtain that the pushout-product

$$(i_*, f_*) : \mathrm{S}_\mathrm{R}(M, B) \bigoplus_{\mathrm{S}_\mathrm{R}(M,A)} \mathrm{S}_\mathrm{R}(N, A) \to \mathrm{S}_\mathrm{R}(N, B)$$

is identified with the pushout-product

$$(k_*, f_*) : \mathrm{S}(K, B) \bigoplus_{\mathrm{S}(K,A)} \mathrm{S}(L, A) \to \mathrm{S}(L, B)$$

of lemma 11.5.1. Thus we can deduce lemma 15.1.1 from the assertion of lemma 11.5.1. □

As a corollary, we obtain:

15.1.2 Lemma.

(a) Let M be a cofibrant object in $\mathcal{M}_\mathrm{R}$. If a morphism of R-algebras $f : A \to B$, where A is $\mathcal{E}$-cofibrant, forms a $\mathcal{E}$-cofibration (respectively, an acyclic $\mathcal{E}$-cofibration), then the induced morphism $\mathrm{S}_\mathrm{R}(M, f) : \mathrm{S}_\mathrm{R}(M, A) \to \mathrm{S}_\mathrm{R}(M, B)$ defines a cofibration (respectively, an acyclic cofibration) in $\mathcal{E}$.

(b) Let A be an $\mathcal{E}$-cofibrant R-algebra. The morphism $\mathrm{S}_\mathrm{R}(i, A) : \mathrm{S}_\mathrm{R}(M, A) \to \mathrm{S}_\mathrm{R}(N, A)$ induced by a cofibration (respectively, an acyclic cofibration) of right R-modules $i : M \to N$ forms a cofibration (respectively, an acyclic cofibration) in $\mathcal{E}$. □

And the claim of theorem 15.1.A follows immediately from Brown's lemma. □

15.2 Homotopy Invariance of Functors for Cofibrant Algebras

In this chapter, we study the homotopy invariance of the functors $\mathrm{S}_\mathrm{R}(M) : {}_\mathrm{R}\mathcal{E} \to \mathcal{E}$ associated to right R-modules M which are not cofibrant. Our main result reads:

Theorem 15.2.A. *Let* R *be an operad. Suppose that* R *is* Σ_*-*cofibrant so that the category of* R-*algebras is equipped with a semi-model structure.*

(a) If A is cofibrant in the category of R-*algebras, then the morphism*

$$S_R(f, A) : S_R(M, A) \to S_R(N, A)$$

induced by a weak-equivalence of right R-*modules* $f : M \xrightarrow{\sim} N$ *forms a weak-equivalence as long as* M, N *are* Σ_*-*cofibrant.*

(b) Let M be a Σ_*-*cofibrant right* R-*module. The morphism*

$$S_R(M, f) : S_R(M, A) \to S_R(M, B)$$

induced by a weak-equivalence of R-*algebras* $f : A \xrightarrow{\sim} B$ *forms a weak-equivalence if A and B are cofibrant in the category of* R-*algebras.*

¶ **Remark.** In the context of a (reduced) symmetric monoidal category with regular tensor powers, we assume that R is a $\mathcal{C}$-cofibrant non-unitary operad. Assertion (a) holds for any weak-equivalence of connected right R-modules $f : M \xrightarrow{\sim} N$ such that M, N are $\mathcal{C}$-cofibrant. Assertion (b) holds for all connected right R-modules M which are $\mathcal{C}$-cofibrant.

¶ **Remark.** Suppose $\mathcal{E} = \mathcal{C} = \mathrm{Sp}^\Sigma$ is the category of symmetric spectra together with the stable positive model structure. Let R be an operad in simplicial sets. If we can ensure that the forgetful functor $U : {}_R\mathcal{E} \to \mathcal{E}$ preserves cofibrations, then our arguments work for right R-modules which are cofibrant in the category of spectra. Hence:

- assertion (a) of theorem 15.2.A holds for any weak-equivalence of right R-modules in spectra $f : M \xrightarrow{\sim} N$ such that M, N are Sp^Σ-cofibrant;
- assertion (b) of theorem 15.2.A holds for all right R-modules in spectra M which are Sp^Σ-cofibrant.

Observe that assertion (a) of theorem 15.2.A has a converse:

Proposition 15.2.B. *Assume* $\mathcal{E} = \mathcal{C}$.

Let $f : M \to N$ *be a morphism of right* R-*modules, where* M, N *are* Σ_*-*cofibrant. If the morphism* $S_R(f, A) : S_R(M, A) \to S_R(N, A)$ *induced by* f *forms a weak-equivalence in* $\mathcal{C}$ *for every cofibrant* R-*algebra* $A \in {}_R\mathcal{E}$, *then* f *is a weak-equivalence as well.*

If $\mathcal{C}$ *is pointed, then this implication holds without the assumption that the modules* M, N *are* Σ_*-*cofibrant.*

Proof. In proposition 11.5.4, we prove a similar assertion for the bifunctor $(M, X) \mapsto S(M, X)$, where $M \in \mathcal{M}$ and $X \in \mathcal{C}$. Proposition 15.2.B is an immediate corollary of this result because the free R-algebras $A = R(X)$, where X is a cofibrant object in $\mathcal{C}$, are cofibrant R-algebras and, by theorem 7.1.1, we have $S_R(M, R(X)) \simeq S(M, X)$ for a free R-algebra $A = R(X)$. $\qquad\square$

Theorem 15.2.A is again a routine consequence of a pushout-product lemma:

Lemma 15.2.C. *The pushout-product*

$$(i_*, f_*) : S_R(M, B) \bigoplus_{S_R(M,A)} S_R(N, A) \to S_R(N, B)$$

induced by $i : M \to N$, *morphism of right* R-*modules, and* $f : A \to B$, *morphism of* R-*algebras, forms a cofibration as long as* i *defines a* Σ_*-*cofibration, the morphism* f *is a cofibration of* R-*algebras, and* A *is a cofibrant* R-*algebra. The pushout-product* (i_*, f_*) *is an acyclic cofibration if* i *or* f *is also acyclic.*

The technical proof of this lemma is postponed to the appendix, §§18-20. The arguments of lemma 15.1.1 can not be carried out in the context of lemma 15.2.C, because we address the image of cofibrations $f : A \to B$ under functors $S_R(M) : {}_R\mathcal{E} \to \mathcal{E}$ which do not preserve colimits.

In the remainder of this section, we survey examples of applications of theorem 15.2.A.

In §16, we prove that the unit morphism $\eta(A) : A \to \psi^* \psi_!(A)$ of the adjunction relation defined by extension and restriction functors

$$\psi_! : {}_R\mathcal{E} \rightleftarrows {}_S\mathcal{E} : \psi^*$$

preserve weak-equivalences. For this purpose, we apply theorem 15.2.A to right R-modules defined by the operads $M = R$ and $N = S$ themselves and we observe that the adjunction unit $\eta(A) : A \to \psi^* \psi_!(A)$ is identified with a morphism of the form

$$S_R(\psi, A) : S_R(R, A) \to S_R(S, A).$$

The homotopy invariance of generalized James's constructions give other instances of applications of theorem 15.2.A. In §5.1.8, we recall that the generalized James construction $J(M, X)$ defined in the literature for a Λ_*-object M is the instance of a functor of the form $S_*(M, X)$, where $*$ is an initial operad in Top_*, the category of pointed topological spaces.

An object $X \in {}_*\mathcal{E}$ is cofibrant in ${}_*\mathcal{E}$ simply if the unit morphism $* : \mathrm{pt} \to X$ forms a cofibration in the underlying model category of topological spaces. Thus the cofibrant algebras over the initial unitary operad are the well-pointed topological spaces and theorem 15.2.A asserts that a weak-equivalence of Σ_*-cofibrant Λ_*-objects, the cofibrant objects for the Reedy model structure of [16], induces a weak-equivalence $S_*(f, X) : S_*(M, X) \xrightarrow{\sim} S_*(N, X)$ for all well-pointed spaces X. In this context, we retrieve a result of [7]. In fact, in this particular example, the space $S_*(M, X)$ can be identified with a coend over the generalized Reedy category Λ_*, considered in [16], and the homotopy invariance of the functor $M \mapsto S_*(M, X)$ can be deduced from general homotopy invariance properties of coends (see [7, Lemma 2.7]).

15.3 ¶ Refinements for the Homotopy Invariance of Relative Composition Products

In §§3.2.9-3.2.10, we identify the category of left R-modules $_R\mathcal{M}$ with the category of R-algebras in Σ_*-objects. Furthermore, we observe in §5.1.5 that the functor $S_R(M) : _R\mathcal{M} \to \mathcal{M}$ defined by a right R-module on this category is identified with the relative composite $N \mapsto M \circ_R N$. As a consequence, theorem 15.1.A and theorem 15.2.A imply that the relative composition products $(M, N) \mapsto M \circ_R N$ satisfy natural homotopy invariance properties.

In this chapter, we prove that stronger homotopy invariance properties hold if we assume that N is a connected left R-module.

Recall that the model category of connected right R-modules is cofibrantly generated for a collection of generating cofibrations of the form

$$\mathcal{I} = \{i \circ R : K \circ R \to L \circ R, \text{ where } i : K \to L \text{ runs over cofibrations in } \mathcal{M}^0 \}.$$

As a consequence, every cofibration of $\mathcal{M}_R^0$ is a retract of a relative $\mathcal{I}$-cell complex. Symmetrically, the semi-model category of connected left R-modules is cofibrantly generated for a collection of generating cofibrations of the form

$$\mathcal{J} = \{R \circ i : R \circ K \to R \circ L, \text{ where } i : K \to L \text{ runs over cofibrations in } \mathcal{M}^0 \}.$$

To identify this set of generating cofibrations, use that the composite $R \circ K$ represents the free R-algebra $R \circ K = R(K)$ and go back to the general definition of the semi-model category of R-algebras in §12.3. As a byproduct, every cofibration of $_R\mathcal{M}^0$ with a cofibrant domain is a retract of a relative $\mathcal{J}$-cell complex.

For our purpose, we consider the collection of morphisms

$$\mathcal{K} = \{i \circ R : K \circ R \to L \circ R, \text{ where } i : K \to L \text{ runs over the } \mathcal{C}\text{-cofibrations in } \mathcal{M}^0 \}$$

in the category of right R-modules and the associated class of relative $\mathcal{K}$-cell complexes. Symmetrically, we consider the collection of morphisms

$$\mathcal{L} = \{R \circ i : R \circ K \to R \circ L, \text{ where } i : K \to L \text{ runs over the } \mathcal{C}\text{-cofibrations in } \mathcal{M}^0 \}$$

in the category of left R-modules and the associated class of relative $\mathcal{L}$-cell complexes. It is not clear that the collection $\mathcal{K}$ (respectively, $\mathcal{L}$) defines the generating cofibrations of a model structure on $\mathcal{M}_R^0$ (respectively, $_R\mathcal{M}^0$) in general.

In the case of a relative composition product $S_R(M, N) = M \circ_R N$, we can improve on theorem 15.1.A and theorem 15.2.A to obtain:

Theorem 15.3.A. *Let R be a $\mathcal{C}$-cofibrant non-unitary operad.*

(a) The morphism

$$f \circ_R A : M \circ_R A \to N \circ_R A$$

induced by a weak-equivalence of connected right R-modules $f : M \xrightarrow{\sim} N$ forms a weak-equivalence as long as M, N are retracts of $\mathcal{K}$-cell complexes, for every $\mathcal{C}$-cofibrant object $A \in {}_R\mathcal{M}^0$.

(b) If a connected right R-module M is a retract of a $\mathcal{K}$-cell complex, then the morphism

$$M \circ_R g : M \circ_R A \to M \circ_R B$$

induced by a weak-equivalence of connected left R-modules $g : A \xrightarrow{\sim} B$ forms a weak-equivalence as long as A, B are $\mathcal{C}$-cofibrant.

Theorem 15.3.B. *Let R be a $\mathcal{C}$-cofibrant non-unitary operad.*

(a) *If a connected left R-module A is a retract of an $\mathcal{L}$-cell complex, then the morphism*

$$f \circ_R A : M \circ_R A \to N \circ_R A$$

induced by a weak-equivalence of connected right R-modules $f : M \xrightarrow{\sim} N$ forms a weak-equivalence as long as M, N are $\mathcal{C}$-cofibrant.

(b) *The morphism*

$$M \circ_R g : M \circ_R A \to M \circ_R B$$

induced by a weak-equivalence of connected left R-modules $g : A \xrightarrow{\sim} B$ forms a weak-equivalence as long as A, B are retracts of $\mathcal{L}$-cell complexes, for every $\mathcal{C}$-cofibrant object $M \in \mathcal{M}_R^0$.

One can observe further that theorem 15.3.A holds for any category of algebras in a reduced category with regular tensor powers, and not only for connected left R-modules, equivalent to algebras in the category of connected Σ_*-objects.

We mention these results as remarks and we give simply a sketch of the proof of theorem 15.3.A and theorem 15.3.B. In the case $\mathcal{C} = \mathrm{dg}\,\Bbbk\,\mathrm{Mod}$, the category of dg-modules over a ring $\Bbbk$, another proof can be obtained by spectral sequence techniques (see the comparison theorems in [14, §2]).

The proof of theorem 15.3.A and theorem 15.3.B is based on the following lemma:

15.3.1 Lemma.

(a) *Let $i : M \to N$ be a relative $\mathcal{K}$-cell complex in the category of connected right R-modules. Let $f : A \to B$ be any morphism of connected left R-modules. Suppose A is $\mathcal{C}$-cofibrant. If f is an (acyclic) $\mathcal{C}$-cofibration, then so is the pushout-product*

$$(i_*, f_*) : M \circ_R B \bigoplus_{M \circ_R A} N \circ_R A \to N \circ_R B$$

associated to i and f.

(b) Let $i : M \to N$ be any morphism of connected right R-modules. Let $f : A \to B$ be a relative $\mathcal{L}$-cell complex in the category of connected left R-modules. Suppose A is an $\mathcal{L}$-cell complex. If i is an (acyclic) $\mathcal{C}$-cofibration, then so is the pushout-product

$$(i_*, f_*) : M \circ_R B \bigoplus_{M \circ_R A} N \circ_R A \to N \circ_R B$$

induced by i.

Proof. These assertions improve on results of §§15.1-15.2 in the case where we consider R-algebras in the category of connected Σ_*-objects $\mathcal{M}_R^0$. Assertion (a) corresponds to the case where i is a cofibration and f is an (acyclic) $\mathcal{C}$-cofibration in lemma 15.1.1 and is proved in by the same argument. Assertion (b) arises from the extension of lemma 15.2.C to reduced categories with regular tensor powers, in the case where i is an (acyclic) $\mathcal{C}$-cofibration and f is a cofibration between cofibrant objects. To prove this latter assertion, check indications given in §15.2.   □

As a corollary, we obtain:

15.3.2 Lemma.
(a) If M is a $\mathcal{K}$-cell complex in the category of connected right R-modules, then the functor $M \circ_R -$ maps (acyclic) $\mathcal{C}$-cofibrations between $\mathcal{C}$-cofibrant objects in the category of connected left R-modules to weak-equivalences.

(b) If A is an $\mathcal{L}$-cell complex in the category of connected left R-modules, then the functor $- \circ_R A$ maps (acyclic) $\mathcal{C}$-cofibrations in the category of connected right R-modules to weak-equivalences.   □

And Brown's lemma implies again:

15.3.3 Lemma.
(a) If M is a $\mathcal{K}$-cell complex in the category of connected right R-modules, then the functor $M \circ_R -$ maps weak-equivalences between $\mathcal{C}$-cofibrant objects in $_R\mathcal{M}^0$ to weak-equivalences.

(b) If A is an $\mathcal{L}$-cell complex in the category of connected left R-modules, then the functor $- \circ_R A$ maps weak-equivalences between $\mathcal{C}$-cofibrant objects in $\mathcal{M}_R^0$ to weak-equivalences.   □

Observe simply that $\mathcal{K}$-cell (respectively, $\mathcal{L}$-cell) complexes are $\mathcal{C}$-cofibrant and check the argument of [28, Lemma 1.1.12].

Obviously, we can extend this lemma to the case where M (respectively, A) is a retract of a $\mathcal{K}$-cell complex (respectively, of an $\mathcal{L}$-cell complex).

Thus lemma 15.3.3 implies assertion (b) of theorem 15.3.A and assertion (a) of theorem 15.3.B.

To achieve the proof of theorem 15.3.A and theorem 15.3.B, we observe:

Claim.

(c) *In theorem 15.3.A, assertion (b) implies assertion (a).*

(d) *In theorem 15.3.B, assertion (a) implies assertion (b).*

Proof. Let $f : M \xrightarrow{\sim} N$ be a weak-equivalence between retracts of $\mathcal{K}$-cell complexes in $\mathcal{M}_R^0$. Let A be a $\mathcal{C}$-cofibrant object of $_R\mathcal{M}^0$.

Let $g : Q_A \xrightarrow{\sim} A$ be a cofibrant replacement of A in $_R\mathcal{M}^0$. The object Q_A is Σ_*-cofibrant by proposition 12.3.2, and hence is $\mathcal{C}$-cofibrant, and forms a (retract of) a $\mathcal{K}$-cell complex as well.

We have a commutative diagram

$$
\begin{array}{ccc}
M \circ_R A & \xrightarrow{\;f \circ_R A\;} & N \circ_R A \\[2pt]
{\scriptstyle M \circ_R g}\Big\uparrow {\scriptstyle \sim} & & {\scriptstyle \sim}\Big\uparrow {\scriptstyle N \circ_R g} \\[2pt]
M \circ_R Q_A & \xrightarrow[\;f \circ_R A\;]{\sim} & N \circ_R Q_A
\end{array}
$$

in which vertical morphisms are weak-equivalences by assertion (b) of theorem 15.3.A. Since Q_A is assumed to be cofibrant, theorem 15.2.A (use the suitable generalization in the context of reduced categories with regular tensor powers – see remarks below the statement of theorem 15.2.A) implies that the lower horizontal morphism is also a weak-equivalence.

We conclude that

$$
f \circ_R A : M \circ_R A \to N \circ_R A
$$

defines a weak-equivalence as well and this proves assertion (c) of the claim.

We prove assertion (c) by symmetric arguments.   □

This verification achieves the proof of theorem 15.3.A and theorem 15.3.B.

□

Chapter 16
Extension and Restriction Functors and Model Structures

Introduction

In §3.3.5, we recall that an operad morphism $\phi : P \to Q$ induces adjoint extension and restriction functors

$$\phi_! : {}_P\mathcal{E} \rightleftarrows {}_Q\mathcal{E} : \phi^*$$

In §7.2, we observe that an operad morphism $\psi : R \to S$ induces similar adjoint extension and restriction functors on module categories:

$$\psi_! : \mathcal{M}_R \rightleftarrows \mathcal{M}_S : \psi^*$$

In this chapter, we study the functors on model categories defined by these extension and restriction functors. Our goal is to prove:

Theorem 16.A. *Let $\phi : P \to Q$ be an operad morphism. Suppose that the operad P (respectively, Q) is Σ_*-cofibrant and use proposition 12.3.A to equip the category of P-algebras (respectively, Q-algebras) with a semi model structure.*

The extension and restriction functors

$$\phi_! : {}_P\mathcal{E} \rightleftarrows {}_Q\mathcal{E} : \phi^*$$

define Quillen adjoint functors. If $\phi : P \to Q$ is a weak-equivalence, then these functors define Quillen adjoint equivalences.

Theorem 16.B. *Let $\psi : R \to S$ be an operad morphism. Assume that the operad R (respectively, S) is C-cofibrant and use proposition 14.1.A to equip the category of right R-modules (respectively, right S-modules) with a model structure.*

The extension and restriction functors

$$\psi_! : \mathcal{M}_R \rightleftarrows \mathcal{M}_S : \psi^*$$

B. Fresse, *Modules over Operads and Functors*, Lecture Notes in Mathematics 1967, 235
DOI: 10.1007/978-3-540-89056-0_16, © Springer-Verlag Berlin Heidelberg 2009

define Quillen adjoint functors. If $\psi : R \to S$ is a weak-equivalence, then these functors define Quillen adjoint equivalences.

¶ Remark. In the context of a reduced category with regular tensor powers, theorem 16.A holds for all C-cofibrant non-unitary operads. In the context of symmetric spectra, theorem 16.A holds for all operads by [23] (see also [21, Theorem 1.2.4]).

The claim of theorem 16.A about the existence of a Quillen equivalence is proved in [4] in the case where the underlying category $\mathcal{E}$ forms a proper model category.

The crux of the proof is to check that the adjunction unit $\eta A : A \to \psi^*\psi_! A$ defines a weak-equivalence for all cofibrant P-algebras A. In our proof, we use the relation $\psi_! A = S_P(Q, A)$ and we apply the homotopy invariance theorems of §§15.1-15.2 to check this property for all cofibrantly generated symmetric monoidal model categories $\mathcal{E}$. The remainder of the proof does not change.

In the next section, we give a detailed proof of theorem 16.A. The proof of theorem 16.B is strictly parallel and we give only a few indications.

In §16.2, we survey applications of our results to categories of bimodules over operads, or equivalently algebras in right modules over operads.

16.1 Proofs

Throughout this section, we suppose given an operad morphism $\phi : P \to Q$, where P, Q are Σ_*-cofibrant operads (as assumed in theorem 16.A), and we study the extension and restriction functors

$$\phi_! : {}_P\mathcal{E} \rightleftarrows {}_Q\mathcal{E} : \phi^*$$

determined by $\phi : P \to Q$.

By definition, the restriction functor $\phi^* : {}_Q\mathcal{E} \to {}_P\mathcal{E}$ reduces to the identity functor $\phi^*(B) = B$ if we forget operad actions. On the other hand, by definition of the semi-model category of algebras over an operad, the forgetful functor creates fibrations and weak-equivalences. Accordingly, we have:

16.1.1 Fact. *The restriction functor $\phi^* : {}_Q\mathcal{E} \to {}_P\mathcal{E}$ preserves fibrations and acyclic fibrations.*

And we obtain immediately:

16.1.2 Proposition (first part of theorem 16.A). *The extension and restriction functors*

$$\phi_! : {}_P\mathcal{E} \rightleftarrows {}_Q\mathcal{E} : \phi^*$$

define Quillen adjoint functors.   □

In the remainder of this chapter, we check that these functors define Quillen adjoint equivalences if $\phi : P \to Q$ is a weak-equivalence of operads. First, we have:

16.1.3 Lemma. *If $\phi : P \to Q$ is a weak-equivalence of Σ_*-cofibrant operads, then, for any cofibrant P-algebra A, the adjunction unit $\eta(A) : A \to \phi^*\phi_!(A)$ is a weak-equivalence.*

Proof. This assertion is a corollary of the next observation and of theorem 15.2.A.   □

16.1.4 Lemma. *Recall that Q forms an algebra over itself in the category of right modules over itself and a P-algebra in right P-modules by restriction on the left and on the right.*

We have $\phi^*\phi_!(A) = S_P(Q, A)$, the image of A under the functor $S_P(Q) : {}_P\mathcal{E} \to {}_P\mathcal{E}$ associated to Q, and the adjunction unit is identified with the morphism

$$A = S_P(P, A) \xrightarrow{S_P(\phi, A)} S_P(Q, A)$$

induced by ϕ.

Proof. The identification $\phi^*\phi_!(A) = S_P(Q, A)$ is a corollary of propositions 9.3.1 and 9.3.3. It is straightforward to check that $S_P(\phi, A)$ represents the adjunction unit.   □

To achieve the proof of theorem 16.A, we use the same arguments as the authors of [4].

Note that lemma 16.1.3 implies condition E1' of Quillen equivalences (see §12.1.8), because ϕ^* preserves all weak-equivalences. Thus it remains to check condition E2' of §12.1.8.

Previously, we observe that the restriction functor $\phi^* : {}_Q\mathcal{E} \to {}_P\mathcal{E}$ preserves fibrations and acyclic fibrations, because, on one hand, the restriction functor reduces to the identity functor if we forget operad actions, and, on the other hand, the forgetful functors creates fibrations and weak-equivalences. In fact, these observations show that the restriction functor $\phi^* : {}_Q\mathcal{E} \to {}_P\mathcal{E}$ creates fibrations and weak-equivalences. As a corollary:

16.1.5 Fact. *The restriction functor $\phi^* : {}_Q\mathcal{E} \to {}_P\mathcal{E}$ reflects weak-equivalences.*

This assertion together with lemma 16.1.3 suffices to obtain:

Claim. *Let $\phi : P \to Q$ be a weak-equivalence of Σ_*-cofibrant operads. Let B be any Q-algebra. Let A be any cofibrant replacement of $\phi^* B$ in the category of P-algebras. The composite*

$$\phi_!(A) \to \phi_!\phi^*(B) \to B$$

is a weak-equivalence.

Proof. In the commutative diagram

$$
\begin{array}{ccc}
A & \xrightarrow{\sim} & \phi^*(B) \\
\downarrow {\scriptstyle \eta(A)} & & \downarrow {\scriptstyle \eta(\phi^*(B))} \\
\phi^*\phi_!(A) & \longrightarrow & \phi^*\phi_!\phi^*(B) \\
& & \downarrow {\scriptstyle \phi^*(\epsilon(B))} \\
& & \phi^*(B)
\end{array}
$$

the left-hand side vertical morphism is a weak-equivalence by lemma 16.1.3 and the right-hand side composite is the identity by adjunction. By the two out of three axiom, we obtain that the image of the composite

$$
\phi_!(A) \to \phi_!\phi^*(B) \to B
$$

under the restriction functor ϕ^* forms a weak-equivalence. Therefore, by the assertion of §16.1.5, this morphism forms a weak-equivalence itself.   □

This claim achieves the proof of:

16.1.6 Proposition (Second part of theorem 16.A). *If* $\phi : \mathsf{P} \to \mathsf{Q}$ *is a weak-equivalence of* Σ_**-cofibrant operads, then the extension and restriction functors*

$$
\phi_! : {}_{\mathsf{P}}\mathcal{E} \rightleftarrows {}_{\mathsf{Q}}\mathcal{E} : \phi^*
$$

define Quillen adjoint equivalences.   □

We prove theorem 16.B about the extension and restriction of right modules

$$
\psi_! : \mathcal{M}_{\mathsf{R}} \rightleftarrows \mathcal{M}_{\mathsf{S}} : \psi^*,
$$

along the same lines of arguments. To prove the analogue of lemma 16.1.3, use the next observation, symmetric to the observation of lemma 16.1.4, and use theorem 15.1.A in place of theorem 15.2.A.

16.1.7 Observation. *Recall that* S *forms an algebra over itself in the category of right modules over itself and an* R*-algebra in right* R*-modules by restriction of structures on the left and on the right.*

For a right R*-module* M*, we have* $\psi^*\psi_!(M) = S_{\mathsf{R}}(M, \mathsf{S})$*, the image of* S *under the functor* $S_{\mathsf{R}}(M) : {}_{\mathsf{R}}\mathcal{M}_{\mathsf{R}} \to \mathcal{M}_{\mathsf{R}}$ *determined by* $M \in \mathcal{M}_{\mathsf{R}}$*, and the adjunction unit* $\eta(M) : M \to \psi^*\psi_!(M)$ *is identified with the morphism*

$$
M = S_{\mathsf{R}}(M, \mathsf{R}) \xrightarrow{S_{\mathsf{R}}(M, \psi)} S_{\mathsf{R}}(M, \mathsf{S})
$$

induced by ψ*.*

Formally, the identification $\psi^*\psi_!(M) = S_R(M, S)$ is a corollary of proposition 9.3.2 and of the functoriality of $S_R(M) : \mathcal{E}_R \to \mathcal{E}$ with respect to the functor of symmetric monoidal categories $\psi^* : \mathcal{M}_S \to \mathcal{M}_R$. Equivalently, use the identifications $\psi^*\psi_!(M) = M \circ_R S$ (see §7.2.1) and $S_R(M, S) = M \circ_R S$ (see §5.1.5).

16.2 ¶ Applications to Bimodules over Operads

In this section, we apply theorem 16.A and theorem 16.B to bimodules over operads. Recall that, in the context of a bimodule category $_P\mathcal{M}_R$, where P, R are operads, we have extension and restriction functors on the left $\phi_! : {}_P\mathcal{M}_R \rightleftarrows {}_Q\mathcal{M}_S : \phi^*$ and on the right $\psi_! : {}_P\mathcal{M}_R \rightleftarrows {}_P\mathcal{M}_S : \psi^*$. Recall further that the category $_P\mathcal{M}_R$ of P-R-bimodules, is isomorphic to the category of P-algebras in right R-modules. Therefore, we use theorems 16.A and 16.B to obtain that these extension and restriction functors define Quillen adjunctions and Quillen equivalences in the case of operad equivalences.

To obtain better results, we can restrict ourself to non-unitary operads and to the subcategory $_P\mathcal{M}_R^0$ formed by connected P-R-bimodules, or equivalent by P-algebras in connected right R-modules. In this context, we obtain:

Theorem 16.2.A.

(a) Let $\phi : P \to Q$ be an operad morphism, where P, Q are $\mathcal{C}$-cofibrant non-unitary operads. Let R be any $\mathcal{C}$-cofibrant non-unitary operad.

The extension and restriction functors

$$\phi_! : {}_P\mathcal{M}_R^0 \rightleftarrows {}_Q\mathcal{M}_R^0 : \phi^*$$

define Quillen adjoint functors. If $\phi : P \to Q$ is a weak-equivalence, then these functors define Quillen adjoint equivalences as well.

(b) Let P be a $\mathcal{C}$-cofibrant non-unitary operad. Let $\psi : R \to S$ be an operad morphism, where R, S are $\mathcal{C}$-cofibrant non-unitary operads.

The extension and restriction functors

$$\psi_! : {}_P\mathcal{M}_R^0 \rightleftarrows {}_P\mathcal{M}_S^0 : \psi^*$$

define Quillen adjoint functors. If $\psi : R \to S$ is a weak-equivalence, then these functors define Quillen adjoint equivalences as well.

Proof. To obtain assertion (a), apply a generalization of theorem 16.A for reduced categories with regular tensor powers to $\mathcal{E}^0 = \mathcal{M}_R^0$.

To obtain assertion (b), apply theorem 16.B to obtain a Quillen adjunction (respectively, a Quillen equivalence) at the module level $\psi_! : \mathcal{M}_R^0 \rightleftarrows \mathcal{M}_S^0 : \psi^*$. Then apply to this adjunction a generalization of proposition 12.3.3 for reduced categories with regular tensor powers. $\square$

Chapter 17
Miscellaneous Applications

Introduction

To conclude the book, we sketch applications of theorems of §15 to the homotopy theory of algebras over operads.

Usual constructions in the semi model category of algebras over an operad R can be realized in the category of right R-modules.

In §17.1 we prove that the functor $S_R(N):_R \mathcal{E} \to {}_R\mathcal{E}$ associated to a cofibrant replacement of the operad R in the category of R-algebras in right R-modules yields functorial cofibrant replacements in the category of R-algebras. The usual cofibrant replacements of the theory of operads (operadic bar constructions, Koszul constructions) are associated to particular cofibrant replacements of R.

In §17.2 we study the cotriple construction as an example of a functor associated to a right R-module.

In §17.3 we study applications of the theorems of §15 to the homology of algebras over operads. In §13 we recall that any category of algebras over an operad R has a natural homology theory derived from the functor of Kähler differentials $A \mapsto \Omega^1_R(A)$. Since we observe in §10.3 that $\Omega^1_R(A)$ is the functor associated to a right R-module Ω^1_R, we can use theorems of §15 to obtain significant results for the homology of R-algebras. To illustrate this principle, we prove that the homology of R-algebras is defined by a Tor-functor (and the cohomology by an Ext-functor) if Ω^1_R and the shifted object R[1] form projective right R-modules (when R is an operad in $\Bbbk$-modules).

As examples, we address the case of the operad A of non-unitary associative algebras, the case of the operad L of Lie algebras, and the case of the operad C of commutative algebras. In §10.3, we proved that Ω^1_R is free (or close to be free) as a right R-module for R = A, L. In these cases, we retrieve that the classical Hochschild homology of associative algebras and the classical Chevalley-Eilenberg homology of Lie algebras are given by Tor-functors (respectively, Ext-functors for the cohomology). On the other hand, one can

B. Fresse, *Modules over Operads and Functors*, Lecture Notes in Mathematics 1967, 241
DOI: 10.1007/978-3-540-89056-0_17, © Springer-Verlag Berlin Heidelberg 2009

check that the Harrison homology of commutative algebras (respectively, the André-Quillen homology in positive characteristic) is not given by a Tor-functor (respectively, an Ext-functor). This observation reflects the fact that Ω_C^1 is not projective as a right C-module.

In §10 we observed that universal operads $U_R(A)$ are associated to right modules over operads A. In §17.4, we use this representation to obtain homotopy invariance results for enveloping operads in the case where R is a cofibrant operad.

17.1 Functorial Cofibrant Replacements for Algebras over Operads

In §12.3 we recall that the category of P-algebras forms a cofibrantly generated semi model category if P is a Σ_*-cofibrant operad. Accordingly, for any P-algebra A, there exists a cofibrant P-algebra Q_A (naturally associated to A) together with a weak-equivalence $Q_A \xrightarrow{\sim} A$. Constructions of the theory of operads (for instance, operadic bar constructions, Koszul constructions in the dg-context) aim to return explicit cofibrant replacements for all P-algebras A, or at least for all P-algebras A which are cofibrant in the underlying category. These particular functorial cofibrant replacements are all functors associated to P-algebras in right P-modules. This observation illustrates the following general proposition:

17.1.1 Proposition. *Assume* P *is a* Σ_*-*cofibrant operad so that the category of* P-*algebras forms a model category.*

If N *forms a cofibrant replacement of* P *in* $_P\mathcal{M}_P$, *then the functor* $S_P(N)$: $_P\mathcal{E} \to {}_P\mathcal{E}$ *associates to every* $\mathcal{E}$-*cofibrant* P-*algebra* $A \in {}_P\mathcal{E}$ *a cofibrant replacement of* A *in the category of* P-*algebras* $_P\mathcal{E}$.

Proof. By definition N is a cofibrant object in $_P\mathcal{M}_P$ together with a weak-equivalence $\epsilon : N \xrightarrow{\sim} P$. Observe that the operad itself P forms a cofibrant object in the category $\mathcal{M}_P$ of right P-modules (but P is not cofibrant as a P-algebra in right P-modules). Thus, by theorem 15.1.A, and since the forgetful functor creates weak-equivalences in the category of P-algebras, we obtain that the morphism $\epsilon : N \xrightarrow{\sim} P$ induces a weak-equivalence of P-algebras

$$S_P(N, A) \xrightarrow{\sim} S_P(P, A) = A$$

as long as A is $\mathcal{E}$-cofibrant.

The next lemma implies that $S_P(N, A)$ forms a cofibrant P-algebra. Thus we conclude that $S_P(N, A)$ defines a cofibrant replacement of A in $_P\mathcal{E}$. □

17.1.2 Lemma. *Let* P *be a* Σ_*-*cofibrant operad. Let* R *be any* C-*cofibrant operad. Let* A *be any* R-*algebra in a category* $\mathcal{E}$ *over* C. *If* A *is* $\mathcal{E}$-*cofibrant, then the functor* $S_R(-, A) : {}_P\mathcal{M}_R \to {}_P\mathcal{E}$ *preserves cofibrations (respectively, acyclic cofibrations) between cofibrant objects.*

Proof. Let $\mathcal{I}$ (respectively, $\mathcal{J}$) denote the class of generating cofibrations (respectively, acyclic cofibrations) in $\mathcal{E}$. Let $\mathcal{K}$ (respectively, $\mathcal{L}$) denote the class of generating cofibrations (respectively, acyclic cofibrations) in $\mathcal{M}_R$. To prove the lemma, we check that the functor $S_R(-, A)$ maps retracts of relative $P(\mathcal{K})$-cell (respectively, $P(\mathcal{L})$-cell) complexes to retracts of relative $P(\mathcal{I})$-cell (respectively, $P(\mathcal{J})$-cell) complexes. Then the conclusion follows from the definition of (acyclic) cofibrations in categories of P-algebras.

By lemma 15.1.2, the morphism $S_R(i, A) : S_R(K, A) \to S_R(L, A)$ induced by a cofibration (respectively, acyclic cofibration) of right R-modules $i : K \rightarrowtail L$ forms a cofibration in $\mathcal{E}$ provided that A is $\mathcal{E}$-cofibrant. As a consequence, the morphism $S_R(i, A) : S_R(K, A) \to S_R(L, A)$ forms a retract of a relative $\mathcal{I}$-cell (respectively, $\mathcal{J}$-cell) complex in $\mathcal{E}$. By proposition 9.2.1, the morphism $S_R(P(i), A) : S_R(P(K), A) \to S_R(P(L), A)$ induced by the morphism of free P-algebras in right R-modules $P(i) : P(K) \to P(L)$ is identified with the morphism of free P-algebras

$$P(S_R(i, A)) : P(S_R(K, A)) \to P(S_R(L, A)).$$

Since the free P-algebra $P(-)$ preserves colimits, we obtain that $R(S_R(i, A))$ forms a retract of a relative $P(\mathcal{I})$-cell (respectively, $P(\mathcal{J})$-cell) complex if $S_R(i, A)$ forms itself a retract of a relative $\mathcal{I}$-cell (respectively, $\mathcal{J}$-cell) complex in $\mathcal{E}$.

Since the functor $S_R : {}_R\mathcal{M}_R \to {}_R\mathcal{F}_R$ preserves colimits too, we deduce easily from these observations that $S_R(-, A)$ maps $P(\mathcal{K})$-cell (respectively, $P(\mathcal{L})$-cell) complexes to retracts of $P(\mathcal{I})$-cell (respectively, $P(\mathcal{J})$-cell) complexes and the conclusion follows immediately. $\qquad\square$

17.1.3 The Operadic Bar and Koszul Constructions in the dg-Context. In [14] we use a two-sided operadic bar construction $B(M, R, N)$ defined for an operad R in dg-modules $\mathcal{C} = dg\,\Bbbk\,\mathrm{Mod}$ with coefficients in a right R-module M and in a left R-module N. For $M = N = R$, the object $B(R, R, R)$ forms a cofibrant replacement of R in ${}_R\mathcal{M}_R$ as long as R is a Σ_*-cofibrant operad. For a right R-module M, we have $B(M, R, R) = M \circ_R B(R, R, R)$, and for an R-algebra A, we have $B(M, R, A) = S_R(B(M, R, R), A)$. The object $B(R, R, A)$ defines a cofibrant replacement of A in the category of R-algebras provided that A is cofibrant in the underlying category.

For certain operads, the Koszul operads, we have a small subcomplex $K(M, R, N) \subset B(M, R, N)$, the Koszul construction, so that the embedding $i : K(M, R, N) \hookrightarrow B(M, R, N)$ defines a weak-equivalence when M, R, N are connected and cofibrant in dg-modules (see [14, §5]). In particular, for $M = N = R$, we obtain weak-equivalences $K(R, R, R) \xrightarrow{\sim} B(R, R, R) \xrightarrow{\sim} R$ in ${}_R\mathcal{M}_R$. For this construction, we also have identities $K(M, R, R) = M \circ_R K(R, R, R)$ and $K(M, R, A) = S_R(K(M, R, R), A)$. But in positive characteristic the Koszul construction is applied to operads R which are usually not Σ_*-cofibrant, like the commutative or the Lie operad.

Nevertheless one can observe that $K(M, R, R)$ (respectively, $K(R, R, N)$) forms a $\mathcal{K}$-cell complex (respectively, an $\mathcal{L}$-cell complex) in the sense of the definition of §15.3 if M, R, N are connected. Therefore, by theorems 15.3.A-15.3.B, we have at least a weak-equivalence $i : K(M, R, N) \xrightarrow{\sim} B(M, R, N)$ for connected objects M, R, N. In [14] we prove this assertion by a spectral sequence argument.

17.2 The Operadic Cotriple Construction

The cotriple construction $B_\Delta(P, P, A)$, whose definition is recalled in §13.3.1, forms a simplicial resolution of A in the sense that the normalized complex of this simplicial P-algebra comes equipped with a weak-equivalence

$$\epsilon : N_*(B_\Delta(P, P, A)) \xrightarrow{\sim} A.$$

In §13.3 we recall the definition of $B_\Delta(P, P, A)$ and of the associated normalized complex $N_*(B_\Delta(P, P, A))$ in the basic case where A is a P-algebra in dg-modules, but the construction makes sense for a P-algebra in Σ_*-objects, respectively for a P-algebra in right modules over an operad R. In particular, we can apply the construction to the P-algebra in right P-modules formed by the operad itself to obtain a simplicial resolution of P in the category of P-algebras in right P-modules. The first purpose of this section is to observe that $B_\Delta(P, P, A)$ is identified with the functor associated to this universal simplicial resolution:

$$B_\Delta(P, P, A) = S_P(B_\Delta(P, P, P), A),$$

and similarly as regards the normalized complex $N_*(B_\Delta(P, P, A))$.

In addition, we review the definition of a cotriple construction with coefficients in a right P-module on a left hand side. The cotriple construction with coefficients satisfies formal identities:

$$B_\Delta(M, P, A) = S_P(B_\Delta(M, P, P), A) = S_P(M, B_\Delta(P, P, A))$$
$$\text{and} \quad N_*(B_\Delta(M, P, A)) = S_P(N_*(B_\Delta(M, P, P)), A),$$

but $S_P(M, N_*(B_\Delta(P, P, A)))$ differs from $N_*(B_\Delta(M, P, A)) = N_*(S_P(M, B_\Delta(P, P, A)))$ because the functor $S_P(M, -)$ does not preserves colimits. Nevertheless, we observe that $N_*(B_\Delta(M, P, A))$ determines the evaluation of $S_P(M, -)$ on a cofibrant replacement of A, under the assumption that the operad P and the right P-module M are Σ_*-cofibrant objects. For this purpose, we prove that $N_*(B_\Delta(M, P, P))$ represents a cofibrant replacement of M in the category of right P-modules and we use the theorems of §15 about the homotopy invariance of functors associated to modules over operads.

17.2.1 Cotriple Constructions with Coefficients. The simplicial cotriple construction $B_\Delta(P, P, A)$ of §13.3.1 is defined for an operad P in any base category $\mathcal{C}$ and for a P-algebra A in any category $\mathcal{E}$ over $\mathcal{C}$. To form a simplicial cotriple construction with coefficients in a right P-module M, we replace simply the first factor $S(P)$ in the definition of §13.3.1 by the functor $S(M)$ associated to M

$$B_\Delta(M, P, A)_n = S(M) \circ \underset{1}{S(P)} \circ \cdots \circ \underset{n}{S(P)}(A)$$

and we use the morphism

$$S(M) \circ S(P) = S(M \circ P) \xrightarrow{\rho} S(M)$$

induced by the right P-action on M to define the face $d_0 : B_\Delta(M, P, A)_n \to B_\Delta(M, P, A)_{n-1}$. The construction returns a simplicial object in $\mathcal{E}$, the underlying category of the P-algebra A.

The construction can be applied to a P-algebra N in the category of Σ_*-objects $\mathcal{E} = \mathcal{M}$, respectively in the category of right modules over an operad $\mathcal{E} = \mathcal{M}_R$. In this context, we have an identity

$$B_\Delta(M, P, N)_n = M \circ P^{\circ n} \circ N$$

since the bifunctor $(M, N) \mapsto S(M, N)$ is identified with the composition product of Σ_*-objects. In the case of a P-algebra in right R-modules, the right R-action on $B_\Delta(M, P, N)_n$ is identified with the obvious morphism

$$M \circ P^{\circ n} \circ N \circ R \xrightarrow{M \circ P^{\circ n} \circ \rho} M \circ P^{\circ n} \circ N,$$

where $\rho : N \circ R \to N$ defines the right R-action on N.

In the case $P = R = N$, the right R-action $\rho : M \circ R \to M$ determines a morphism of right R-modules $\epsilon : B_\Delta(M, R, R)_0 \to M$ so that $\epsilon d_0 = \epsilon d_1$. Accordingly, the simplicial right R-module $B_\Delta(M, R, R)$ comes equipped with a natural augmentation

$$\epsilon : B_\Delta(M, R, R) \to M,$$

where M is identified with a constant simplicial object.

If we also have $M = R$, then the construction returns a simplicial R-algebra in right R-modules $B_\Delta(R, R, R)$ together with an augmentation $\epsilon : B_\Delta(R, R, R) \to R$ in the category of R-algebras in right R-modules. In this case, we retrieve the construction of §13.3.1 applied to the R-algebra in right R-modules formed by the operad itself $A = R$.

In the remainder of this section, we assume that the base category is the category of dg-modules $\mathcal{C} = \mathrm{dg}\,\Bbbk\,\mathrm{Mod}$. In this setting, we can use normalized complexes to produce a resolution of M from the simplicial object $B_\Delta(M, R, R)$. First of all, we check that the normalized complex functor has a natural generalization in the context of Σ_*-objects and right modules over an operad.

17.2.2 Normalized Complexes in Σ_*-Objects and Right Modules over Operads. In §13.3.2, we define the normalized complex of a simplicial dg-module C as the total complex of a bicomplex $N_*(C)$ such that

$$N_m(C) = C_m/s_0 C_{m-1} + \cdots + s_{m-1} C_{m-1}.$$

For a simplicial Σ_*-object in dg-modules M, the collection $N_*(M) = \{N_*(M(n))\}_{n \in \mathbb{N}}$ defines clearly a Σ_*-object in dg-modules associated to M, the normalized complex of the Σ_*-object M.

The functor $N_* : \mathcal{M}_\Delta \to \mathcal{M}$ does not preserves the composition structure (and the symmetric monoidal structure) of the category Σ_*-objects, just like the normalized complex of simplicial dg-modules does not preserves symmetric monoidal structures. But the Eilenberg-Mac Lane equivalence $\nabla : N_*(C) \otimes N_*(D) \xrightarrow{\sim} N_*(C \otimes D)$ gives rise to a natural transformation

$$\nabla : N_*(K) \circ N_*(L) \to N_*(K \circ L),$$

for any $K, L \in \mathcal{M}_\Delta$. In the case $L \in \mathcal{M}$, this natural transformation gives an isomorphism $\nabla : N_*(K) \circ L \xrightarrow{\sim} N_*(K \circ L)$. If we assume symmetrically $K \in \mathcal{M}$, then we have still a natural transformation $\nabla : K \circ N_*(L) \to N_*(K \circ L)$ only, because the composite $K \circ L$ involve tensor products of copies of L.

If M is a simplicial right R-module, where R is any operad in dg-modules, then the object $N_*(M)$ inherits a natural right R-action

$$N_*(M) \circ R \xrightarrow{\sim} N_*(M \circ R) \xrightarrow{N_*(\rho)} N_*(M),$$

so that $N_*(M)$ forms a right R-module. If M is a simplicial P-algebra in Σ_*-objects, where P is any operad in dg-modules, then the object $N_*(M)$ inherits a natural left P-action

$$P \circ N_*(M) \xrightarrow{\nabla} N_*(P \circ M) \xrightarrow{N_*(\lambda)} N_*(M),$$

so that $N_*(M)$ forms a P-algebra in Σ_*-objects. If M is a simplicial P-algebra in right R-modules, then the normalized complex $N_*(M)$ returns a P-algebra in right R-modules.

In the particular example of the cotriple construction, we obtain that $N_*(B_\Delta(M, R, R))$ forms a right R-module and $N_*(B_\Delta(R, R, R))$ forms an R-algebra in right R-modules.

The next assertion is symmetric to the result of lemma 13.3.3 about the cotriple resolution of a P-algebra:

17.2.3 Lemma. *The augmentation* $B_\Delta(M, \mathsf{R}, \mathsf{R}) \to M$ *induces a weak-equivalence*

$$\epsilon : N_*(B_\Delta(M, \mathsf{R}, \mathsf{R})) \xrightarrow{\sim} M$$

in $\mathcal{M}_{\mathsf{R}}$.

Proof. The arguments are symmetrical to the case of the cotriple resolution $B_\Delta(\mathsf{P}, \mathsf{P}, A)$ of an algebra A over an operad P.

The augmentation $\epsilon : B_\Delta(M, \mathsf{R}, \mathsf{R}) \to M$ has a section $\eta : M \to B_\Delta(M, \mathsf{R}, \mathsf{R})$ in the underlying category of Σ_*-objects. This section is given by the morphism

$$M = M \circ I \xrightarrow{M \circ \eta} M \circ \mathsf{R} = B_\Delta(M, \mathsf{R}, \mathsf{R})_0$$

induced by the operad unit $\eta : I \to \mathsf{R}$. Since $\epsilon \eta = \mathrm{id}$, we have $N_*(\epsilon) \cdot N_*(\eta) = \mathrm{id}$ at the realization level.

The object $B_\Delta(M, \mathsf{R}, \mathsf{R})$ comes also equipped with an extra degeneracy

$$s_{n+1} : B_\Delta(M, \mathsf{R}, \mathsf{R})_{n-1} \to B_\Delta(M, \mathsf{R}, \mathsf{R})_n$$

given by the insertion of an operad unit

$$M \circ \mathsf{R}^{\circ n-1} \circ \mathsf{R} \xrightarrow{M \circ \mathsf{R}^{\circ n} \circ \eta} M \circ \mathsf{R}^{\circ n} \circ \mathsf{R}.$$

This extra degeneracy determines a morphisms

$$N_*(B_\Delta(M, \mathsf{R}, \mathsf{R})) \otimes N_*(\Delta^1) \xrightarrow{h} N_*(B_\Delta(M, \mathsf{R}, \mathsf{R}))$$

which gives a chain homotopy $N_*(\eta) \cdot N_*(\epsilon) \simeq \mathrm{id}$. Hence we conclude that $N_*(\epsilon)$ forms a weak-equivalence and this achieves the proof of the lemma. $\square$

The normalized complex functor does not preserves tensor products and free algebras over operads. As a byproduct, the cotriple resolution of a P-algebra does not return a cofibrant P-algebra in dg-modules though $B_\Delta(\mathsf{P}, \mathsf{P}, A)$ forms a cofibrant object in the category of simplicial P-algebras under mild assumptions on A. In contrast, for right modules over operads, we obtain:

17.2.4 Lemma. *Suppose* R *is a C-cofibrant operad. If the right R-module M is Σ_*-cofibrant, then $N_*(B_\Delta(M, \mathsf{R}, \mathsf{R}))$ is cofibrant as a right R-module.*

Proof. The normalized complex of a simplicial object C has a natural filtration

$$0 = N_{*\leq -1}(C) \subset \cdots \subset N_{*\leq n}(C) \subset \cdots \subset \operatorname*{colim}_n N_{*\leq n}(C) = N_*(C),$$

where $N_{*\leq n}(C)$ consists of components $N_m(C)_*$ so that $m \leq n$. Furthermore, the embedding $j_n : N_{*\leq n-1}(C) \hookrightarrow N_{*\leq n}(C)$ fits a pushout

$$B^{n-1} \otimes C_n \xrightarrow{\ f\ } N_{*\leq n-1}(C)$$

$$i^n \otimes C_n \downarrow \qquad\qquad \vdots$$

$$E^n \otimes C_n \dashrightarrow_{\ g\ } N_{*\leq n-1}(C)$$

where $i^n : B^{n-1} \rightarrow E^n$ is the generating cofibration in degree n of the category of dg-modules. Recall that E^n is spanned by a homogeneous element $e = e_n$ in degree n and a homogeneous element $b = b_{n-1}$ in degree $n-1$, together with the differential such that $\delta(e) = b$. The dg-module B^{n-1} is the submodule of E^n spanned by b. The morphism $f : B^{n-1} \otimes C_n \rightarrow N_{*\leq n-1}(C)$ is yielded by the homogeneous map

$$C_n \xrightarrow{\partial} C_{n-1} \rightarrow N_{n-1}(C)$$

where $\partial = \sum_i \pm d_i$. The morphism $g : E^n \otimes C_n \rightarrow N_{*\leq n-1}(C)$ is given by f on the summand $\Bbbk\, b \otimes C_n$ and by the canonical projection of $\pi : C_n \rightarrow N_n(C)$ on the summand $\Bbbk\, e \otimes C_n$.

In the case $C = B_\Delta(M, \mathrm{R}, \mathrm{R})$, we have $C_n = M \circ \mathrm{R}^{\circ n} \circ \mathrm{R}$ and $i^n \otimes C_n$ is identified with the morphism of free right R-modules:

$$(i^n \otimes M \circ \mathrm{R}^{\circ n}) \circ \mathrm{R} : (B^{n-1} \otimes M \circ \mathrm{R}^{\circ n}) \circ \mathrm{R} \rightarrow (E^n \otimes M \circ \mathrm{R}^{\circ n}) \circ \mathrm{R}.$$

The assumptions imply that $M \circ \mathrm{R}^n$ forms a Σ_*-cofibrant object, from which we deduce that $(i^n \otimes M \circ \mathrm{R}^{\circ n}) \circ \mathrm{R}$ defines a cofibration in the category of right R-modules. The lemma follows readily. $\square$

Lemmas 17.2.3-17.2.4 imply:

17.2.5 Theorem. *Let* R *be a* C-*cofibrant operad. Let* M *be a right* R-*module. If* M *is* Σ_*-*cofibrant, then* $N_*(B_\Delta(M, \mathrm{R}, \mathrm{R}))$ *defines a natural cofibrant replacement of* M *in the category of right* R-*modules.* $\square$

The next observation, announced in the introduction of this section, is an easy consequence of the definition of the cotriple construction:

17.2.6 Observation. *We have:*

$$B_\Delta(M, \mathrm{R}, A) = S_{\mathrm{R}}(B_\Delta(M, \mathrm{R}, \mathrm{R}), A) = S_{\mathrm{R}}(M, B_\Delta(\mathrm{R}, \mathrm{R}, A))$$

$$and \quad N_*(B_\Delta(M, \mathrm{R}, A)) = S_{\mathrm{R}}(N_*(B_\Delta(M, \mathrm{R}, \mathrm{R})), A).$$

The relation $B_\Delta(M, \mathrm{R}, A) = S_{\mathrm{R}}(B_\Delta(M, \mathrm{R}, \mathrm{R}), A)$ of the observation is a consequence of identities

$$S_{\mathrm{R}}(B_\Delta(M, \mathrm{R}, \mathrm{R})_n, A) = S_{\mathrm{R}}(M \circ \mathrm{R}^{\circ n} \circ \mathrm{R}, A) = S(M \circ \mathrm{R}^{\circ n}, A) = B_\Delta(M, \mathrm{R}, A)_n$$

deduced from assertion (a) of theorem 7.1.1. From this relation, we deduce an identity of normalized complexes

$$S_R(N_*(B_\Delta(M, R, R)), A) = N_*(B(M, R, A)),$$

because the functor $S_R(-, A)$ preserves all colimits. The relation $B_\Delta(M, R, A)$ = $S_R(M, B_\Delta(R, R, A))$ arise from symmetric identities

$$S_R(M, B_\Delta(R, R, A)_n) = S_R(M, R(S(R^{\circ n}, A))) = S(M, S(R^{\circ n}, A)) = B_\Delta(M, R, A)_n,$$

deduced from assertion (b) of theorem 7.1.1. Since the functor $S_R(M, -)$ does not preserves all colimits, we can not deduce an identity of normalized complexes from these relations. Nevertheless:

17.2.7 Theorem. *Let* R *be any operad in dg-modules. Let* M *be any right* R*-module. Let* A *be any* R*-algebra in a category* $\mathcal{E}$ *over dg-modules.*

Suppose that the operad R *is* Σ_**-cofibrant so that the category of* R*-algebras in* $\mathcal{E}$ *forms a semi model category. If* M *is* Σ_**-cofibrant and* A *is* $\mathcal{E}$*-cofibrant, then we have weak-equivalences*

$$N_*(B_\Delta(M, R, A)) \xleftarrow{\sim} \cdot \xrightarrow{\sim} S_R(M, Q_A),$$

where Q_A *is any cofibrant replacement of* A *in the category of* R*-algebras.*

Proof. By theorem 15.2.A, the augmentation $\epsilon : N_*(B_\Delta(M, R, R)) \xrightarrow{\sim} M$ induces a weak-equivalence

$$S_R(N_*(B_\Delta(M, R, R)), Q_A) \xrightarrow{\sim} S_R(M, Q_A),$$

since the R-algebra Q_A is a cofibrant by assumption, the cofibrant right R-module $N_*(B_\Delta(M, R, R))$ is Σ_*-cofibrant by proposition 14.1.1, and the right R-module M is Σ_*-cofibrant by assumption. By theorem 15.2.A, the augmentation $\epsilon : Q_A \xrightarrow{\sim} A$ induces a weak-equivalence

$$S_R(N_*(B_\Delta(M, R, R)), Q_A) \xrightarrow{\sim} S_R(N_*(B_\Delta(M, R, R)), A),$$

since the right R-module $N_*(B_\Delta(M, R, R))$ is cofibrant, the cofibrant R-algebra Q_A is cofibrant in the underlying category $\mathcal{E}$, and the R-algebra A is also $\mathcal{E}$-cofibrant by assumption. The conclusion follows since we have an identity $N_*(B_\Delta(M, R, A)) = S_R(N_*(B_\Delta(M, R, R)), A)$ according to the observation above the theorem. □

17.3 The (Co)homology of Algebras over Operads Revisited

Throughout this section, we assume that P is any Σ_*-cofibrant operad in dg-modules. The purpose of this section is to revisit the definition of the (co)homology of algebras over operads, and to prove the statements which have been put off in §13.1. Throughout this section, we use the short notation $U_P(A) = U_P(A)(1)$ to refer to the enveloping algebra of a P-algebra A.

In §13.1, we define the cohomology of a P-algebra A with coefficients in a representation E by a formula of the form

$$H^*_P(A, E) = H^*(\mathrm{Der}_P(Q_A, E)),$$

where Q_A refers to a cofibrant replacement of A in the category of P-algebras. Recall that representations of A are equivalent to left modules the enveloping algebra $U_P(A)$. Dually to the cohomology, we define the homology of A with coefficients in a right $U_P(A)$-module E by the formula

$$H_*^P(A, E) = H^*(E \otimes_{U_P(Q_A)} \Omega^1(Q_A)),$$

where $\Omega^1_P(-)$ refers to the functor of Kähler differentials on the category of P-algebras.

To distinguish the role of coefficients, we have introduced the object

$$T^1_P(Q_A) = U_P(A) \otimes_{U_P(Q_A)} \Omega^1(Q_A)),$$

which determines the homology of A with coefficients in the enveloping algebra $E = U_P(A)$. We can use the formula

$$H^*_P(A, E) = H^*(Hom_{U_P(A)}(T^1(Q_A), E),$$
$$\text{respectively} \quad H_*^P(A, E) = H^*(E \otimes_{U_P(A)} T^1(Q_A)),$$

to determine the cohomology $H^*_P(A, E)$, respectively the homology $H_*^P(A, E)$.

To give a sense to these definitions, we have to prove that the cohomology $H^*_P(A, E)$, respectively the homology $H_*^P(A, E)$, does not depend on the choice of the cofibrant replacement Q_A. In §13.1, we prove that this homotopy invariance assertion reduces to the next statement:

17.3.1 Lemma. *The morphism* $U_P(f) : U_P(Q_A) \to U_P(Q_B)$, *induced by a weak-equivalence of P-algebras* $f : Q_A \xrightarrow{\sim} Q_B$ *forms a weak-equivalence if* Q_A *and* Q_B *are cofibrant, and similarly as regards the morphism* $\Omega^1_P(f) :$ $\Omega^1_P(Q_A) \to \Omega^1_P(Q_B)$,

In §§10.1-10.2, we observed that $U_P(-)$ is the instance of a functor associated to a right module over the operad P, namely the shifted object P[1]. Observe that P[1] is Σ_*-cofibrant if the operad P is so. Accordingly, theorem 15.2.A implies immediately that the morphism $U_P(f) : U_P(Q_A) \to U_P(Q_B)$ induced by a weak-equivalence of P-algebras $f : Q_A \xrightarrow{\sim} Q_B$ forms a weak-equivalence if Q_A and Q_B are cofibrant.

In §10.3, we observe that $\Omega^1_P(-)$ is also a functor associated to a right module over the operad P, the module of Kähler differentials Ω^1_P. Moreover, we have $\Omega^1_P(n) = \Sigma_n \otimes_{\Sigma_{n-1}} P[1](n-1)$ if we forget the right P-action. Consequently, the right P-module Ω^1_P is Σ_*-cofibrant if the operad P is so and we conclude again that the morphism $\Omega^1_P(f) : \Omega^1_P(Q_A) \to \Omega^1_P(Q_B)$ induced by a weak-equivalence of P-algebras $f : Q_A \xrightarrow{\sim} Q_B$ forms a weak-equivalence if Q_A and Q_B are cofibrant.

These observations complete the arguments of §13.1 to prove that the cohomology $H^*_P(A, E)$, respectively the homology $H^P_*(A, E)$, does not depend on the choice of the cofibrant replacement Q_A. $\qquad\square$

In the next paragraphs, we exploit the representations

$$U_P(-) = S_P(P[1], -) \quad \text{and} \quad \Omega^1_P(-) = S_P(\Omega^1_P, -)$$

further to prove that the cohomology of P-algebras is defined by an Ext-functor (and the homology by a Tor-functor) in good cases. Theorem 15.1 implies:

17.3.2 Lemma. *If* $P[1]$*, respectively* Ω^1_P*, forms a cofibrant object in the category of right P-modules, then the morphism* $U_P(f) : U_P(Q_A) \to U_P(Q_B)$*, respectively* $\Omega^1_P(f) : \Omega^1_P(Q_A) \to \Omega^1_P(Q_B)$*, induced by a weak-equivalence of P-algebras* $f : Q_A \xrightarrow{\sim} Q_B$ *forms a weak-equivalence as long as the P-algebras* Q_A *and* Q_B *are* $\mathcal{E}$*-cofibrant.* $\qquad\square$

Recall that a P-algebra is $\mathcal{E}$-cofibrant if the initial morphism $\eta : P(0) \to Q_A$ defines a cofibration in the underlying category.

In the situation of lemma 17.3.2 the arguments of lemma 13.1.12 give the following result:

17.3.3 Lemma. *Let* A *be any* $\mathcal{E}$*-cofibrant P-algebra. If* $P[1]$ *and* Ω^1_P *form cofibrant object in the category of right P-modules, then the object*

$$T^1(Q_A) = U_P(A) \otimes_{U_P(Q_A)} \Omega^1(Q_A)$$

associated to a cofibrant replacement of A *in the category of P-algebras forms a cofibrant replacement of* $\Omega^1(A)$ *in the category of left* $U_P(A)$*-modules.*

Proof. By lemma 17.3.2, we have a weak-equivalence of dg-algebras $U_P(\epsilon) : U_P(Q_A) \xrightarrow{\sim} U_P(A)$, and a weak-equivalence of left $U_P(Q_A)$-modules $\Omega^1_P(f) : \Omega^1_P(Q_A) \xrightarrow{\sim} \Omega^1_P(A)$. By lemma 13.1.9, the object $\Omega^1_P(Q_A)$ forms a cofibrant $U_P(Q_A)$-modules, whenever Q_A is a cofibrant algebra. By Quillen equivalence (as in the proof of lemma 13.1.12), we deduce from these assertions that the morphism

$$\Omega^1_P(f)_\sharp : U_P(A) \otimes_{U_P(Q_A)} \Omega^1_P(Q_A) \xrightarrow{\sim} \Omega^1_P(A),$$

adjoint to $\Omega^1_P(f)$, defines a weak-equivalence of left $U_P(A)$-modules. Moreover, the object $T^1(Q_A) = U_P(A) \otimes_{U_P(Q_A)} \Omega^1(Q_A)$ is also cofibrant as a left $U_P(A)$-module. $\qquad\square$

This lemma gives immediately:

17.3.4 Theorem. *Let* P *be any* Σ_**-cofibrant operad in dg-modules. If* Ω^1_P *and* $P[1]$ *form cofibrant right P-modules, then we have*

$$H_{\mathsf{P}}^{*}(A, E) = \mathrm{Ext}_{U_{\mathsf{P}}(A)}^{*}(\Omega_{\mathsf{P}}^{1}(A), E),$$

$$respectively \; H_{*}^{\mathsf{P}}(A, E) = \mathrm{Tor}_{*}^{U_{\mathsf{P}}(A)}(E, \Omega_{\mathsf{P}}^{1}(A)),$$

for every $\mathcal{E}$-cofibrant P-algebra $A \in {}_{\mathsf{P}}\mathcal{E}$ and every $E \in U_{\mathsf{P}}(A) \, \mathrm{Mod}$, respectively $E \in \mathrm{Mod} \, U_{\mathsf{P}}(A)$.   □

If the operad P has a trivial differential, then the objects Ω_{P}^{1} and $\mathsf{P}[1]$ have a trivial differential as well. In this context, the right P-modules Ω_{P}^{1} and $\mathsf{P}[1]$ are cofibrant objects if and only if they are projective (retracts of free objects) in the category of right P-modules. By immediate applications of the five lemma, we can extend theorem §17.3.4 to operads P such that the module of Kähler differentials Ω_{P}^{1} is equipped with a filtration whose subquotients consist of projective right P-modules.

17.3.5 Classical Examples. In §13.1.5, we recall that the (co)homology associated to the usual operads $\mathsf{P} = \mathsf{A}, \mathsf{L}, \mathsf{C}$ agrees with:

– the Hochschild (co)homology in the case of the associative operad $\mathsf{P} = \mathsf{A}$,
– the Chevalley-Eilenberg (co)homology in the case of Lie operad $\mathsf{P} = \mathsf{L}$,
– the Harrison (co)homology in the case of the commutative operad $\mathsf{P} = \mathsf{C}$.

Classical theorems assert that Hochschild and Chevalley-Eilenberg cohomology (respectively, homology) are given by Ext-functors (respectively, Tor-functors). According to theorem 17.3.4, these assertions are implied by the deeper property, checked in §10.3, that the module Ω_{P}^{1} forms a free right P-module for $\mathsf{P} = \mathsf{A}, \mathsf{L}$. In the case $\mathsf{P} = \mathsf{C}$, the non-vanishing of $H_{\mathsf{C}}^{*}(A, A)$ for non-smooth algebras (recall that $U_{\mathsf{C}}(A) = A$ for commutative operads) implies that Ω_{C}^{1} does not form a projective right C-module and gives an obstruction for this property.

17.3.6 Remark. In positive characteristic, one has to use simplicial objects to address the case of commutative algebras and Lie algebras because the commutative and Lie operads are not Σ_{*}-cofibrant. In this context, the operadic (co)homology agrees with the Chevalley-Eilenberg (co)homology for the Lie operad $\mathsf{R} = \mathsf{L}$, with the André-Quillen (co)homology for the commutative operad $\mathsf{R} = \mathsf{C}$,

For the Lie operad L, both objects Ω_{L}^{1} and $\mathsf{L}[1]$ are not projective as right L-modules, but have only a filtration by subquotients of the form $M \circ \mathsf{L}$, where $M \in \mathcal{M}$. This property is sufficient to obtain that the functors $U_{\mathsf{L}}(\mathsf{G}) = S_{\mathsf{L}}(\mathsf{L}[1], \mathsf{G})$ and $\Omega_{\mathsf{L}}^{1}(\mathsf{G}) = S_{\mathsf{L}}(\Omega_{\mathsf{L}}^{1}, \mathsf{G})$ preserve weak-equivalences between all Lie algebras which are cofibrant in the underlying category.

In the remainder of this section, we use theorems of §17.2 to compare the cohomology $H_{*}^{\mathsf{P}}(A, E)$, respectively homology $H_{\mathsf{P}}^{*}(A, E)$, with the cohomology, respectively homology, of cotriple constructions. For short, we set $B_{\Delta}(Q_A) = B_{\Delta}(\mathsf{P}, \mathsf{P}, Q_A)$, for any P-algebra Q_A.

In the case $M = \mathsf{P}[1]$, respectively $M = \Omega_{\mathsf{P}}^{1}$, theorem 17.2.7 returns:

17.3.7 Lemma. *Suppose A is an $\mathcal{E}$-cofibrant P-algebra. Let Q_A be any cofibrant replacement of A in the category of P-algebras. The natural morphisms*

$$N_*(U_P(B_\Delta(A))) \leftarrow N_*(U_P(B_\Delta(Q_A))) \rightarrow U_P(Q_A)$$
$$and \quad N_*(\Omega^1_P(B_\Delta(A))) \leftarrow N_*(\Omega^1_P(B_\Delta(Q_A))) \rightarrow \Omega^1_P(Q_A)$$

are weak-equivalences.

The Eilenberg-Mac Lane morphism implies that $N_*(U_P(B_\Delta(Q_A))$ forms a dg-algebra, for any P-algebra Q_A, and $N_*(\Omega^1_P(B_\Delta(Q_A))$ forms a module over this dg-algebra $N_*(U_P(B_\Delta(Q_A)))$. The natural morphisms

$$N_*(U_P(B_\Delta(A)) \leftarrow N_*(U_P(B_\Delta(Q_A)) \rightarrow U_P(Q_A)$$

are morphisms of dg-algebras and

$$N_*(\Omega^1_P(B_\Delta(A)) \leftarrow N_*(\Omega^1_P(B_\Delta(Q_A)) \rightarrow \Omega^1_P(Q_A)$$

are morphism of $N_*(U_P(B_\Delta(Q_A)))$-modules. Our idea is to adapt the arguments of §§13.1.7-13.1.13 to deduce from the equivalences of lemma 17.3.7 that $H^P_*(A, E)$, respectively $H_P^*(A, E)$, can be determined by the cohomology, respectively homology, of a cotriple construction.

Lemma 13.1.9 asserts that $\Omega^1_P(Q_A)$ forms a cofibrant $U_P(Q_A)$-module. Similarly:

17.3.8 Lemma. *The simplicial object $\Omega^1_P(B_\Delta(A))$, respectively $\Omega^1_P(B_\Delta(Q_A))$, forms dimensionwise a free module over the simplicial dg-algebra $U_P(B_\Delta(A))$, respectively $U_P(B_\Delta(Q_A))$.* □

To adapt the arguments of §§13.1.7-13.1.13 in the simplicial context, we use the bar construction of algebras $B(E, R, F)$, defined for a dg-algebra R, with coefficients in a right R-module E and a left R-module F. The definition of $B(E, R, F)$ makes sense for simplicial objects. In any situation, we have a right-half-plane homological spectral sequence $(E^0, d^0) \Rightarrow H_*(B(E, R, F))$ such that d^0 is determined by internal differentials. If we can assume the existence of a Künneth isomorphism, then we have $E^1 = B(H_*(E), H_*(R), H_*(F))$, where we set $H_*(-) = H_*(N_*(-))$. In any case, a weak-equivalence on E, R, F yields a weak-equivalence at the E^1-level when R and F (or R and E) are cofibrant in the underlying category of simplicial dg-modules. Dually, we have a left-half-plane cohomological spectral sequence $(E^0, d^0) \Rightarrow H^*(Hom_{R\,Mod}(B(R, R, F), T))$ to compute the cohomology of Hom-objects.

By routine applications of spectral sequences, we obtain:

17.3.9 Lemma. *Let $\phi : R \rightarrow S$ be a morphism of simplicial dg-algebras so that $N_*(\phi) : N_*(R) \rightarrow N_*(S)$ forms a weak-equivalence of dg-algebras. Let $f : E \rightarrow F$ be a morphism of simplicial dg-modules, so that $N_*(f) : N_*(E) \rightarrow$*

$N_*(F)$ *forms a weak-equivalence. Suppose E is a left R-module, the object F is a left S-module and the morphism f is a morphism of R-modules.*
   Then:

(a) For any right S-module T, the morphism

$$N_*(B(T,R,E)) \xrightarrow{N_*(\phi,\phi,f)_*} N_*(B(T,S,F))$$

forms a weak-equivalence.

(b) For any left S-module T, the morphism

$$N^*(Hom_{S\,Mod}(B(S,S,F),T)) \xrightarrow{N^*(\phi,\phi,f)^*} N^*(Hom_{R\,Mod}(B(R,R,E),T))$$

forms a weak-equivalence. $\square$

   Moreover:

17.3.10 Lemma. *Let R be a simplicial dg-algebra. Let E be a right R-module. Suppose E is dimensionwise cofibrant over R. Then:*

(a) For any right R-module T, the natural morphism

$$N_*(B(T,R,E)) \to N_*(T \otimes_R E)$$

forms a weak-equivalence.

(b) For any left R-module T, the natural morphism

$$N^*(Hom_{R\,Mod}(E,T)) \to N^*(Hom_{R\,Mod}(B(R,R,E),T))$$

forms a weak-equivalence.

Proof. Use that $B(T,R,E) \to T \otimes_R E$ defines a weak-equivalence dimensionwise, and similarly as regards the morphism $Hom_{R\,Mod}(E,T) \to Hom_{R\,Mod}(B(R,R,E),T)$. $\square$

   These general lemmas imply immediately:

17.3.11 Theorem. *We have*

$$H_{\mathrm{P}}^*(A,E) = H^*N^*\big(Hom_{U_{\mathrm{P}}(A)}(T_{\mathrm{P}}^1(B_\Delta(\mathrm{P},\mathrm{P},A)),E)\big)$$
$$and \qquad H^{\mathrm{P}}_*(A,E) = H_*N_*\big(E \otimes_{U_{\mathrm{P}}(A)} T_{\mathrm{P}}^1(B_\Delta(\mathrm{P},\mathrm{P},A))\big),$$

where on the right hand side we use a dimensionwise evaluation of functors to form a cosimplicial (respectively, simplicial) object in dg-modules and we take the cohomology (respectively, homology) of the associated conormalized (respectively, normalized) chain complex. $\square$

   Thus we obtain the result announced by theorem 13.3.4.

17.4 Homotopy Invariance of Enveloping Operads and Enveloping Algebras

In this section we study applications of the homotopy invariance results of §§15.1-15.2 to enveloping operads.

For this purpose, we use that the enveloping operad $U_R(A)$ of an R-algebra A is identified with the functor associated to the shifted object $R[\,\cdot\,]$. By §10.1.5, an operad morphism $\psi : R \to S$ gives rise to a morphism of operads in right S-modules $\psi_! : \psi_! R[\,\cdot\,] \to S[\,\cdot\,]$, which, at the functor level, represents a natural transformation $\psi_\flat : U_R(\psi^*B) \to U_S(B)$ on enveloping operads.

The shifted Σ_*-objects $M[r]$ associated to any cofibrant Σ_*-object M are clearly cofibrant. Hence, theorem 15.1.A, implies immediately:

Theorem 17.4.A.

(a) Let $\psi : R \xrightarrow{\sim} S$ be a weak-equivalence between Σ_*-cofibrant operads. The morphism $\psi_\flat : U_R(\psi^*B) \to U_S(B)$ induced by ψ defines a weak-equivalence if B is a cofibrant S-algebra.

(b) Suppose R is a Σ_*-cofibrant operad. The morphism $U_R(f) : U_R(A) \to U_R(B)$ induced by a weak-equivalence of R-algebras $f : A \xrightarrow{\sim} B$ is a weak-equivalence if A and B are cofibrant R-algebras.   □

See also [5] for another proof of this proposition.

The main purpose of this section is to prove that better results hold if we deal with cofibrant operads:

Theorem 17.4.B.

(a) If $\psi : R \to S$ is a weak-equivalence between cofibrant operads, then $\psi_\flat : U_R(\psi^*B) \to U_S(B)$ defines a weak-equivalence for every S-algebra B which is $\mathcal{E}$-cofibrant.

(b) Suppose R is a cofibrant operad. The morphism $U_R(f) : U_R(A) \to U_R(B)$ induced by a weak-equivalence of R-algebras $f : A \xrightarrow{\sim} B$ defines a weak-equivalence as long as A and B are $\mathcal{E}$-cofibrant.

By theorem 15.1.A, this result is a corollary of the following theorem:

Theorem 17.4.C.

(a) If R is a cofibrant operad, then $R[\,\cdot\,]$ forms a cofibrant right R-module.

(b) If $\psi : R \to S$ is a weak-equivalence between cofibrant operads, then the morphism $\psi_\flat : \psi_! R[\,\cdot\,] \to S[\,\cdot\,]$ defines a weak-equivalence between cofibrant right S-modules.

The proof of proposition 17.4.C is deferred to a series of lemma. The result holds in any cofibrantly generated symmetric monoidal model category $\mathcal{C}$, but we assume for simplicity that $\mathcal{C}$ is the category of dg-modules. We determine the structure of the shifted object $R[\,\cdot\,]$ associated to a free operad $R = F(M)$

and we use that cofibrant cell operads are represented by quasi-free objects to deduce our result.

By definition, the shifted object $M[\,\cdot\,]$ of a Σ_*-object M forms a Σ_*-object in Σ_*-objects, and hence comes equipped with an internal grading $M[\,\cdot\,] = M[\,\cdot\,](-)$. To distinguish the role of gradings, we call external (respectively, internal) grading the first (respectively, second) grading of any Σ_*-object in Σ_*-objects $T(\,\cdot\,,-) = \{T(m,n)\}_{m,n\in\mathbb{N}}$ and we use always dots $\cdot$ (respectively, hyphens $-$) to refer to external (respectively, internal) gradings. Recall that the shifted operad $R[\,\cdot\,]$ forms an operad in right R-modules. The external grading gives the operadic grading of $R[\,\cdot\,](-)$. The internal grading gives the right R-module grading.

By definition, we have $R[m](n) = R(m+n)$. To have an intuitive representation of the shifted operad $R[\,\cdot\,]$, we identified the elements of $R[m](n) = R(m+n)$ with operations $p \in R(m+n)$ in two distinguished sets of variables $p = p(x_1,\ldots,x_m,y_1,\ldots,y_n)$, referred to as the external and internal variables (or inputs) of p. The operad structure is given by operadic composites on external variables x_i, $i = 1,\ldots,m$. The right R-module structure is given by operadic composites on internal variables y_j, $j = 1,\ldots,n$.

The next lemma can easily be deduced from this intuitive representation:

17.4.1 Lemma. *For a free operad $R = F(M)$, we have the identity of right $F(M)(-)$-modules*

$$F(M)[\,\cdot\,](-) = F(M)(-) \oplus F(T(\,\cdot\,,-)) \circ F(M)(-),$$

where $F(T(\,\cdot\,,-))$ is a free operad in Σ_*-objects generated by the Σ_*-object in Σ_*-objects $T(\,\cdot\,,-)$ such that

$$T(m,n) = \begin{cases} 0, & \text{if } m = 0, \\ M[m](n) = M(m+n), & \text{otherwise.} \end{cases}$$

The composition product $\circ$ on the right-hand side is applied to the internal grading $-$.

For a quasi-free operad $R = (F(M),\partial)$, the shifted object $R[\,\cdot\,]$ forms a quasi-free right $R(-)$-module such that

$$R[\,\cdot\,](-) = R(-) \oplus (F(T(\,\cdot\,,-)) \circ R(-), \partial). \qquad \square$$

The twisting cochain $\partial : F(T(\,\cdot\,,-)) \to F(T(\,\cdot\,,-)) \circ F(M)(-)$ is determined as follows. Let $p \in F(T)(m,n)$. Forget external and internal labeling of variables to identify p with an element of $F(M)$ and take the image of p under the twisting cochain of $R = (F(M),\partial)$. The inputs of $\partial p \in F(M)(m+n)$ are naturally in bijection with inputs of $p \in F(T)(m,n)$, and hence correspond either to external or internal variables. Thus we can distinguish external and internal variables in inputs of ∂p.

The operation $\partial p \in \mathrm{F}(M)(m+n)$ is identified with a formal composite of elements $\xi_v \in M$, whose inputs are either connected to inputs of ∂p, or to other factors $\xi_w \in M$. The inputs of $\xi_v \in M$ which are connected to internal variables of ∂p are identified with internal variables attached to the factor ξ_v. The inputs of $\xi_v \in M$ which are connected to external variables of ∂p and to other factors $\xi_w \in M$ are identified with external variables. In this way, we identify ∂p with a composite of elements $\xi_v \in M[r_v](s_v)$. The factors $\xi_v \in M[r_v](s_v)$ such that $r_v = 0$ can be moved outside the composite so that we can identify ∂p with an element of $\mathrm{F}(T(\,\cdot\,, -)) \circ \mathrm{F}(M)(-)$.

The next lemma is a straightforward consequence of this explicit description of the twisting cochain:

17.4.2 Lemma. *If* R *is a cofibrant cell operad, then the shifted objects* $\mathrm{R}[m]$, $m \in \mathbb{N}$, *forms a cofibrant cell right* $\mathrm{R}(-)$-*module.*

Proof. Recall that a cofibrant cell operad is equivalent to a quasi-free operad $\mathrm{R} = (\mathrm{F}(L), \partial)$ where L is a free Σ_*-object in graded $\Bbbk$-modules $L = \bigoplus_\alpha \Bbbk\, e_{d_\alpha} \otimes F_{r_\alpha}$ together with a basis filtration $L_\lambda = \bigoplus_{\alpha < \lambda} \Bbbk\, e_{d_\alpha} \otimes F_{r_\alpha}$ such that $\partial(L_\lambda) \subset \mathrm{F}(L_\lambda)$.

From our description of $\mathrm{R}[m] = (\mathrm{F}(L)[m], \partial)$, we obtain readily that $\mathrm{R}[m]$ has a filtration by subobjects

$$\mathrm{R}_\lambda[m](-) = \mathrm{R}(-) \oplus K_\lambda \circ \mathrm{R}(-), \partial)$$

so that $\partial(K_\lambda) \subset K_{\lambda-1} \subset K_\lambda \circ \mathrm{R}(-)$: take the Σ_*-object

$$T_\lambda(m, n) = \begin{cases} 0, & \text{if } m = 0, \\ L_\lambda(m+n), & \text{otherwise,} \end{cases}$$

and set $K_\lambda(-) = \mathrm{F}(T_\lambda)(m, -)$. Check that K_λ form free Σ_*-objects in graded $\Bbbk$-modules to conclude. $\qquad\square$

Since the cofibrant objects in a cofibrantly generated (semi-)model category are retracts of cofibrant cell complexes, we conclude:

17.4.3 Lemma. *If* R *is a cofibrant operad, then the shifted objects* $\mathrm{R}[m]$, $m \in \mathbb{N}$, *form cofibrant right* $\mathrm{R}(-)$-*modules.* $\qquad\square$

This lemma achieves the proof of assertion (a) of theorem 17.4.C. To prove the other assertion of theorem 17.4.C, we observe further:

17.4.4 Lemma. *If* $\psi : \mathrm{R} \to \mathrm{S}$ *is a weak-equivalence and* R *is a cofibrant operad, then the morphism* $\psi_b : \psi_! \mathrm{R}[\,\cdot\,] \to \mathrm{S}[\,\cdot\,]$ *forms a weak-equivalence as well. Moreover the object* $\psi_! \mathrm{R}[\,\cdot\,]$ *is cofibrant as a right* S-*module.*

Proof. By theorem 16.B, the extension and restriction functors $\psi_! : \mathcal{M}_{\mathrm{R}} \leftrightarrows \mathcal{M}_{\mathrm{S}} : \psi^*$ define Quillen adjoint equivalences if $\psi : \mathrm{R} \to \mathrm{S}$ is a weak-equivalence of cofibrant (and hence $\mathcal{C}$-cofibrant) operads.

The morphism $\psi : R \to S$ defines obviously a weak-equivalence of right S-modules $\psi_\sharp : R[m] \to \psi^* S[m]$, for every $m \in \mathbb{N}$, and the object $R[m]$ is cofibrant as a right R-module by lemma 17.4.3. By Quillen equivalence, we conclude that the morphism $\psi_\flat : \psi_! R[\cdot] \to S[\cdot]$ adjoint to $\psi_\sharp$ is also a weak-equivalence and $\psi_! R[\cdot]$ is cofibrant as a right S-module.   □

This lemma achieves the proof of theorem 17.4.C and theorem 17.4.B.   □

We give a another example of applications of the realization of the enveloping operad $U_R(A)$ by a right module over an operad. We study rather the enveloping algebra $U_R(A) = U_R(A)(1)$.

Recall that an E_∞-operad is a Σ_*-cofibrant operad E equipped with an acyclic fibration $\epsilon : E \to C$ where C refers to the operad of commutative algebras. An A_∞-operad is a Σ_*-cofibrant operad K equipped with an acyclic fibration $\epsilon : K \to A$ where A refers to the operad of associative algebras. An E_∞-algebra is an algebra over some E_∞-operad. An A_∞-algebra is an algebra over some A_∞-operad.

We have an operad morphism $\epsilon : A \to C$ so that the restriction functor $\epsilon^* : {}_C\mathcal{E} \to {}_A\mathcal{E}$ represent the obvious category embedding from commutative algebras to associative algebras. We can take a cofibrant A_∞-operad K and pick a lifting

$$
\begin{array}{ccc}
K & \dashrightarrow & E \\
\sim \downarrow & & \downarrow \sim \\
A & \longrightarrow & C
\end{array}
$$

to equip every E-algebra A with an A_∞-algebra structure.

The morphism $\phi_\sharp : E[\cdot] \to \phi^* C[\cdot]$ forms clearly a weak-equivalence. We use this equivalence to transport properties of the enveloping algebra of commutative algebras to algebras over E_∞-operads.

For a commutative algebra A, we have $U_C(A) = A$. For algebras over E_∞-operads, we obtain:

17.4.5 Theorem. *Let* E *be a cofibrant* E_∞-*operad. Pick a lifting*

$$
\begin{array}{ccc}
K & \dashrightarrow & E \\
\sim \downarrow & & \downarrow \sim \\
A & \longrightarrow & C
\end{array}
$$

to equip every E-*algebra* A *with the structure of an algebra over some cofibrant* A_∞-*operad* K.

If A *is an* $\mathcal{E}$-*cofibrant* E-*algebra, then we have functorial weak-equivalences of* K-*algebras*

$$U_E(A) \xleftarrow{\sim} \cdot \xrightarrow{\sim} A$$

that connect the enveloping algebra of A *to the* A_∞-*algebra underlying* A.

Proof. By theorem 17.4.C, the object $E[1]$ forms a cofibrant right E-module.

For the commutative operad C, we have an isomorphism of associative algebras in right C-modules $C[1] \simeq C$ that reflects the relation $U_C(A) \simeq A$ for commutative algebras. The operad morphism $\phi : E \to C$ defines a weak-equivalence of associative algebras in right E-modules $\phi_\sharp : E[1] \to \phi^* C[1]$.

The operad E forms an K-algebra in right E-modules and the morphism $\phi : E \to C$ defines a weak-equivalence of K-algebras in right E-modules $\phi_\sharp : E \to \phi^* C$.

Finally we have weak-equivalences of K-algebras in right E-modules

$$E[1] \xrightarrow{\sim} \phi^* C[1] \simeq \phi^* C \xleftarrow{\sim} E$$

from which we deduce the existence of weak-equivalences

$$E[1] \xleftarrow{\sim} \cdot \xrightarrow{\sim} E,$$

where the middle term consists of a cofibrant K-algebra in right E-modules. Since cofibrant K-algebras in right E-modules are cofibrant in the underlying category, we deduce from theorem §15.1.A that these weak-equivalences yield weak-equivalences at the functor level. Hence we obtain functorial weak-equivalences of K-algebras

$$U_E(A) \xleftarrow{\sim} \cdot \xrightarrow{\sim} A$$

as stated. $\square$

Bibliographical Comments on Part IV

This part gives the first systematic and comprehensive study of the relationship between the homotopy of modules over operads and functors. Nevertheless, we note that another model structure for right modules and bimodules over operads is defined in [54] in the context of simplicial sets and simplicial modules. A version of theorem 15.1.A also occurs in [54], but the other unifying homotopy invariance result, theorem 15.2.A, is completely new for modules over operads. Recall however that theorem 15.2.A has already been proved for certain very particular instances of modules (see §15.2). The homotopy invariance of the relative composition product is proved in the dg-context in [14] by using spectral sequence arguments, and in the context of spectra in [23].

The homotopy of extension and restriction functor has already been studied in various situations: in the context of simplicial sets and simplicial modules, [54, §3.6]; for certain operads in differential graded modules, [26]; in the context of spectra, [21, Theorem 1.2.4] and [23]; under the assumption that the underlying model category is left proper, [4].

The ideas of §17 are original, though certain results already occur in the literature. In particular, theorem 17.4.A, the homotopy invariance of enveloping operads (algebras) on cofibrant algebras, is proved in [26] and [5] by other methods. On the other hand, the parallel case of cofibrant operads, stated in theorem 17.4.B, is completely new.

Part V
Appendix: Technical Verifications

Foreword

The goal of this part is to achieve technical verifications of §12.3 and §15.2:

- we check condition (2) of theorem 12.1.4 to prove that the category of algebras over an operad R is equipped with a semi model structure,
- we prove that the bifunctor $(M, A) \mapsto S_R(M, A)$ associated to an operad R satisfies an analogue of the pushout-product axiom of tensor products in symmetric monoidal model categories.

As usual, we address the case of R-algebras in any symmetric monoidal category $\mathcal{E}$ over the base category $\mathcal{C}$.

The bifunctor $(M, A) \mapsto S_R(M, A)$ does not preserves all colimits in A. Therefore we can not apply the argument of lemma 11.3.2 to prove the required pushout-product axiom.

To handle the difficulty we introduce right R-modules $S_R[M, A]$ such that $S_R(M, A) = S_R[M, A](0)$ and we study pushout-products on the bifunctor $(M, A) \mapsto S_R[M, A]$. The same method is used in the proof that R-algebras form a semi model category.

Chapter 18
Shifted Modules over Operads and Functors

Introduction

The purpose of this chapter is to define the functor $S_R[M] : A \mapsto S_R[M, A]$ used to determine the image of pushouts of R-algebras under the functor $S_R(M) : A \mapsto S(M, A)$ associated to a right module over an operad.

The explicit construction of the functor $S_R[M] : A \mapsto S_R[M, A]$ is carried out in §18.1. The categorical features of the construction are studied in §18.2.

In this chapter we address only categorical constructions and we do not use any model structure.

18.1 Shifted Modules and Functors to Modules over Operads

To define $S_R[M, A]$, we use the shifted objects $M[n]$ associated to M and the associated functors $S[M, -](n) = S(M[n], -)$, defined in §4.1.4. The idea is to substitute $S[M, -]$ to the functor $S(M, -)$ in the definition of $S_R(M, A)$.

The purpose of the next paragraph is to review the definition of the shifted objects $M[n]$ and to study the structure of the functor $S[M, -]$ in the case where M is a right R-module. The definition of $S_R[M, A]$ comes immediately after these recollections.

18.1.1 Shifted Modules over Operads. The shifted objects $M[m]$ are defined in §4.1.4 by $M[m](n) = M(m+n)$. We use the morphism $\Sigma_n \to \Sigma_{m+n}$ yielded by the natural action of Σ_n on $\{m+1, \ldots, m+n\} \subset \{1, \ldots, m+n\}$ to determine a Σ_n-action on the object $M[m](n)$, for each $m, n \in \mathbb{N}$, so that every collection $M[m] = \{M[m](n)\}_{n \in \mathbb{N}}$ forms a Σ_*-object.

B. Fresse, *Modules over Operads and Functors*, Lecture Notes in Mathematics 1967, 267
DOI: 10.1007/978-3-540-89056-0_18, © Springer-Verlag Berlin Heidelberg 2009

For any object $X \in \mathcal{E}$, we set $S[M, X](m) = S(M[m], X)$. For short, we also use the notation $M[X](m) = S[M, X](m)$. By definition, we have the relation $S[M, X](0) = S(M, X)$.

By §4.1.4, the symmetric group Σ_m acts on the collection $M[m] = \{M[m](n)\}_{n \in \mathbb{N}}$ by automorphisms of Σ_*-objects: we use simply the morphism $\Sigma_m \to \Sigma_{m+n}$ yielded by the natural action of Σ_m on $\{1, \ldots, m\} \subset \{1, \ldots, m + n\}$ to determine the Σ_m-action on the object $M[m](n)$. As a byproduct, we obtain that the collection $S[M, X] = \{S(M[m], X)\}_{m \in \mathbb{N}}$ forms a Σ_*-object, for every $X \in \mathcal{E}$. Note that $S[M, X]$ forms a Σ_*-object in $\mathcal{E}$ and not in the base category $\mathcal{C}$.

For a right R-module M, the operadic composites at last positions

$$M(r + s) \otimes R(n_1) \otimes \cdots \otimes R(n_s) \to M(r + n_1 + \cdots + n_s)$$

provide each $M[r]$ with the structure of a right R-module. This definition is an obvious generalization of the right R-module structure of the shifted operad $R[\cdot]$ introduced in §10.1.1. At the level of the functor $S[M, X]$, the morphism

$$S(M[r], R(X)) = S(M[r] \circ R, X) \xrightarrow{S(\rho[r], X)} S(M[r], X),$$

induced by the right R-action on $M[r]$ gives rise to a natural morphism

$$S[M, R(X)] \xrightarrow{d_0} S[M, X],$$

for all $X \in \mathcal{E}$.

On the other hand, we can use operadic composites at first positions to form morphisms

$$M(r + s) \otimes R(m_1) \otimes \cdots \otimes R(m_r) \to M(m_1 + \cdots + m_r + s).$$

These morphisms provide the collection $M[\cdot] = \{M[r]\}_{r \in \mathbb{N}}$ with the structure of a right R-module in the category of right R-modules. At the level of the functor $S[M, X]$, we obtain a natural right R-module structure on $S[M, X]$, for every $X \in \mathcal{E}$, so that the composition products

$$S[M, X](r) \otimes R(m_1) \otimes \cdots \otimes R(m_r) \to S[M, X](m_1 + \cdots + m_r)$$

commute with the morphisms $d_0 : S[M, R(X)] \to S[M, E]$. Equivalently, we obtain that d_0 defines a morphism of right R-modules.

18.1.2 The Construction. For an R-algebra A, the action of the operad on A induces a morphism of right R-modules

$$S[M, R(A)] \xrightarrow{d_1} S[M, A],$$

parallel to the morphism

$$S[M, R(A)] \xrightarrow{d_0} S[M, A]$$

defined in §18.1.1. In the converse direction, the morphism $\eta(A) : A \to \mathrm{R}(A)$ induced by the operad unit $\eta : I \to \mathrm{R}$ gives rise to a morphism of right R-modules

$$S[M, A] \xrightarrow{s_0} S[M, \mathrm{R}(A)]$$

so that $d_0 s_0 = d_1 s_0$. The object $S_\mathrm{R}[M, A]$ is defined by the reflexive coequalizer

$$S[M, \mathrm{R}(A)] \overset{\overset{\displaystyle s_0}{\overset{\displaystyle d_0}{\underset{\displaystyle d_1}{\rightrightarrows}}}}{} S[M, A] \longrightarrow S_\mathrm{R}[M, A]$$

in the category of right R-modules.

For a fixed component $S_\mathrm{R}[M, A](r)$, this coequalizer is identified with the coequalizer of the object $S_\mathrm{R}(M[r], A)$ associated to the right R-module $M[r]$. Hence we have the relation $S_\mathrm{R}[M, A](r) = S_\mathrm{R}(M[r], A)$. In particular, for $r = 0$, we have $S_\mathrm{R}[M, A](0) = S_\mathrm{R}(M, A)$.

Note that $S_\mathrm{R}[M, X]$ forms a right R-module in $\mathcal{E}$ and not in the base category $\mathcal{C}$.

18.2 Shifted Functors and Pushouts

In this section, we study the image of pushouts

$$\begin{array}{ccc}
\mathrm{R}(X) & \xrightarrow{\ u\ } & A \\
{\scriptstyle \mathrm{R}(i)} \downarrow & & \downarrow {\scriptstyle f} \\
\mathrm{R}(Y) & \dashrightarrow[v]{} & B
\end{array}$$

under the functor $S_\mathrm{R}[M] : A \mapsto S_\mathrm{R}[M, A]$ associated to a right module M over an operad R. Our aim is to decompose the morphism $S_\mathrm{R}[M, f] : S_\mathrm{R}[M, A] \to S_\mathrm{R}[M, B]$ into a composite

$$S[M, A] = S_\mathrm{R}[M, B]_0 \xrightarrow{j_1} \cdots$$

$$\cdots \to S_\mathrm{R}[M, B]_{n-1} \xrightarrow{j_n} S_\mathrm{R}[M, B]_n \to \cdots$$

$$\cdots \to \operatorname*{colim}_n S_\mathrm{R}[M, B]_n = S_\mathrm{R}[M, B]$$

so that each morphism j_n is obtained by a pushout

$$\begin{array}{ccc}
S_\mathrm{R}[L_n M[Y/X], A] & \longrightarrow & S_\mathrm{R}[M, B]_{n-1} \\
\downarrow & & \downarrow \\
S_\mathrm{R}[T_n M[Y/X], A] & \dashrightarrow & S_\mathrm{R}[M, B]_n
\end{array}$$

in the category of right R-modules. In §19, we use this construction to ana-
lyze the image of relative cell complexes under functors associated to right
R-modules.

The objects $L_n M[Y/X]$ and $T_n M[Y/X]$ are defined in §§18.2.6-18.2.10.

A preliminary step is to reduce the construction of pushouts to reflexive
coequalizers of free R-algebras, because we have immediately:

18.2.1 Proposition.

*(a) For a free R-algebra $A = R(X)$, we have a natural isomorphism
$S_R[M, R(X)] = S[M, X]$.*

*(b) The functor $S_R[M, -] : {}_R\mathcal{E} \to \mathcal{M}_R$ preserves the coequalizers which are
reflexive in $\mathcal{E}$.*

Proof. Since we have an identity $S_R[M, A](n) = S_R(M[n], A)$, assertion (a) is
a corollary of assertion (b) of theorem 7.1.1, assertion (b) is a corollary of
proposition 5.2.2.   □

In any category, a pushout

$$\begin{array}{ccc} S & \longrightarrow & A \\ \downarrow & & \downarrow \\ T & \longrightarrow & B \end{array}$$

is equivalent to a reflexive coequalizer such that:

$$T \vee S \vee A \underset{d_1}{\overset{d_0}{\rightrightarrows}} T \vee A \longrightarrow B.$$

Thus, by assertion (b) of proposition 18.2.1, the object $S_R[M, B]$ that we aim
to understand is determined by a reflexive coequalizer of the form:

$$S_R[M, R(Y) \vee R(X) \vee A] \underset{d_1}{\overset{d_0}{\rightrightarrows}} S_R[M, R(Y) \vee A] \longrightarrow S_R[M, B].$$

The motivation to introduce refined functors $S_R[M, -]$ appears in the next
observations:

18.2.2 Lemma. *For any Σ_*-object M and any sum of objects $X \oplus Y$ in $\mathcal{E}$,
we have identities*

$$S[M, X \oplus Y] \simeq S[M[X], Y] \simeq S[M, Y][X].$$

Proof. Easy consequence of the decomposition of the tensor power:

$$(X \oplus Y)^{\otimes n} = \bigoplus_{p+q=n} \Sigma_n \otimes_{\Sigma_p \times \Sigma_q} X^{\otimes p} \otimes Y^{\otimes q}$$

(we use a similar splitting in the proof of lemma 2.3.9).   □

Recall that $M[X]$ is a short notation for $M[X] = S[M, X]$.

18.2.3 Lemma. *For all $M \in \mathcal{M}_R$, $A \in {}_R\mathcal{E}$ and $X \in \mathcal{E}$, we have natural isomorphisms*

$$S_R[M, R(X) \vee A] \simeq S_R[M[X], A]$$
$$\simeq S_R[M, A][X],$$

where $\vee$ denotes the coproduct in the category of R-algebras.

Proof. Observe that the coproduct $R(X) \vee A$ is realized by a reflexive coequalizer of the form

$$R(X \oplus R(A)) \underset{d_1}{\overset{d_0}{\rightrightarrows}} \overset{s_0}{\underset{}{\overgroup{}}} R(X \oplus A) \longrightarrow R(X) \vee A.$$

The morphism d_0 is defined by the composite

$$R(X \oplus R(A)) \xrightarrow{R(\eta(X) \oplus \mathrm{id})} R(R(X) \oplus R(A)) \xrightarrow{R(R(i_X), R(i_A))} R(R(X \oplus A)) \xrightarrow{\mu(X \oplus A)} R(X \oplus A),$$

where $\mu : R \circ R \to R$ denotes the composition product of the operad R. The morphism d_1 is the morphism of free R-algebras

$$R(X \oplus R(A)) \xrightarrow{R(X \oplus \lambda)} R(X \oplus A)$$

induced by the operad action $\lambda : R(A) \to A$. The reflection s_0 is the morphism of free R-algebras

$$R(X \oplus A) \xrightarrow{R(X \oplus \eta(A))} R(X \oplus R(A)),$$

where $\eta : I \to R$ denotes the unit of the operad R.

by proposition 18.2.1, the functor $S_R[M, -]$ maps this reflexive coequalizer to a coequalizer of the form

$$S[M, X \oplus R(A)] \underset{d_1}{\overset{d_0}{\rightrightarrows}} \overset{s_0}{\underset{}{\overgroup{}}} S[M, X \oplus A] \longrightarrow S_R[M, R(X) \vee A].$$

The natural isomorphism $S[M, X \oplus Y] \simeq S[M[X], Y]$ of lemma 18.2.2 transports our coequalizer to a coequalizer of the form

$$S[M[X], R(A)] \underset{d_1}{\overset{d_0}{\rightrightarrows}} S[M[X], A] \dashrightarrow S_R[M, R(X) \vee A] .$$

with s_0 above.

A straightforward verification shows that this coequalizer is identified with the coequalizer that determines $S_R[M[X], A]$. Hence we obtain the relation $S_R[M, R(X) \vee A] \simeq S_R[M[X], A]$.

Similarly, we use the natural isomorphism $S[M, X \oplus Y] \simeq S[M, Y][X]$ to obtain $S_R[M, R(X) \vee A] \simeq S_R[M, A][X]$. $\square$

18.2.4 The Decomposition of $S_R[M, B]$. Thus we are reduced to study a reflexive coequalizer of the form:

$$S_R[M, A][X \oplus Y] \underset{d_1}{\overset{d_0}{\rightrightarrows}} S_R[M, A][Y] \longrightarrow S_R[M, B] .$$

with s_0 above.

The morphism

$$S_R[M, A][X \oplus Y] \xrightarrow{S_R[M,A][(i,\mathrm{id})]} S_R[M, A][Y]$$

induced by $i : X \to Y$ gives d_0. The morphism

$$S_R[M, A][X \oplus Y] \simeq S_R[M, A \vee R(X)][Y] \xrightarrow{S_R[M,(\mathrm{id},u)][Y]} S_R[M, A][Y]$$

determined by $u : R(X) \to A$ gives d_1. The canonical embedding

$$S_R[M, A][Y] \xrightarrow{S_R[M,A][(0,\mathrm{id})]} S_R[M, A][X \oplus Y]$$

gives s_0.

The functor $S(N) : \mathcal{E} \to \mathcal{E}$ associated to any Σ_*-object N has a canonical filtration by subfunctors $S(N)_n : Y \mapsto S(N, Y)_n$ defined by:

$$S(N, Y)_n = \bigoplus_{m=0}^{n} N(m) \otimes_{\Sigma_m} Y^{\otimes m}.$$

The shifted functor $S[N, Y] = S(N[\cdot], Y)$ inherits a filtration from $S(N[\cdot], Y)$:

$$S[N, Y]_n = S(N[\cdot], Y)_n = \bigoplus_{m=0}^{n} N(m + \cdot) \otimes_{\Sigma_m} Y^{\otimes m}.$$

If N is a right R-module, then $S[N, Y]_n$ forms a subobject of $S[N, Y]$ in the category of right R-modules.

To form the decomposition $S_R[M, B] = \mathrm{colim}_n\, S_R[M, B]_n$, we use the sequence of reflexive coequalizers

$$
\begin{array}{ccc}
\cdots \longrightarrow S_R[M, A][X \oplus Y]_{n-1} & \longrightarrow S_R[M, A][X \oplus Y]_n & \longrightarrow \cdots \\
\Big\downarrow\Big\downarrow\Big) & \Big\downarrow\Big\downarrow\Big) & \\
\cdots \longrightarrow S_R[M, A][Y]_{n-1} & \longrightarrow S_R[M, A][Y]_n & \longrightarrow \cdots \\
\Big\downarrow & \Big\downarrow & \\
\cdots \cdots\!\!\succ S_R[M, B]_{n-1} & \underset{j_n}{\cdots\cdots\!\!\succ} S_R[M, B]_n & \cdots\!\!\succ \cdots
\end{array}
$$

determined by the filtration of $S_R[M, A][X \oplus Y]$ and $S_R[M, A][Y]$ (note that the morphisms d_0, d_1, s_0 preserve filtrations). We have $S_R[M, B] = \mathrm{colim}_n\, S_R[M, B]_n$ by interchange of colimits. We have moreover $S_R[M, A][X \oplus Y]_0 = S_R[M, A][Y]_0 = S_R[M, B]_0 = S_R[M, A]$.

The coequalizers of $S_R[M, B]_{n-1}$ and $S_R[M, B]_n$ can be reorganized to give:

18.2.5 Lemma. *The morphism $j_n : S_R[M, B]_{n-1} \to S_R[M, B]_n$ fits a natural pushout of the form:*

$$
\begin{array}{ccc}
\bigoplus_{\substack{p+q=n \\ q<n}} N(n + \cdot) \otimes_{\Sigma_p \times \Sigma_q} X^{\otimes p} \otimes Y^{\otimes q} & \longrightarrow & S_R[M, B]_{n-1} \\
\Big\downarrow & & \Big\downarrow{\scriptstyle j_n} \\
N(n + \cdot) \otimes_{\Sigma_n} Y^{\otimes n} & \cdots\cdots\cdots\longrightarrow & S_R[M, B]_n,
\end{array}
$$

where we set $N = S_R[M, A]$.

Proof. Set

$$
S = S_R[M, A][X \oplus Y]_{n-1}, \quad T = S_R[M, A][Y]_{n-1},
$$
$$
\text{and} \quad U = \bigoplus_{\substack{p+q=n \\ q<n}} N(n + \cdot) \otimes_{\Sigma_p \times \Sigma_q} X^{\otimes p} \otimes Y^{\otimes q},
$$
$$
V = N(n + \cdot) \otimes_{\Sigma_n} Y^{\otimes n}.
$$

We have by definition

$$
S_R[M, A][X \oplus Y]_n = S_R[M, A][X \oplus Y]_{n-1} \oplus U \oplus V = S \oplus U \oplus V,
$$
$$
S_R[M, A][Y]_n = S_R[M, A][Y]_{n-1} \oplus V = T \oplus V,
$$

and the coequalizers $C = S_R[M, B]_{n-1}$ and $D = S_R[M, B]_n$ fit a diagram of the form:

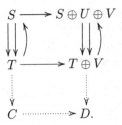

The morphisms $d_0, d_1 : S \oplus U \oplus V \rightrightarrows T \oplus V$ are the identity on V. The summand U is mapped into V by the morphism d_0, into T by the morphism d_1. Accordingly, we have a commutative square

$$
\begin{array}{ccc}
U & \xrightarrow{\;d_1\;} & C \\
{\scriptstyle d_0}\downarrow & & \downarrow \\
V & \cdots\!\!\rightarrow & D.
\end{array}
$$

By an easy inspection of the categorical constructions, we check that this square forms a pushout and the lemma follows.   □

18.2.6 Iterated Pushout-Products. Let $T_0 = X$, $T_1 = Y$. To improve the result of lemma 18.2.5 we use the cubical diagram formed by the tensor products $T_{\epsilon_1} \otimes \cdots \otimes T_{\epsilon_n}$ on vertices and the morphisms

$$
T_{\epsilon_1} \otimes \cdots \otimes T_0 \otimes \cdots \otimes T_{\epsilon_n} \xrightarrow{\;T_{\epsilon_1} \otimes \cdots \otimes i \otimes \cdots \otimes T_{\epsilon_n}\;} T_{\epsilon_1} \otimes \cdots \otimes T_1 \otimes \cdots \otimes T_{\epsilon_n}
$$

on edges. The terminal vertex of the cube is associated to the tensor power

$$
T_n(Y/X) = T_1 \otimes \cdots \otimes T_1 = Y^{\otimes n}.
$$

The latching morphism $\lambda : L_n(Y/X) \to T_n(Y/X)$ associated to the diagram $T_{\epsilon_1} \otimes \cdots \otimes T_{\epsilon_n}$ is the canonical morphism from the colimit

$$
L_n(Y/X) = \operatorname*{colim}_{(\epsilon_1, \ldots, \epsilon_n) < (1, \ldots, 1)} T_{\epsilon_1} \otimes \cdots \otimes T_{\epsilon_n}
$$

to the terminal vertex $T_n(Y/X) = T_1 \otimes \cdots \otimes T_1$. The next observation is used in §19.1 to determine the class of a latching morphism whenever i is a cofibration (respectively, an acyclic cofibration) in a symmetric monoidal model category:

18.2.7 Observation. *The latching morphism*

$$
L_n(Y/X) \xrightarrow{\;\lambda\;} T_n(Y/X) = Y^{\otimes n}
$$

is identified with the pushout-product

$$
L_{n-1}(Y/X) \otimes Y \bigoplus_{L_{n-1}(Y/X) \otimes X} Y^{\otimes n-1} \otimes X \xrightarrow{\;(\lambda_*, i_*)\;} Y^{\otimes n-1} \otimes Y
$$

of the previous latching morphism $\lambda : L_n(Y/X) \to Y^{\otimes n-1}$ *with* $i : X \to Y$.

This observation follows from an immediate interchange of colimits. Because of this observation, we also refer to the morphism $\lambda : L_n(Y/X) \to T_n(Y/X)$ as an n-fold pushout-product of $i : X \to Y$.

18.2.8 Iterated Pushout-Products and Functor Decomposition. The symmetric group Σ_n acts naturally (on the right) on $T_n(Y/X)$ and $L_n(Y/X)$. Moreover the latching morphism $\lambda : L_n(Y/X) \to T_n(Y/X)$ is equivariant. For a Σ_*-object N, we set

$$T_n N[Y/X] = N(n+\,\cdot\,) \otimes_{\Sigma_n} T_n(Y/X) \text{ and } L_n N[Y/X] = N(n+\,\cdot\,) \otimes_{\Sigma_n} L_n(Y/X).$$

Note that $T_n N[Y/X]$ (respectively, $L_n N[Y/X]$) inherits the structure of a right R-module if N has this structure. By definition, we have

$$T_n N[Y/X] = N(n+\,\cdot\,) \otimes_{\Sigma_n} Y^{\otimes n}$$

and the morphism of lemma 18.2.5 has an obvious factorization

$$
\begin{array}{ccc}
 & \bigoplus_{\substack{p+q=n \\ q<n}} N(n+\,\cdot\,) \otimes_{\Sigma_p \times \Sigma_q} X^{\otimes p} \otimes Y^{\otimes q} & \\
 & & \downarrow \\
N(n+\,\cdot\,) \otimes_{\Sigma_n} Y^{\otimes n} & \longleftarrow\cdots\cdots\cdots\cdots & L_n N[Y/X],
\end{array}
$$

Furthermore:

18.2.9 Lemma. *In lemma 18.2.5, the base of the pushout can be replaced by factorizations:*

$$
\begin{array}{ccc}
 & \bigoplus_{\substack{p+q=n \\ q<n}} N(n+\,\cdot\,) \otimes_{\Sigma_p \times \Sigma_q} X^{\otimes p} \otimes Y^{\otimes q} & \\
 & \downarrow & \\
N(n+\,\cdot\,) \otimes_{\Sigma_n} Y^{\otimes n} \longleftarrow\cdots\cdots L_n N[Y/X] & \cdots\cdots\longrightarrow & S_R[M,B]_{n-1}.
\end{array}
$$

Consequently, we have a pushout of the form:

$$
\begin{array}{ccc}
L_n N[Y/X] & \longrightarrow & S_R[M,B]_{n-1} \\
\downarrow & & \downarrow{\scriptstyle j_n} \\
T_n N[Y/X] & \dashrightarrow & S_R[M,B]_n,
\end{array}
$$

where we set again $N = S_R[M,A]$.

Proof. The existence of factorizations follows from a straightforward inspection. To prove the identity of the pushout of lemma 18.2.5 with the pushout based on $L_n N[X/Y]$, use that the morphism $\bigoplus_{\substack{p+q=n \\ q<n}} \Sigma_n \otimes_{\Sigma_p \times \Sigma_q} X^{\otimes p} \otimes Y^{\otimes q} \to L_n(Y/X)$ is epi, as well as the induced morphism

$$N(n + \cdot) \otimes_{\Sigma_p \times \Sigma_q} X^{\otimes p} \otimes Y^{\otimes q}$$
$$= N(n + \cdot) \otimes_{\Sigma_n} \Big\{ \bigoplus_{\substack{p+q=n \\ q<n}} \Sigma_n \otimes_{\Sigma_p \times \Sigma_q} X^{\otimes p} \otimes Y^{\otimes q} \Big\}$$
$$\to N(n + \cdot) \otimes_{\Sigma_n} L_n(Y/X) = L_n N[X/Y].$$

$\square$

In §19.2, we also use:

18.2.10 Lemma. *For the functor $N = S_R[M, A]$, we have identities $L_n S_R$ $[M, A][Y/X] = S_R[L_n M[Y/X], A]$ and $T_n S_R[M, A][Y/X] = S_R[T_n M[Y/X], A]$.*

Proof. Immediate by permutation of tensors and interchange of colimits. $\square$

To summarize:

18.2.11 Proposition. *Let M be a right module over an operad* R. *Let*

$$
\begin{array}{ccc}
R(X) & \xrightarrow{\ u\ } & A \\
{\scriptstyle R(i)}\downarrow & & \downarrow{\scriptstyle f} \\
R(Y) & \dashrightarrow[v] & B
\end{array}
$$

be a pushout in the category of R-*algebras. The morphism $S_R[M, f] : S_R[M, A]$* $\to S_R[M, B]$ *has a canonical decomposition*

$$S[M, A] = S_R[M, B]_0 \xrightarrow{\ j_1\ } \cdots$$
$$\cdots \to S_R[M, B]_{n-1} \xrightarrow{\ j_n\ } S_R[M, B]_n \to \cdots$$
$$\cdots \to \operatorname*{colim}_n S_R[M, B]_n = S_R[M, B]$$

so that each morphism j_n is obtained by a pushout

$$
\begin{array}{ccc}
S_R[L_n M[Y/X], A] & \longrightarrow & S_R[M, B]_{n-1} \\
{\scriptstyle S_R[\lambda, A]}\downarrow & & \downarrow{\scriptstyle j_n} \\
S_R[T_n M[Y/X], A] & \dashrightarrow & S_R[M, B]_n
\end{array}
$$

in the category of right R-*modules, where $\lambda : L_n M[Y/X] \to T_n M[Y/X]$ is defined by the tensor product $M(n + \cdot) \otimes_{\Sigma_n} -$ with a certain latching morphism $\lambda : L_n(Y/X) \to Y^{\otimes n}$.* $\square$

Chapter 19
Shifted Functors
and Pushout-Products

Introduction

In this chapter we assume that the base category $\mathcal{C}$ is equipped with a model structure and satisfies axioms MM0-MM1 of §11.3.3. Let R be any operad in $\mathcal{C}$. In principle we consider R-algebras in any symmetric monoidal model category $\mathcal{E}$ over $\mathcal{C}$ such that axioms MM0-MM1 hold in $\mathcal{E}$ and the canonical functor $\eta : \mathcal{C} \to \mathcal{E}$ preserves cofibrations (see §11.3.3).

But we use constructions of §18 which return Σ_*-objects in $\mathcal{E}$ and right R-modules in $\mathcal{E}$. For this reason we do not assume necessarily that Σ_*-objects and right R-modules belong to the base category $\mathcal{C}$. To simplify we use the canonical functor $\eta : \mathcal{C} \to \mathcal{E}$ associated to $\mathcal{E}$ to transport all objects in $\mathcal{E}$, and we assume $\mathcal{C} = \mathcal{E}$. This operation makes sense because the canonical functor $\eta : \mathcal{C} \to \mathcal{E}$ is supposed to preserves cofibrations and acyclic cofibrations.

¶ Our results can be improved in the context of (reduced symmetric monoidal categories) with regular tensor powers. These refinements are addressed in the note apparatus.

Say that a morphism of R-algebras $f : A \to B$ is a Σ_*-flat cofibration if the pushout-product

$$(i_*, f_*) : S_R[M, B] \bigoplus_{S_R[M, A]} S_R[N, A] \to S_R[N, B]$$

forms a Σ_*-cofibration of right R-modules (respectively, an acyclic Σ_*-cofibration) whenever i is so, for every morphism of right R-modules $i : M \to N$ in $\mathcal{E}$. Say that $f : A \to B$ is a Σ_*-flat acyclic cofibration if the same pushout-product (i_*, f_*) forms an acyclic Σ_*-cofibration of right R-modules for every Σ_*-cofibration of right R-modules $i : M \to N$ in $\mathcal{E}$.

Suppose that the operad R is Σ_*-cofibrant. The goal of this chapter is to prove that certain relative cell extensions of R-algebras are Σ_*-flat (acyclic) cofibrations. The result is used in §12.3 to prove that R-algebras form a semi

B. Fresse, *Modules over Operads and Functors*, Lecture Notes in Mathematics 1967, 277
DOI: 10.1007/978-3-540-89056-0_19, © Springer-Verlag Berlin Heidelberg 2009

model category and implies that any (acyclic) cofibration of R-algebras such that A is cofibrant is a Σ_*-flat (acyclic) cofibration (see §20.1).

But we can not use the axiom of semi model categories for the moment. For this reason we consider the class $R(\mathcal{E}_c)$ (respectively, $R(\mathcal{E}_a)$) of morphisms of free R-algebras $R(i) : R(X) \to R(Y)$ such that $i : X \to Y$ is a cofibration in $\mathcal{E}$ (respectively an acyclic cofibration). The main result of this chapter reads:

Lemma 19.A. *Every relative $R(\mathcal{E}_c)$-cell (respectively, $R(\mathcal{E}_a)$) complex f : $A \to B$ whose domain A is an $R(\mathcal{E}_c)$-cell complex is a Σ_*-flat cofibration (respectively, a Σ_*-flat acyclic cofibration).*

¶ If $\mathcal{E}$ has regular tensor powers with respect to a class of $\mathcal{B}$-cofibrations. then the lemma holds for (acyclic) $\mathcal{B}$-cofibrations of Σ_*-objects.

We prove lemma 19.A by induction on the cell decompositions, using that:

(1) the pushout of a morphism of free R-algebras $R(j) : R(X) \to R(Y)$ forms a Σ_*-flat cofibration (respectively, a Σ_*-flat acyclic cofibration) if j is a cofibration (respectively, an acyclic cofibration),
(2) the class of Σ_*-flat cofibrations (respectively, Σ_*-flat acyclic cofibrations) is stable under (possibly transfinite) composites.

We check assertion (1) in §19.2, assertion (2) in §19.3. We recapitulate the results and conclude in §19.4.

The verification of (1) use the decomposition of §18.2 and the latching morphism $\lambda : L_n M[Y/X] \to T_n M[Y/X]$ associated to morphisms $j : X \to Y$ in $\mathcal{E}$, for a Σ_*-object M. To prove (1), we need first to determine the class of the latching morphism $\lambda : L_n M[Y/X] \to T_n M[Y/X]$ when j is a cofibration (respectively, an acyclic cofibration). This preliminary step is addressed in the next section.

19.1 Preliminary Step

The goal of this section is to prove:

Lemma 19.1.A. *Let $i : M \to N$ be any morphism in $\mathcal{E}^{\Sigma_*}$, the category of Σ_*-objects in $\mathcal{E}$. Let $j : X \to Y$ be a morphism in $\mathcal{E}$. The pushout-product*

$$(i_*, \lambda_*) : T_n M[Y/X] \bigoplus_{L_n M[Y/X]} L_n N[Y/X] \to T_n N[Y/X]$$

forms a cofibration in $\mathcal{E}^{\Sigma_*}$ if i is so and j is a cofibration in $\mathcal{E}$, an acyclic cofibration if i or j is also acyclic.

19.1.1 ¶ Remark. In the context of categories with regular tensor powers, we replace (acyclic) cofibrations in $\mathcal{E}^{\Sigma_*}$ by (acyclic) $\mathcal{B}$-cofibrations. The statement is nothing but axiom R1.

Proof. By lemma 11.3.2, we can assume that $i : M \to N$ is a generating (acyclic) cofibration of the category of Σ_*-objects, and hence is given by a tensor products $k \otimes F_r : K \otimes F_r \to L \otimes F_r$, where $k : K \to L$ is a generating (acyclic) cofibration of $\mathcal{E}$,

For a Σ_*-object M, the latching morphism $\lambda : L_n M[Y/X] \to T_n M[Y/X]$ is defined by a tensor product

$$M(n + \cdot) \otimes_{\Sigma_n} \lambda : M(n + \cdot) \otimes_{\Sigma_n} L_n(Y/X) \to M(n + \cdot) \otimes_{\Sigma_n} T_n(Y/X)$$

where $T_n(Y/X) = Y^{\otimes n}$ and $\lambda : L_n(Y/X) \to T_n(Y/X)$ is an n-fold pushout-product of the morphism $j : X \to Y$. Note that Σ_n operates on $L_n(Y/X)$ and $T_n(Y/X)$ on the right.

Let T be any object equipped with a right Σ_n-action. Recall that

$$F_r(s) = \begin{cases} 1[\Sigma_r], & \text{if } r = s, \\ 0, & \text{otherwise.} \end{cases}$$

According to this definition, we have $F_r(n + s) \otimes_{\Sigma_n} T = 0$ if $n + s \neq r$ and

$$\begin{aligned} F_r(n + r - n) \otimes_{\Sigma_n} T &= 1[\Sigma_r] \otimes_{\Sigma_n} T \\ &\simeq \Sigma_n \backslash \Sigma_r \otimes T \\ &\simeq \Sigma_n \times \Sigma_{r-n} \backslash \Sigma_r \otimes T \otimes F_{r-n}(r - n) \end{aligned}$$

otherwise. Hence we obtain an isomorphism

$$(F_r(n + \cdot) \otimes T)_{\Sigma_n} \simeq T \otimes (\Pi_{r,n} \otimes F_{r-n}),$$

where $\Pi_{r,n} = \Sigma_n \times \Sigma_{r-n} \backslash \Sigma_r$, and the latching morphism $\lambda : L_n F_r[Y/X] \to T_n F_r[Y/X]$ is identified with a tensor product

$$\lambda \otimes (\Pi_{r,n} \otimes F_{r-n}) : L_n(Y/X) \otimes (\Pi_{r,n} \otimes F_{r-n}) \to T_n(Y/X) \otimes (\Pi_{r,n} \otimes F_{r-n}).$$

For a morphism of the form

$$i = k \otimes F_r : K \otimes F_r \to L \otimes F_r,$$

the pushout-product

$$(i_*, j_*) : S[K \otimes F_r, Y] \bigoplus_{S[K \otimes F_r, X]} S[L \otimes F_r, X] \to S[L \otimes F_r, Y]$$

is identified with the tensor product of the object $\Pi_{r,n} \otimes F_{r-n} \in \mathcal{M}$ with the pushout product in $\mathcal{E}$

$$(k_*, \lambda_*) : K \otimes T_n(Y/X) \bigoplus_{K \otimes L_n(Y/X)} L \otimes L_n(Y/X) \to L \otimes T_n(Y/X).$$

The pushout-product (k_*, λ_*) forms a cofibration in $\mathcal{E}$ if k and λ are so, an acyclic cofibration if k or λ is also an acyclic cofibration. Since λ : $L_n M[Y/X] \to T_n M[Y/X]$ is defined by an n-fold pushout-product in $\mathcal{E}$, we obtain that λ forms a cofibration (respectively, an acyclic cofibration) if j is so. The conclusion follows. $\square$

19.2 Pushouts

In this section, we study the morphism $f : A \to B$ obtained by a pushout

$$\begin{array}{ccc} R(X) & \xrightarrow{u} & A \\ \scriptstyle R(j) \downarrow & & \downarrow \scriptstyle f \\ R(Y) & \dashrightarrow[v] & B, \end{array}$$

where $R(j) : R(X) \to R(Y)$ is a morphism of free R-algebras induced by a morphism $j : X \to Y$ in $\mathcal{E}$. Our goal is to prove:

Lemma 19.2.A. *Let A be an R-algebra. Suppose that the initial R-algebra morphism $\eta : R(0) \to A$ forms a Σ_*-flat cofibration. If j is a cofibration (respectively, an acyclic cofibration) in $\mathcal{E}$, then the morphism $f : A \to B$ obtained by a pushout of $R(j) : R(X) \to R(Y)$ is a Σ_*-flat cofibration (respectively, a Σ_*-flat acyclic cofibration).*

We begin the proof of lemma 19.2.A by an observation:

19.2.1 Lemma. *If $\eta : R(0) \to A$ is a Σ_*-flat cofibration, then the morphism $S_R[i, A] : S_R[M, A] \to S_R[N, A]$ forms a Σ_*-cofibration (respectively, an acyclic Σ_*-cofibration) whenever $i : M \to N$ is so.*

Proof. For the initial algebra $R(0)$, proposition 18.2.1 gives an identity $S_R[M, R(0)] = S[M, 0] = M$. Hence the pushout-product (i_*, η_*) is identified with the morphism $S_R[M, A] \bigoplus_M N \to S_R[N, A]$ induced by $S_R[i, A]$ and the natural morphism $N \to S_R[N, A]$. The pushout

$$\begin{array}{ccc} M & \longrightarrow & S_R[M, A] \\ \downarrow & & \downarrow \\ N & \dashrightarrow & S_R[M, A] \bigoplus_M N \end{array}$$

returns a Σ_*-cofibration (respectively, an acyclic Σ_*-cofibration) if i is so. Hence, if $\eta : R(0) \to A$ is a Σ_*-flat cofibration, then we obtain that the composite

$$S_R[M, A] \to S_R[M, A] \bigoplus_M N \to S_R[N, A]$$

forms a Σ_*-cofibration (respectively, an acyclic Σ_*-cofibration) as well. $\square$

In the next proofs, we use the following independent lemma repeatedly:

19.2.2 Lemma. *For all $S, T \in \mathcal{M}_R$, the functor $-\bigoplus_S T : \mathcal{M}_R \to \mathcal{M}_R$ defined by the relative sum $M \mapsto M \bigoplus_S T$ in the category of right R-modules preserves (acyclic) Σ_*-cofibrations.*

Proof. The operation $-\bigoplus_S T : \mathcal{A} \to \mathcal{A}$ preserves (acyclic) cofibrations in any model category $\mathcal{A}$.

The lemma is an immediate corollary of this assertion and of the fact that the forgetful functor $U : \mathcal{M}_R \to \mathcal{M}$ from right R-modules to Σ_*-objects creates colimits. □

For the remainder of this section, we suppose given a pushout

$$
\begin{array}{ccc}
R(X) & \longrightarrow & A \\
{\scriptstyle R(j)}\downarrow & & \downarrow{\scriptstyle f} \\
R(Y) & \dashrightarrow & B
\end{array}
$$

and we assume that $\eta : R(0) \to A$ forms a Σ_*-flat cofibration. We aim to prove:

Claim (claim of lemma 19.2.A). *The pushout-product*

$$
(i_*, f_*) : S_R[M, B] \bigoplus_{S_R[M, A]} S_R[N, A] \to S_R[N, B]
$$

forms a Σ_*-cofibration if i is so and j is a cofibration in $\mathcal{E}$, an acyclic Σ_*-cofibration if i or j is also acyclic.

Hence the morphism f forms a Σ_*-flat cofibration (respectively, Σ_*-flat acyclic cofibration) if j is a cofibration (respectively, an acyclic cofibration), as asserted by lemma 19.2.A.

Our arguments are based on the decomposition of proposition 18.2.11. Explicitly, for any right R-module M, we have a natural decomposition

$$S[M, A] = S_R[M, B]_0 \xrightarrow{j_1} \cdots$$

$$\cdots \to S_R[M, B]_{n-1} \xrightarrow{j_n} S_R[M, B]_n \to \cdots$$

$$\cdots \to \operatorname*{colim}_n S_R[M, B]_n = S_R[M, B]$$

of the morphism $S_R[M, f] : S_R[M, A] \to S_R[M, B]$ so that each morphism j_n is obtained by a pushout

$$S_R[L_n M[Y/X], A] \longrightarrow S_R[M, B]_{n-1}$$

$$S_R[\lambda, A] \downarrow \qquad\qquad \downarrow j_n$$

$$S_R[T_n M[Y/X], A] \dashrightarrow S_R[M, B]_n$$

in the category of right R-modules. By naturality of the decomposition, any morphism of right R-modules $i : M \to N$, gives rise to a morphism $S_R[i, B]_n :$ $S_R[M, B]_n \to S_R[N, B]_n$ which can also be determined by the pushout of morphisms

$$S_R[T_n M[Y/X], A] \longleftarrow S_R[L_n M[Y/X], A] \longrightarrow S_R[M, B]_{n-1}$$

$$S_R[T_n f[Y/X], A] \downarrow \qquad S_R[T_n f[Y/X], A] \downarrow \qquad\qquad \downarrow S_R[f, B]_{n-1}$$

$$S_R[T_n N[Y/X], A] \longleftarrow S_R[L_n N[Y/X], A] \longrightarrow S_R[N, B]_{n-1}$$

induced by f.

The proof of the claim is achieved by the next lemmas. The idea is use the categorical decomposition to split the pushout-product of the claim into (acyclic) cofibrations.

19.2.3 Lemma. *The pushout-product*

$$((S_R[i, B]_n)_*, (j_n)_*) : S_R[M, B]_n \bigoplus_{S_R[M,B]_{n-1}} S_R[N, B]_{n-1} \to S_R[N, B]_n$$

forms a Σ_*-cofibration if i is so and j is a cofibration in $\mathcal{E}$, an acyclic Σ_*-cofibration if i or j is also acyclic.

Proof. We use a categorical decomposition of the pushout-product

$$((S_R[i, B]_n)_*, (j_n)_*) : S_R[M, B]_n \bigoplus_{S_R[M,B]_{n-1}} S_R[N, B]_{n-1} \xrightarrow{(1)} S_R[N, B]_n$$

to check that $((S_R[i, B]_n)_*, (j_n)_*)$ forms a Σ_*-cofibration (respectively an acyclic Σ_*-cofibration) under the assumptions of the lemma.

Throughout this proof we adopt the short notation

$$L_n[M, A] = L_n S_R[M, A][Y/X], \qquad T_n[M, A] = T_n S_R[M, A][Y/X],$$
$$L_n M[Y/X] = L_n M, \qquad\qquad T_n M[Y/X] = T_n M,$$

and so on. Hence proposition 18.2.11 gives natural isomorphisms

$$S_R[M, B]_n = T_n[M, A] \bigoplus_{L_n[M,A]} S_R[M, B]_{n-1}$$

$$\text{and} \qquad S_R[N, B]_n = T_n[N, A] \bigoplus_{L_n[N,A]} S_R[N, B]_{n-1}$$

and morphism (1) is identified with the morphism of pushouts

$$T_n[M, A] \bigoplus_{L_n[M,A]} S_R[N, B]_{n-1} \xrightarrow{(2)} T_n[N, A] \bigoplus_{L_n[N,A]} S_R[N, B]_{n-1}$$

induced by the morphisms

$$\begin{array}{ccccc}
T_n[M, A] & \longleftarrow & L_n[M, A] & \longrightarrow & S_R[N, B]_{n-1} \\
{\scriptstyle T_n[i,A]}\downarrow & & {\scriptstyle L_n[i,A]}\downarrow & & \downarrow{\scriptstyle =} \\
T_n[N, A] & \longleftarrow & L_n[N, A] & \longrightarrow & S_R[N, B]_{n-1}.
\end{array}$$

Since

$$T_n[M, A] \bigoplus_{L_n[M,A]} S_R[N, B]_{n-1} \simeq \left\{ T_n[M, A] \bigoplus_{L_n[M,A]} L_n[N, A] \right\} \bigoplus_{L_n[N,A]} S_R[N, B]_{n-1}$$

morphism (2) is also equal to the image of the natural morphism

$$T_n[M, A] \bigoplus_{L_n[M,A]} L_n[N, A] \xrightarrow{(3)} T_n[N, A]$$

under the relative sum operation

$$- \bigoplus_{L_n[N,A]} S_R[N, B]_{n-1}.$$

Recall that we use short notation $T_n M = T_n M[Y/X]$, $T_n[M, A] = T_n S_R[M, A][Y/X]$, and so on. Since the functor $S_R[-, A]$ preserves colimits, lemma 18.2.10 implies that morphism (3) is the image of the morphism

$$T_n M \bigoplus_{L_n M} L_n N \xrightarrow{(4)} T_n N$$

under the functor $S_R[-, A]$.

By lemma 19.1.A, morphism (4) forms a Σ_*-cofibration if i is so and j is a cofibration in $\mathcal{E}$, an acyclic Σ_*-cofibration if i or j is also acyclic. So does morphism (3) by lemma 19.2.1, morphism (2) by lemma 19.2.2, and the conclusion regarding morphism (1) follows. $\square$

19.2.4 Lemma. *If each pushout-product*

$$((S_R[i, B]_n)_*, (j_n)_*) : S_R[M, B]_n \bigoplus_{S_R[M,B]_{n-1}} S_R[N, B]_{n-1} \to S_R[N, B]_n$$

forms a Σ_*-cofibration (respectively, an acyclic Σ_*-cofibration), then so does the pushout-product

$$(i_*, f_*) : \mathrm{S_R}[M, B] \bigoplus_{\mathrm{S_R}[M,A]} \mathrm{S_R}[N, A] \to \mathrm{S_R}[N, B].$$

Proof. The pushout-product

$$(i_*, f_*) : \mathrm{S_R}[M, B] \bigoplus_{\mathrm{S_R}[M,A]} \mathrm{S_R}[N, A] \xrightarrow{(1)} \mathrm{S_R}[N, B]$$

can be decomposed into:

$$\mathrm{S_R}[M, B] \bigoplus_{\mathrm{S_R}[M,A]} \mathrm{S_R}[N, A] = \mathrm{S_R}[M, B] \bigoplus_{\mathrm{S_R}[M,B]_0} \mathrm{S_R}[N, B]_0 \to \cdots$$

$$\cdots \to \mathrm{S_R}[M, B] \bigoplus_{\mathrm{S_R}[M,B]_{n-1}} \mathrm{S_R}[N, B]_{n-1} \to \mathrm{S_R}[M, B] \bigoplus_{\mathrm{S_R}[M,B]_n} \mathrm{S_R}[N, B]_n \to \cdots$$

$$\cdots \to \operatorname*{colim}_{n}\left\{ \mathrm{S_R}[M, B] \bigoplus_{\mathrm{S_R}[M,B]_n} \mathrm{S_R}[N, B]_n \right\} \simeq \mathrm{S_R}[N, B].$$

The morphism

$$\mathrm{S_R}[M, B] \bigoplus_{\mathrm{S_R}[M,B]_{n-1}} \mathrm{S_R}[N, B]_{n-1} \xrightarrow{(2)} \mathrm{S_R}[M, B] \bigoplus_{\mathrm{S_R}[M,B]_n} \mathrm{S_R}[N, B]_n$$

is also identified with the image of

$$\mathrm{S_R}[M, B]_n \bigoplus_{\mathrm{S_R}[M,B]_{n-1}} \mathrm{S_R}[N, B]_{n-1} \xrightarrow{(3)} \mathrm{S_R}[N, B]_n$$

under the relative sum operation

$$\mathrm{S_R}[M, B] \bigoplus_{\mathrm{S_R}[M,B]_n} -.$$

By lemma 19.2.2, morphism (2) forms a Σ_*-cofibration (respectively, an acyclic cofibration) if morphism (3) is so. Since the class of (acyclic) cofibrations in a model category is closed under composites, we conclude that morphism (1) forms a Σ_*-cofibration (respectively, an acyclic cofibration) if every morphism (3) is so.   □

This proof achieves the proof of lemma 19.2.A.   □

19.3 Third Step: Composites

In this section, we prove:

Lemma 19.3.A. *A morphism $f : A \to B$ that decomposes into a (possibly transfinite) composite of Σ_*-flat (acyclic) cofibrations*

$$A = B_0 \xrightarrow{f_0} \cdots \to B_{\lambda-1} \xrightarrow{f_\lambda} B_\lambda \to \cdots \to \operatorname*{colim}_\lambda B_\lambda = B$$

forms itself a Σ_*-flat (acyclic) cofibration.

We deduce this lemma from formal categorical constructions.

Proof. Let $i : M \to N$ be a fixed morphism in the category of right R-modules.

Recall that functors of the form $S_R(M) : A \mapsto S_R(M, A)$ preserve filtered colimits in A. So do the extended functors $S_R[M, -] : A \mapsto S_R[M, A]$ since we have $S_R[M, A](m) = S_R(M[m], A)$, for any $m \in \mathbb{N}$.

As a consequence, for any sequential colimit

$$A = B_0 \xrightarrow{f_0} \cdots \to B_\lambda \xrightarrow{i_\lambda} B_\lambda \to \cdots \to \operatorname*{colim}_\lambda B_\lambda = B,$$

we have

$$S_R[M, B] = \operatorname*{colim}_\lambda S_R[M, B_\lambda], \qquad S_R[N, B] = \operatorname*{colim}_\lambda S_R[N, B_\lambda]$$

and the pushout-product

$$(i_*, f_*) : S_R[M, B] \bigoplus_{S_R[M,A]} S_R[N, A] \xrightarrow{(1)} S_R[N, B]$$

can be decomposed into a colimit

$$S_R[M, B] \bigoplus_{S_R[M,A]} S_R[N, A] = S_R[M, B] \bigoplus_{S_R[M,B_0]} S_R[N, B_0] \to \cdots$$

$$\cdots \to S_R[M, B] \bigoplus_{S_R[M,B_{\lambda-1}]} S_R[N, B_{\lambda-1}] \to S_R[M, B] \bigoplus_{S_R[M,B_\lambda]} S_R[N, B_\lambda] \to \cdots$$

$$\cdots \to \operatorname*{colim}_\lambda \left\{ S_R[M, B] \bigoplus_{S_R[M,B_\lambda]} S_R[N, B_\lambda] \right\} \simeq S_R[N, B].$$

The morphism

$$S_R[M, B] \bigoplus_{S_R[M,B_{\lambda-1}]} S_R[N, B_{\lambda-1}] \xrightarrow{(2)} S_R[M, B] \bigoplus_{S_R[M,B_\lambda]} S_R[N, B_\lambda]$$

is also identified with the image of

$$S_R[M, B_\lambda] \underset{S_R[M, B_{\lambda-1}]}{\bigoplus} S_R[N, B_{\lambda-1}] \xrightarrow{(3)} S_R[N, B_\lambda]$$

under the relative sum operation

$$S_R[M, B] \underset{S_R[M, B_\lambda]}{\bigoplus} -.$$

If $i : M \to N$ is a Σ_*-cofibration and $f_\lambda : B_{\lambda-1} \to B_\lambda$ is a Σ_*-flat cofibration, then (3) forms a Σ_*-cofibration, and so does morphism (2) by lemma 19.2.2. Since the class of cofibrations in a model category is closed under composites, we obtain that morphism (1) forms a Σ_*-cofibration whenever $i : M \to N$ is a Σ_*-cofibration and every $f_\lambda : B_{\lambda-1} \to B_\lambda$ is a Σ_*-flat cofibration. We prove similarly that morphism (1) forms an acyclic Σ_*-cofibration if i or every f_λ is also acyclic. $\square$

19.4 Recapitulation and Conclusion

By definition, a relative $R(\mathcal{E}_c)$-cell (respectively, $R(\mathcal{E}_a)$-cell) complex is a morphism of R-algebras $f : A \to B$ which can be decomposed into a composite

$$A = B_0 \xrightarrow{f_0} B_1 \to \cdots \to B_{\lambda-1} \xrightarrow{f_\lambda} B_\lambda \to \cdots \to \operatorname*{colim}_\lambda B_\lambda = B$$

so that the morphisms $f_\lambda : B_{\lambda-1} \to B_\lambda$ are obtained by cobase extensions

$$
\begin{array}{ccc}
R(X_\lambda) & \longrightarrow & B_{\lambda-1} \,, \\
{\scriptstyle R(i_\lambda)} \downarrow & & \downarrow {\scriptstyle f_\lambda} \\
R(Y_\lambda) & \dashrightarrow & B_\lambda
\end{array}
$$

where $i_\lambda : X_\lambda \to Y_\lambda$ is a cofibration (respectively, an acyclic cofibration).

Lemmas 19.2.A-19.3.A imply by induction that a relative $R(\mathcal{E}_c)$-cell complex $f : A \to B$ is a Σ_*-flat cofibration as long as the initial morphism $\eta : R(0) \to A$ is so. Observe that the identity morphism $\eta : R(0) \to R(0)$ is trivially a Σ_*-flat cofibration to conclude that any relative $R(\mathcal{E}_c)$-cell complex $f : A \to B$, where A is an $R(\mathcal{E}_c)$-cell complex, is a Σ_*-flat cofibration.

Lemmas 19.2.A-19.3.A imply further that a relative $R(\mathcal{E}_a)$-cell complex $f : A \to B$ is a Σ_*-flat acyclic cofibration as long as the initial morphism $\eta : R(0) \to A$ is a Σ_*-flat cofibration. We conclude that any relative $R(\mathcal{E}_a)$-cell complex $f : A \to B$, where A is an $R(\mathcal{E}_c)$-cell complex, is a Σ_*-flat acyclic cofibration.

This conclusion achieves the proof of lemma 19.A. $\square$

Chapter 20
Applications of the Pushout-Products of Shifted Functors

Introduction

The purpose of this chapter is to check lemma 12.3.1 and lemma 15.2.C whose verification was put off in §12.3 and §15.2. Both statements are applications of the results of §19.

20.1 The Semi Model Category of Algebras over an Operad

The purpose of this section is to check the following lemma announced in §12.3:

Lemma 20.1.A (see lemma 12.3.1). *Suppose* P *is a* Σ_*-*cofibrant operad in a base model category* $\mathcal{C}$. *Let* $\mathcal{E}$ *be a symmetric monoidal category over* $\mathcal{C}$. *Let* $P(\mathcal{E}_c)$ *denote the class of morphisms* $P(i) : P(C) \to P(D)$ *so that* i *is a cofibration in* $\mathcal{E}_c$.

For any pushout

$$
\begin{array}{ccc}
P(X) & \longrightarrow & A \\
{\scriptstyle P(j)}\downarrow & & \downarrow{\scriptstyle f} \\
P(Y) & \dashrightarrow & B
\end{array}
$$

such that A *is a* $P(\mathcal{E}_c)$-*cell complex, the morphism* f *forms a cofibration (respectively, an acyclic cofibration) in the underlying category* $\mathcal{E}$ *if* $i : X \to Y$ *is so.*

In the context of a (reduced) symmetric monoidal category with regular tensor powers, the proposition holds if the operad is $\mathcal{C}$-cofibrant.

B. Fresse, *Modules over Operads and Functors*, Lecture Notes in Mathematics 1967, 287
DOI: 10.1007/978-3-540-89056-0_20, © Springer-Verlag Berlin Heidelberg 2009

Proof. The proof of lemma 20.1.A relies on an easy applications of the results of §19.2.

The initial morphism of the P-algebra A is a Σ_*-flat cofibration since we assume that A is a $P(\mathcal{E}_c)$-cell complex and lemma 19.2.A implies that the morphism $f : A \to B$ forms a Σ_*-flat cofibration (respectively, a Σ_*-flat acyclic cofibration) if j is a cofibration (respectively, an acyclic cofibration) in $\mathcal{E}$. The trivial morphism $0 : 0 \to P$ forms a Σ_*-cofibration by assumption. Therefore the pushout-product

$$(0_*, f_*) : S_P[0, B] \bigoplus_{S_P[0,A]} S_P[P, A] \to S_P[P, B]$$

forms a Σ_*-cofibration (respectively, an acyclic Σ_*-cofibration) if j is a cofibration (respectively, an acyclic cofibration) in $\mathcal{E}$.

Since $S_P[0, -] = 0$, the pushout-product $(0_*, f_*)$ is identified with the morphism $S_P[P, f] : S_P[P, A] \to S_P[P, B]$ induced by f. Since $S_P(P, -) = S_P[P, -](0)$ is identified with the forgetful functor $U : {}_P\mathcal{E} \to \mathcal{E}$, we conclude that the morphism $f : A \to B$ forms a cofibration (respectively, an acyclic cofibration) in $\mathcal{E}$ if j is so. □

The verification of lemma 20.1.A achieves the proof of theorem 12.3.A: the category of P-algebras in $\mathcal{E}$ forms a semi model category. □

20.2 Applications: Homotopy Invariance of Functors for Cofibrant Algebras

Lemma 19.A gives as a corollary:

Lemma 20.2.A (see lemma 15.2.C). *Suppose R is a Σ_*-cofibrant operad in a base model category C. Let $\mathcal{E}$ be a symmetric monoidal category over C.*

Let $i : M \to N$ be a morphism of right R-modules. Let $f : A \to B$ be a morphism of R-algebras in $\mathcal{E}$. The pushout-product

$$(i_*, f_*) : S_R(M, B) \bigoplus_{S_R(M,A)} S_R(N, A) \to S(N, B)$$

forms a cofibration in $\mathcal{E}$ if i defines a Σ_*-cofibration, the morphism f is a cofibration of R-algebras, and A is a cofibrant R-algebra, an acyclic cofibration if i or f is also acyclic.

Proof. As in §19, we use the notation $\mathcal{E}_c$ (respectively, $\mathcal{E}_a$) to refer to the class of cofibrations (respectively, an acyclic cofibrations) in $\mathcal{E}$. Let $\mathcal{I} \subset \mathcal{E}_c$ (respectively, $\mathcal{J} \subset \mathcal{E}_a$) be the set of generating cofibrations (respectively, acyclic cofibrations) of $\mathcal{E}$. By construction, the semi model category of R-algebras is

cofibrantly generated, with $R(\mathcal{I}) = \{R(i), i \in \mathcal{I}\}$ as a set of generating cofibrations and $R(\mathcal{J}) = \{R(i), i \in \mathcal{I}\}$ as a set of generating acyclic cofibrations.
Any cofibrant R-algebra A forms a retract of an $R(\mathcal{I})$-cell complex in the
category of R-algebras. Any cofibration of R-algebras $f : A \to B$ such that A
is cofibrant forms a retract of a relative $R(\mathcal{I})$-cell complex.

Lemma 19 asserts that the pushout-product

$$(i_*, f_*) : S_R[M, B] \bigoplus_{S_R[M, A]} S_R[N, A] \to S[N, B]$$

forms a Σ_*-cofibration if i defines a Σ_*-cofibration, the object A is an $R(\mathcal{E}_c)$-
cell complex and the morphism f is a relative $R(\mathcal{E}_c)$-complex, an acyclic Σ_*-
cofibration if i is also acyclic or f is a relative $R(\mathcal{E}_a)$-complex. The result
can be extended to morphisms f which are retracts of relative $R(\mathcal{E}_c)$-cell
(respectively, $R(\mathcal{E}_a)$-cell) complexes with a domain A which forms a retract
of a $R(\mathcal{E}_c)$-cell complex.

By applying this result to the term $S_R(-,-) = S_R[-,-](0)$ of the Σ_*-
objects $S_R[-,-]$, we obtain that the morphism

$$(i_*, f_*) : S_R(M, B) \bigoplus_{S_R(M, A)} S_R(N, A) \to S(N, B)$$

forms a cofibration (respectively, an acyclic cofibrations) in $\mathcal{E}$ under the same
assumptions on i and f. The conclusion follows.   □

The verification of this statement, put off in §15.2, achieves the proof of
theorem 15.2.A.   □

References

[1] D. Balavoine, *Homology and cohomology with coefficients, of an algebra over a quadratic operad*, J. Pure Appl. Algebra **132** (1998), 221–258.

[2] M. Barratt, *Twisted Lie algebras*, *in* Lecture Notes in Math. **658**, Springer-Verlag (1978), 9–15.

[3] J. Beck, *Triples, algebras and cohomology*, Dissertation, Columbia University, 1967.

[4] C. Berger, I. Moerdijk, *Axiomatic homotopy theory for operads*, Comment. Math. Helv. **78** (2003), 805–831.

[5] C. Berger, I. Moerdijk, *On the derived category of an algebra over an operad*, preprint `arXiv:0801.2664` (2008).

[6] J. Boardman, R. Vogt, *Homotopy invariant algebraic structures on topological spaces*, Lecture Notes in Mathematics **347**, Springer-Verlag, 1973.

[7] F. Cohen, P. May, L. Taylor, *Splitting of certain spaces CX*, Math. Proc. Camb. Phil. Soc. **84** (1978), 465–496.

[8] A. Connes, *Non-commutative differential geometry*, Publ. Math. IHES **62** (1985), 257–360.

[9] A. Dold, D. Puppe, *Homologie nicht-additiver funktoren*, Ann. Inst. Fourier **11** (1961), 201–312.

[10] A. Elmendorf, I. Kriz, I, M. Mandell, P. May, *Rings, modules, and algebras in stable homotopy theory*, with an appendix by M. Cole, Mathematical Surveys and Monographs **47**, American Mathematical Society, 1997.

[11] B. Fresse, *Cogroupes dans les algèbres sur une opérade*, Thèse de Doctorat, Université Louis Pasteur, Strasbourg, 1996.

[12] B. Fresse, *Cogroups in algebras over operads are free algebras*, Comment. Math. Helv. **73** (1998), 637–676.

[13] B. Fresse, *Lie theory of formal groups over an operad*, J. Algebra **202** (1998), 455–511.

[14] B. Fresse, *Koszul duality of operads and homology of partition posets*, *in* "Homotopy theory: relations with algebraic geometry, group

cohomology, and algebraic K-theory", Contemp. Math. **346**, Amer. Math. Soc. (2004), 115–215.

[15] B. Fresse, *The bar complex of an E-infinity algebra*, preprint arXiv:math.AT/0601085, (2006).

[16] B. Fresse, *The bar complex as an E-infinity Hopf algebra*, in preparation.

[17] E. Getzler, J. Jones, *Operads, homotopy algebra and iterated integrals for double loop spaces*, preprint arXiv:hep-th/9403055 (1994).

[18] V. Ginzburg, M. Kapranov, *Koszul duality for operads*, Duke Math. J. **76** (1995), 203–272.

[19] P. Goerss, *Barratt's desuspension spectral sequence and the Lie ring analyzer*, Quart. J. Math. Oxford **44** (1993), 43–85.

[20] P. Goerss, M. Hopkins, *André-Quillen (Co-)homology for simplicial algebras over simplicial operads*, in "Une dégustation topologique [Topological morsels]: homotopy theory in the Swiss Alps", Contemp. Math. **265** (2000), 41–85.

[21] P.   Goerss,   M.   Hopkins,   *Moduli   problems   for   structured   ring   spectra*   preprint   (2005).   Available   at: http://www.math.northwestern.edu/~pgoerss/spectra/obstruct.pdf.

[22] J. E. Harper, *Homotopy theory of modules over operads and non-Sigma operads in monoidal model categories*, preprint arXiv:0801.0191 (2007).

[23] J. E. Harper, *Homotopy theory of modules over operads in symmetric spectra*, preprint arXiv:0801.0193 (2007).

[24] J. E. Harper, *Bar constructions and Quillen homology of modules over operads*, preprint arXiv:0802.2311 (2008).

[25] K. Hess, P.-E. Parent, J. Scott, *Co-rings over operads characterize morphisms*, preprint arXiv:math.AT/0505559 (2005).

[26] V. Hinich, *Homological algebra of homotopy algebras*, Comm. Algebra **25** (1997), 3291–3323.

[27] P. Hirschhorn, *Model categories and their localizations*, Mathematical Surveys and Monographs **99**, American Mathematical Society, 2003.

[28] M. Hovey, *Model categories*, Mathematical Surveys and Monographs **63**, American Mathematical Society, 1999.

[29] M. Hovey, *Monoidal model categories*, preprint arXiv:math.AT/9803002 (1998).

[30] M. Hovey, B. Shipley, J. Smith, *Symmetric spectra*, J. Amer. Math. Soc. **13** (2000), 149–208.

[31] I. James, *Reduced product spaces*, Ann. Math. **62** (1955), 170–197.

[32] A. Joyal, *Une théorie combinatoire des séries formelles*, Adv. Math. **42** (1981), 1–82.

[33] M. Kapranov, Y. Manin, *Modules and Morita theorem for operads*, Amer. J. Math. **123** (2001), 811–838.

[34] M. Karoubi, *Homologie cyclique et K-théorie*, Astérisque **149**, Société Mathématique de France, 1987.

[35] M. Kontsevich, *Operads and motives in deformation quantization*, Lett. Math. Phys. **48** (1999), 35–72.

[36] T. A. Kro, *Model structure on operads in orthogonal spectra*, Homology, Homotopy and Applications **9** (2007), 397–412.

[37] M. Lazard, *Lois de groupes et analyseurs*, Ann. Sci. Ecole Norm. Sup. **62** (1955), 299–400.

[38] M. Livernet, *From left modules to algebras over an operad: application to combinatorial Hopf algebras*, preprint arXiv:math/0607427 (2001).

[39] M. Livernet, F. Patras, *Lie theory for Hopf operads*, J. Algebra **319** (2008), 4899–4920.

[40] J.-L. Loday, *Cyclic homology*, Grundlehren der mathematischen Wissenschaften **301**, Springer-Verlag, 1992.

[41] M. Mandell, *E-infinity algebras and p-adic homotopy theory*, Topology **40** (2001), 43–94.

[42] M. Mandell, P. May, S. Schwede, B. Shipley, *model categories of diagram spectra*, Proc. London Math. Soc. **82** (2001), 441–512.

[43] S. Mac Lane, *Categorical algebra*, Bull. Amer. Math. Soc. **71** (1965), 40–106.

[44] S. Mac Lane, *Categories for the Working Mathematician (second edition)*, Graduate Texts in Mathematics **5**, Springer Verlag, 1998.

[45] M. Markl, *Models for operads*, Comm. Algebra **24** (1996), 1471–1500.

[46] M. Markl, S. Shnider, J. Stasheff, *Operads in algebra, topology and physics*, Mathematical Surveys and Monographs **96**, American Mathematical Society, 2002.

[47] P. May, *The geometry of iterated loop spaces*, Lecture Notes in Mathematics **271**, Springer-Verlag, 1972.

[48] P. May, *Pairing of categories and spectra*, J. Pure Appl. Algebra **19** (1980), 299–346.

[49] F. Patras, M. Schocker, *Twisted descent algebras and the Solomon-Tits algebra*, Adv. Math. **199** (2006), 151–184.

[50] D. Quillen, *Homotopical algebra*, Lecture Notes in Mathematics **43**, Springer-Verlag, 1967.

[51] D. Quillen, *Rational homotopy theory*, Ann. Math. **90** (1969), 205–295.

[52] D. Quillen, *On the (co)-homology of commutative rings*, Proc. Symp. Pure Math. **17**, Amer. Math. Soc. (1970), 65–87.

[53] C. Reutenauer, Free Lie algebras, London Mathematical Society Monographs, New Series **7**, Oxford University Press, 1993.

[54] C. Rezk, *Spaces of algebra structures and cohomology of operads*, PhD Thesis, Massachusetts Institute of Technology, 1996.

[55] B. Shipley, *A convenient model category for commutative ring spectra*, in "Homotopy theory: relations with algebraic geometry, group cohomology, and algebraic K-theory", Contemp. Math. **346**, Amer. Math. Soc. (2004), 473–483.

[56] V. Smirnov, *On the cochain complex of topological spaces*, Math. USSR Sbornik **43** (1982), 133–144.

[57] V. Smirnov, *On the chain complex of an iterated loop space*, Izv. Akad. Nauk SSSR Ser. Mat. **53** (1989), 1108–1119, 1135–1136. Translation in Math. USSR-Izv. **35** (1990), 445–455.

[58] M. Spitzweck, *Operads, Algebras and Modules in General Model Categories*, preprint `arXiv:math.AT/0101102` (2001).

[59] S. Schwede, *An untitled book project about symmetric spectra*, preprint (2007). Available at `http://www.math.uni-bonn.de/people/schwede/`.

[60] J. Stasheff, *Homotopy associativity of H-spaces I, II*, Trans. Amer. Math. Soc. **108** (1963), 275–312.

[61] C. Stover, *The equivalence of certain categories of twisted and Hopf algebras over a commutative ring*, J. Pure Appl. Algebra **86** (1993), 289–326.

[62] A. Robinson, *Gamma homology, Lie representations and E_∞ multiplications*, Invent. Math. **152** (2003), 331–348.

[63] A. Robinson, S. Whitehouse, *Operads and Γ-homology of commutative rings*, Math. Proc. Cambridge Philos. Soc. **132** (2002), 197–234.

[64] J. Millès, *Andre-Quillen cohomology of algebras over an operad*, preprint `arXiv:0806.4405` (2008).

Index and Glossary of Notation

Index

adjunction
   of symmetric monoidal categories over a base, §1.1.8
   Quillen adjunction in a model category, §11.1.15
   Quillen adjunction in a semi-model category, §12.1.8

associative operad, §3.1.8, §3.1.9
   algebras over the associative operad, §3.2.4, §3.2.5
   (co)homology of algebras over the associative operad, §13.1.5, §17.3.5
   enveloping algebra of algebras over the associative operad, §4.3.6, §§10.2.3-
      10.2.4
   Kähler differentials of algebras over the associative operad, §4.4.6, §§10.3.2-
      10.3.3
   representations of algebras over the associative operad, §4.2.4

algebras over an operad, §3.2
   free algebras over an operad, §3.2.13
   in Σ_*-objects, §3.2.9, §3.2.10, §9
   in functors, §3.2.7, §3.2.8
   in right modules over an operad, §9.1
   semi-model category of algebras over an operad, §12.3

bimodules over operads, §9.1
   extension of structures for, §9.3
   functors associated to, §9.2
   restriction of structures for, §9.3

Brown's lemma, §11.1.18, §12.1.6

$\mathcal{C}$-cofibrant objects, §11.1.17
   in the category of Σ_*-objects, §11.4.1
   in the category of right modules over an operad, §14.1.2

Glossary of Notation

A: the associative operad, §3.1.8, §3.1.9

$\mathcal{A}$, $\mathcal{B}$: some (model) categories

$\mathcal{A}_\Delta$: the category of simplicial objects in a category $\mathcal{A}$, §13.3

C: the commutative operad, §3.1.8, §3.1.9

$\mathcal{C}$: the base symmetric monoidal category, §1.1.2

$\mathcal{E}$: a symmetric monoidal category over the base category, §1.1.2

$\mathcal{E}_c$: the class of cofibrations in $\mathcal{E}$, when $\mathcal{E}$ is equipped with a model structure, §19

$\mathcal{E}^0$: a reduced symmetric monoidal category over the base category, §1.1.17

$_P\mathcal{E}$: the category of P-algebras in a category $\mathcal{E}$, for P an operad in the base category, §3.2

End_X: endomorphism operads, §3.4, §3.4.1

$End_{X,Y}$: the endomorphism module of a pair, §2.3.1, §8.1.1

$\mathcal{F}$: the category of functors $F : \mathcal{E} \to \mathcal{E}$, where $\mathcal{E}$ is a fixed symmetric monoidal category over a base category $\mathcal{C}$, §2.1

$\mathcal{F}_R$: the category of functors $F : {}_R\mathcal{E} \to \mathcal{E}$, §5

$_P\mathcal{F}$: the category of functors $F : \mathcal{E} \to {}_P\mathcal{E}$, §3.2.7, §3.2.8

$_P\mathcal{F}_R$: the category of functors $F : {}_R\mathcal{E} \to {}_P\mathcal{E}$, §9

$\mathcal{F}^0$: the category formed by functors $F : \mathcal{E} \to \mathcal{E}$ which vanish on the initial object $0 \in \mathcal{E}$, §2.1.4

F_r: generators of the category of Σ_*-objects, §2.1.12

$\mathcal{F}(\mathcal{A}, \mathcal{X})$: the category of functors $F : \mathcal{A} \to \mathcal{X}$

$F(M)$: the free operad on a Σ_*-object M, §3.1.5

$\Gamma(G)$: the Σ_*-object associated to a functor $G : \mathcal{E} \to \mathcal{E}$, §2.3, §2.3.4

$\Gamma_R(G)$: the right R-module associated to a functor $G : {}_R\mathcal{E} \to \mathcal{E}$, for R an operad, §8, §8.1.4

dg k Mod: the category of differential graded k-modules over a ground ring k, §1.1.4, §11.1.9, §11.2

$Hom_{\mathcal{E}}(-,-)$: the hom-objects of a category $\mathcal{E}$ enriched over the base category $\mathcal{C}$, §1.1.12

$Hom_{\mathcal{F}}(-,-)$: the hom-objects of the category of functors $\mathcal{F}$, §2.1.11

$Hom_{\mathcal{M}}(-,-)$: the hom-objects of the category of Σ_*-objects $\mathcal{M}$, §2.1.11

$Hom_{\mathcal{M}_R}(-,-)$: the hom-objects of the category of right modules over an operad R, §6.3

$\mathcal{I}$: the set of generating cofibrations in a cofibrantly generated (semi-)model category, §§11.1.5-11.1.8

$\mathcal{J}$: the set of generating acyclic cofibrations in a cofibrantly generated (semi-)model category, §§11.1.5-11.1.8

k: a ground ring

k Mod: the category of k-modules

L: the Lie operad, §3.1.8

$\mathcal{M}$: the category of Σ_*-objects, §2.1

$\mathcal{M}_R$: the category of right modules over an operad R, §5.1

${}_P\mathcal{M}$: the category of left modules over an operad P, or, equivalently, the category of P-algebras in Σ_*-objects, §3.2.9, §3.2.10, §9

${}_P\mathcal{M}_R$: the category of bimodules over operads P, R, or, equivalently, the category of P-algebras in right R-modules, §9

$\mathcal{M}^0$: the category of connected Σ_*-objects, §2.1

$\mathcal{M}_R^0$: the category of connected objects in right modules over an operad R, §5.1.4

$M[X]$, same as $S[M, X]$: the Σ_*-object defined by shifting the Σ_*-object M in the construction $(M, X) \mapsto S(M, X)$, §4.1.4, §18.1.1

$M \circ N$: the composition product of Σ_*-objects, §2.2, §2.2.2

$M \circ_R N$: the relative composition product between right and left modules over an operad, §5.1.5

$\mathrm{Mor}_\mathcal{A}(-, -)$: the morphism sets of a category $\mathcal{A}$

$\mathcal{O}$: the category of operads in the base category $\mathcal{C}$, §3.1.1

$\mathcal{O}_\mathcal{E}$: the category of operads in a symmetric monoidal category $\mathcal{E}$, §3.1.1

$\mathcal{O}_R$: the category of operads in right R-modules, §4.1

Ω_R^1: the module of Kähler differentials over an operad, §10.3

$\Omega_R^1(A)$: the module of Kähler differentials of an algebra A over an operad R, §4.4, §10.3

1: the unit object
  in a symmetric monoidal category, §1.1.2
  in Σ_*-objects, §2.1.7
  in right modules over an operad, §6.1

P (and also Q, R, S): any operad in the base category $\mathcal{C}$

$P(X)$: the free algebra over an operad P, §3.2.13

$\phi_!$: the functor of extension of structures associated to an operad morphism ϕ
  for bimodules, §9.3
  for left modules, §3.3.6
  for right modules, §7.2
  for algebras, §3.3.5

ϕ^*: the functor of restriction of structures associated to an operad morphism ϕ
  for bimodules, §9.3
  for left modules, §3.3.6
  for right modules, §7.2
  for algebras, §3.3.5

$\mathcal{R}_P(A)$: the category of representations of an algebra A over an operad P, §4.2

S(M): the functor on a symmetric monoidal category $\mathcal{E}$ associated to a Σ_*-object M, §2, §2.1.1

S(M, X): the image of an object $X \in \mathcal{E}$ under the functor S(M) : $\mathcal{E} \to \mathcal{E}$ associated to a Σ_*-object M, §2, §2.1.1

S[M, X], same as $M[X]$: the Σ_*-object defined by shifting the Σ_*-object M in the construction $(M, X) \mapsto$ S(M, X), §4.1.4, §18.1.1

S_R(M): the functor on R-algebras associated to a right R-module M, for R an operad, §5, §5.1.3

S_R(M, A): the image of an R-algebra A under the functor S_R(M) : $_R\mathcal{E} \to {}_R\mathcal{E}$ associated to a right R-module M, §5, §5.1.3

S_R[M, A]: the right R-module defined by shifting the right R-module M in the construction $(M, A) \mapsto S_R(M, A)$, §18.1

Σ_r: the symmetric group in r letters

Σ_*: the sequence of symmetric groups

U_R(A): the enveloping operad (respectively, algebra) of an algebra over an operad, §4.1, §4.3, §10.1, §10.2, §17.4

$\mathcal{X}$, $\mathcal{Y}$: some (model) categories

$\mathcal{X}_c$: the class of cofibrations in a model category $\mathcal{X}$, §11.1.14, §12.1.4

Lecture Notes in Mathematics

For information about earlier volumes
please contact your bookseller or Springer
LNM Online archive: springerlink.com

Vol. 1825: J. H. Bramble, A. Cohen, W. Dahmen, Multiscale Problems and Methods in Numerical Simulations, Martina Franca, Italy 2001. Editor: C. Canuto (2003)

Vol. 1826: K. Dohmen, Improved Bonferroni Inequalities via Abstract Tubes. Inequalities and Identities of Inclusion-Exclusion Type. VIII, 113 p, 2003.

Vol. 1827: K. M. Pilgrim, Combinations of Complex Dynamical Systems. IX, 118 p, 2003.

Vol. 1828: D. J. Green, Gröbner Bases and the Computation of Group Cohomology. XII, 138 p, 2003.

Vol. 1829: E. Altman, B. Gaujal, A. Hordijk, Discrete-Event Control of Stochastic Networks: Multimodularity and Regularity. XIV, 313 p, 2003.

Vol. 1830: M. I. Gil', Operator Functions and Localization of Spectra. XIV, 256 p, 2003.

Vol. 1831: A. Connes, J. Cuntz, E. Guentner, N. Higson, J. E. Kaminker, Noncommutative Geometry, Martina Franca, Italy 2002. Editors: S. Doplicher, L. Longo (2004)

Vol. 1832: J. Azéma, M. Émery, M. Ledoux, M. Yor (Eds.), Séminaire de Probabilités XXXVII (2003)

Vol. 1833: D.-Q. Jiang, M. Qian, M.-P. Qian, Mathematical Theory of Nonequilibrium Steady States. On the Frontier of Probability and Dynamical Systems. IX, 280 p, 2004.

Vol. 1834: Yo. Yomdin, G. Comte, Tame Geometry with Application in Smooth Analysis. VIII, 186 p, 2004.

Vol. 1835: O.T. Izhboldin, B. Kahn, N.A. Karpenko, A. Vishik, Geometric Methods in the Algebraic Theory of Quadratic Forms. Summer School, Lens, 2000. Editor: J.-P. Tignol (2004)

Vol. 1836: C. Năstăsescu, F. Van Oystaeyen, Methods of Graded Rings. XIII, 304 p, 2004.

Vol. 1837: S. Tavaré, O. Zeitouni, Lectures on Probability Theory and Statistics. Ecole d'Eté de Probabilités de Saint-Flour XXXI-2001. Editor: J. Picard (2004)

Vol. 1838: A.J. Ganesh, N.W. O'Connell, D.J. Wischik, Big Queues. XII, 254 p, 2004.

Vol. 1839: R. Gohm, Noncommutative Stationary Processes. VIII, 170 p, 2004.

Vol. 1840: B. Tsirelson, W. Werner, Lectures on Probability Theory and Statistics. Ecole d'Eté de Probabilités de Saint-Flour XXXII-2002. Editor: J. Picard (2004)

Vol. 1841: W. Reichel, Uniqueness Theorems for Variational Problems by the Method of Transformation Groups (2004)

Vol. 1842: T. Johnsen, A. L. Knutsen, K_3 Projective Models in Scrolls (2004)

Vol. 1843: B. Jefferies, Spectral Properties of Noncommuting Operators (2004)

Vol. 1844: K.F. Siburg, The Principle of Least Action in Geometry and Dynamics (2004)

Vol. 1845: Min Ho Lee, Mixed Automorphic Forms, Torus Bundles, and Jacobi Forms (2004)

Vol. 1846: H. Ammari, H. Kang, Reconstruction of Small Inhomogeneities from Boundary Measurements (2004)

Vol. 1847: T.R. Bielecki, T. Björk, M. Jeanblanc, M. Rutkowski, J.A. Scheinkman, W. Xiong, Paris-Princeton Lectures on Mathematical Finance 2003 (2004)

Vol. 1848: M. Abate, J. E. Fornaess, X. Huang, J. P. Rosay, A. Tumanov, Real Methods in Complex and CR Geometry, Martina Franca, Italy 2002. Editors: D. Zaitsev, G. Zampieri (2004)

Vol. 1849: Martin L. Brown, Heegner Modules and Elliptic Curves (2004)

Vol. 1850: V. D. Milman, G. Schechtman (Eds.), Geometric Aspects of Functional Analysis. Israel Seminar 2002-2003 (2004)

Vol. 1851: O. Catoni, Statistical Learning Theory and Stochastic Optimization (2004)

Vol. 1852: A.S. Kechris, B.D. Miller, Topics in Orbit Equivalence (2004)

Vol. 1853: Ch. Favre, M. Jonsson, The Valuative Tree (2004)

Vol. 1854: O. Saeki, Topology of Singular Fibers of Differential Maps (2004)

Vol. 1855: G. Da Prato, P.C. Kunstmann, I. Lasiecka, A. Lunardi, R. Schnaubelt, L. Weis, Functional Analytic Methods for Evolution Equations. Editors: M. Iannelli, R. Nagel, S. Piazzera (2004)

Vol. 1856: K. Back, T.R. Bielecki, C. Hipp, S. Peng, W. Schachermayer, Stochastic Methods in Finance, Bressanone/Brixen, Italy, 2003. Editors: M. Fritelli, W. Runggaldier (2004)

Vol. 1857: M. Émery, M. Ledoux, M. Yor (Eds.), Séminaire de Probabilités XXXVIII (2005)

Vol. 1858: A.S. Cherny, H.-J. Engelbert, Singular Stochastic Differential Equations (2005)

Vol. 1859: E. Letellier, Fourier Transforms of Invariant Functions on Finite Reductive Lie Algebras (2005)

Vol. 1860: A. Borisyuk, G.B. Ermentrout, A. Friedman, D. Terman, Tutorials in Mathematical Biosciences I. Mathematical Neurosciences (2005)

Vol. 1861: G. Benettin, J. Henrard, S. Kuksin, Hamiltonian Dynamics – Theory and Applications, Cetraro, Italy, 1999. Editor: A. Giorgilli (2005)

Vol. 1862: B. Helffer, F. Nier, Hypoelliptic Estimates and Spectral Theory for Fokker-Planck Operators and Witten Laplacians (2005)

Vol. 1863: H. Führ, Abstract Harmonic Analysis of Continuous Wavelet Transforms (2005)

Vol. 1864: K. Efstathiou, Metamorphoses of Hamiltonian Systems with Symmetries (2005)

Vol. 1865: D. Applebaum, B.V. R. Bhat, J. Kustermans, J. M. Lindsay, Quantum Independent Increment Processes I. From Classical Probability to Quantum Stochastic Calculus. Editors: M. Schürmann, U. Franz (2005)

Vol. 1866: O.E. Barndorff-Nielsen, U. Franz, R. Gohm, B. Kümmerer, S. Thorbjønsen, Quantum Independent Increment Processes II. Structure of Quantum Lévy Processes, Classical Probability, and Physics. Editors: M. Schürmann, U. Franz, (2005)

Vol. 1867: J. Sneyd (Ed.), Tutorials in Mathematical Biosciences II. Mathematical Modeling of Calcium Dynamics and Signal Transduction. (2005)

Vol. 1868: J. Jorgenson, S. Lang, $Pos_n(R)$ and Eisenstein Series. (2005)

Vol. 1869: A. Dembo, T. Funaki, Lectures on Probability Theory and Statistics. Ecole d'Eté de Probabilités de Saint-Flour XXXIII-2003. Editor: J. Picard (2005)

Vol. 1870: V.I. Gurariy, W. Lusky, Geometry of Müntz Spaces and Related Questions. (2005)

Vol. 1871: P. Constantin, G. Gallavotti, A.V. Kazhikhov, Y. Meyer, S. Ukai, Mathematical Foundation of Turbulent Viscous Flows, Martina Franca, Italy, 2003. Editors: M. Cannone, T. Miyakawa (2006)

Vol. 1872: A. Friedman (Ed.), Tutorials in Mathematical Biosciences III. Cell Cycle, Proliferation, and Cancer (2006)

Vol. 1873: R. Mansuy, M. Yor, Random Times and Enlargements of Filtrations in a Brownian Setting (2006)

Recent Reprints and New Editions

LECTURE NOTES IN MATHEMATICS

Springer

Edited by J.-M. Morel, F. Takens, B. Teissier, P.K. Maini

Editorial Policy (for the publication of monographs)

1. Lecture Notes aim to report new developments in all areas of mathematics and their applications - quickly, informally and at a high level. Mathematical texts analysing new developments in modelling and numerical simulation are welcome.

   Monograph manuscripts should be reasonably self-contained and rounded off. Thus they may, and often will, present not only results of the author but also related work by other people. They may be based on specialised lecture courses. Furthermore, the manuscripts should provide sufficient motivation, examples and applications. This clearly distinguishes Lecture Notes from journal articles or technical reports which normally are very concise. Articles intended for a journal but too long to be accepted by most journals, usually do not have this "lecture notes" character. For similar reasons it is unusual for doctoral theses to be accepted for the Lecture Notes series, though habilitation theses may be appropriate.

2. Manuscripts should be submitted either to Springer's mathematics editorial in Heidelberg, or to one of the series editors. In general, manuscripts will be sent out to 2 external referees for evaluation. If a decision cannot yet be reached on the basis of the first 2 reports, further referees may be contacted: The author will be informed of this. A final decision to publish can be made only on the basis of the complete manuscript, however a refereeing process leading to a preliminary decision can be based on a pre-final or incomplete manuscript. The strict minimum amount of material that will be considered should include a detailed outline describing the planned contents of each chapter, a bibliography and several sample chapters.

   Authors should be aware that incomplete or insufficiently close to final manuscripts almost always result in longer refereeing times and nevertheless unclear referees' recommendations, making further refereeing of a final draft necessary.

   Authors should also be aware that parallel submission of their manuscript to another publisher while under consideration for LNM will in general lead to immediate rejection.

3. Manuscripts should in general be submitted in English. Final manuscripts should contain at least 100 pages of mathematical text and should always include

   - a table of contents;
   - an informative introduction, with adequate motivation and perhaps some historical remarks: it should be accessible to a reader not intimately familiar with the topic treated;
   - a subject index: as a rule this is genuinely helpful for the reader.

   For evaluation purposes, manuscripts may be submitted in print or electronic form, in the latter case preferably as pdf- or zipped ps-files. Lecture Notes volumes are, as a rule, printed digitally from the authors' files. To ensure best results, authors are asked to use the LaTeX2e style files available from Springer's web-server at:

   ftp://ftp.springer.de/pub/tex/latex/svmonot1/ (for monographs).

Additional technical instructions, if necessary, are available on request from:
lnm@springer.com.

4. Careful preparation of the manuscripts will help keep production time short besides ensuring satisfactory appearance of the finished book in print and online. After acceptance of the manuscript authors will be asked to prepare the final LaTeX source files (and also the corresponding dvi-, pdf- or zipped ps-file) together with the final printout made from these files. The LaTeX source files are essential for producing the full-text online version of the book (see www.springerlink.com/content/110312 for the existing online volumes of LNM).

The actual production of a Lecture Notes volume takes approximately 12 weeks.

5. Authors receive a total of 50 free copies of their volume, but no royalties. They are entitled to a discount of 33.3% on the price of Springer books purchased for their personal use, if ordering directly from Springer.

6. Commitment to publish is made by letter of intent rather than by signing a formal contract. Springer-Verlag secures the copyright for each volume. Authors are free to reuse material contained in their LNM volumes in later publications: a brief written (or e-mail) request for formal permission is sufficient.

Addresses:
Professor J.-M. Morel, CMLA,
École Normale Supérieure de Cachan,
61 Avenue du Président Wilson, 94235 Cachan Cedex, France
E-mail: Jean-Michel.Morel@cmla.ens-cachan.fr

Professor F. Takens, Mathematisch Instituut,
Rijksuniversiteit Groningen, Postbus 800,
9700 AV Groningen, The Netherlands
E-mail: F.Takens@math.rug.nl

Professor B. Teissier, Institut Mathématique de Jussieu,
UMR 7586 du CNRS, Équipe "Géométrie et Dynamique",
175 rue du Chevaleret
75013 Paris, France
E-mail: teissier@math.jussieu.fr

For the "Mathematical Biosciences Subseries" of LNM:

Professor P.K. Maini, Center for Mathematical Biology,
Mathematical Institute, 24-29 St Giles,
Oxford OX1 3LP, UK
E-mail: maini@maths.ox.ac.uk

Springer, Mathematics Editorial I, Tiergartenstr. 17
69121 Heidelberg, Germany,
Tel.: +49 (6221) 487-8259
Fax: +49 (6221) 4876-8259
E-mail: lnm@springer.com